The HTI+ Cram Sheet

This Cram Sheet contains the distilled, key facts about the Residential Systems and Systems Infrastructure and Integration exams for the Home Technology Integrator (HTI+) certification. Review this information as the last thing you do before you enter the testing center, paying special attention to those areas where you feel you need the most review.

RESIDENTIAL SYSTEMS

Computer Networking Fundamentals

- The HomeRF 2.0 specification supports transmission rates of 10Mbps and operates in the 2.4GHz radio band.
- Bluetooth wireless technology is designed to connect one device to another with a short-range radio link.
- Wireless equipment that meets the 802.11b standard is approved to use the Wi-Fi label. The key points to remember for 802.11b devices are that the standard sets a maximum speed of 11Mbps and is authorized for use in the 2.4GHz unlicensed radio band.
- Items that are needed to control automated systems in a home using a Web page are a modem, Web browser, valid username, and password.
- The standard adopted for using the phone lines for networking is called Home Phone Network Alliance (HomePNA).
- The HPNA acronym is associated with the phone-line networking standard. You should memorize the transmission speed of 10Mbps for HPNA version 2.0.
- HomePNA technology is often listed as a network media solution where homeowners do not want adapter modules plugged in to all their power wiring outlets.
- The home automation products that use power-line technology are also referred to as *power-line carrier (PLC)*.
- Surge protectors must not be used between HPLA adapters and the power-line wall outlets.
- X10 is a widely used open standard protocol for home automation. X10 has an address structure that supports a total of 256 receiver locations. It transmits at a digital bit rate of 60bps.
- Problems can occur if X10 receiver modules are installed on the opposite phase of a three-wire 240v hose wiring system. A signal bridge (also known as a *phase coupler*) is used to provide a signal path across the two sides of the 240v mains to remove typical isolation conditions that exist.
- The CEBus communications protocol was developed by the Electronic Industries Association (EIA) as the ANSI/EIA-600 standard. CEBus is an acronym for consumer electronics bus. The main performance parameter for the CEBus standard is a transmission rate of 8000bps, and addresses are set at the factory. You should be able to identify EIA-600 as the CEBus standard.
- The IEEE Ethernet 10BASE-T standard indicates the maximum recommended length for 10BASE-T Ethernet segments using UTP cable is 100 meters (328 feet).
- The popular type of coaxial cable used for home cabling applications is type RG-6.
- TOSLINK is a type of optoelectronic cable used to connect audio and video components. Also, ST and SC connectors are used with fiber-optic cables.
- UTP cabling is considered to be the most susceptible to electromagnetic interference (EMI) among the various types of communications cable.
- UTP cables are designed to have different twist rates between each pair to minimize crosstalk between adjacent cables. More twists per foot raises the performance and the category rating, as well as increases the immunity to EMI and crosstalk.
- Heat buildup in a location where servers or network workstations are located can cause outages or unreliable operation.
- FireWire, i.Link, and Lynx are brand names for IEEE-1394 standard serial cables.

- The tip (green) and ring (red) are the respective line 1 location color codes for telephone cable on an RJ-11 jack.
- The RJ-31x is a type of telephone jack installed and located near the security system control box.
- The main purpose of a residential gateway is to integrate several functions, including a switch, router, and firewall into a single unit.
- The *UPnP* is an industry standard for allowing UPnP-compliant network devices to be plugged in to a network and automatically start communicating with other UPnP devices.
- The basic purpose of DHCP is to allocate IP addresses from a pool of addresses. On large networks, the task is usually allocated to a dedicated DHCP server.
- Memorize the port numbers for the frequently used protocols, such as HTTP (80), SMTP (25), FTP Control (20), and POP3 (110).
- A firewall is computer program or router used for protecting computers on a private network from intrusion by unauthorized users or hackers using the Internet.
- A modem is required to connect a computer or a home network to the Internet. Modem types vary according to the type of connection used, such as dial-up analog modems, cable modems, or DSL modems.
- In addition to the DSL modem, the splitter is important to proper operation of the DSL data transmission and analog voice service, which share the same phone line.
- The cable modem specification called DOCSIS was developed by a consortium of cable companies. The full name of this cable modem standard is Data Over Cable Service Interface Specification.

Audio and Video Fundamentals

- The Dolby Digital Surround Sound format has specific names and locations for the five surround speakers and the sixth low frequency effects (LFE) channel. The locations are center (C), left (L), right(R), right surround (RS), and left surround (LS). The LS and RS are ideally placed to the side and slightly behind the listener's seating position. There is no designated rear surround sound speaker location. The subwoofer (or LFE) can be located anywhere in the room. The subwoofer does not contribute to stereo imaging; therefore, adding a second subwoofer does not enhance the surround quality.
- A bridged amplifier can be used to increase the power available to a single speaker because the output voltage is effectively doubled, which theoretically increases the power output by a factor of 4.
- When an amplifier is too small for the speaker system, at high volume levels it can cause the amplifier to automatically shut down due to clipping of the output audio signal. This problem can be cured by using a higher-powered amplifier consistent with the power rating of the speakers.
- When connecting a DVD player to a TV receiver that uses component video inputs, three cables are required. Two additional cables are required for analog stereo sound (left and right) connections.
- Personal video recorders (PVRs) should not be confused with video cassette recorders (VCRs).
- The analog commercial television transmission standard for analog TV in the United States is the National Television Systems Committee (NTSC standard). It is a 525-line, 30-frames-per-second interlaced standard.
- Remember the basic requirements for selecting a site and installing a satellite antenna. The height above ground for the antenna is not important; however, the site must have a clear view of the southern sky with no trees or buildings in the receiving path of the antenna. The antenna must also be grounded to protect against lightning damage.

Home Security and Surveillance Systems

- Common sense should guide the selection of a security system password used with a monitoring service. Never use family or pet names, birth dates, street names, or any word found in the dictionary.
- Surveillance cameras can be used in most exterior or residential locations; however, you should remember that a surveillance camera cannot be located where there is a reasonable expectation of privacy.
- Exterior security lighting should be used near landscaped areas around the home such as shrubs, bushes, foliage, and trees.
- The local building codes are the final authority for determining construction and installation requirements for wiring, smoke detector location, and connectivity.
- The ANSI/TIA/EIA 570-A standard requires at least 12" of separation between parallel runs of security wire and AC power wiring.
- All alarm systems that use the telephone lines in a home to aromatically dial a central monitoring station must have an RJ-31x jack installed.

Telecommunications Standards

- A key system supported by a KSU is a popular design choice for an average home office or small business. A non-KSU phone system is the best choice for small businesses that have only two phone lines and no requirement for Web hosting.
- The special tool required for inserting wires into the connectors on a 110 or 66 block is called a *punch-down tool*.

Join the
CRAMMERS CLUB

Get exclusive discount offers on books, practice exams, test vouchers, and other exam prep materials by joining The Crammers Club.

It's easy!

Go to **www.examcram2.com/crammersclub**, answer a few short questions, and you will become eligible for several exciting benefits, including:

- Free books and test vouchers
- Exclusive content
- Free Exam Cram 2 merchandise
- Special prizes and offers

The first 500 readers who register at www.examcram2.com/crammersclub receive a free book of their choice!

Also, don't forget to check out examcram2.com, which features:

- The latest certification news
- Author interviews and articles
- Free practice exams
- Newsletters

and much more!

EXAM CRAM™ 2

EXAM CRAM 2

HTI+

Max Main

HTI+ Exam Cram 2

Copyright © 2004 by Que Certification

All rights reserved. No part of this book shall be reproduced, stored in a retrieval system, or transmitted by any means, electronic, mechanical, photocopying, recording, or otherwise, without written permission from the publisher. No patent liability is assumed with respect to the use of the information contained herein. Although every precaution has been taken in the preparation of this book, the publisher and author assume no responsibility for errors or omissions. Nor is any liability assumed for damages resulting from the use of the information contained herein.

International Standard Book Number: 0-7897-2937-7

Library of Congress Catalog Card Number: 2003100808

Printed in the United States of America

First Printing: December 2003

06   05   04   03   4   3   2   1

Trademarks

All terms mentioned in this book that are known to be trademarks or service marks have been appropriately capitalized. Que Certification cannot attest to the accuracy of this information. Use of a term in this book should not be regarded as affecting the validity of any trademark or service mark.

Warning and Disclaimer

Every effort has been made to make this book as complete and as accurate as possible, but no warranty or fitness is implied. The information provided is on an "as is" basis. The authors and the publisher shall have neither liability nor responsibility to any person or entity with respect to any loss or damages arising from the information contained in this book or from the use of the CD or programs accompanying it.

Publisher
Paul Boger

Executive Editor
Jeff Riley

Development Editor
Susan Brown Zahn

Managing Editor
Charlotte Clapp

Project Editor
Tonya Simpson

Production Editor
Megan Wade

Indexer
Chris Barrick

Proofreader
Wendy Ott

Technical Editor
Brian Alley

Team Coordinator
Pamalee Nelson

Page Layout
Stacey DeRome
Ron Wise

Que Certification • 800 East 96th Street • Indianapolis, Indiana 46240

A Note from Series Editor Ed Tittel

You know better than to trust your certification preparation to just anybody. That's why you, and more than two million others, have purchased an Exam Cram book. As Series Editor for the new and improved Exam Cram 2 series, I have worked with the staff at Que Certification to ensure you won't be disappointed. That's why we've taken the world's best-selling certification product—a finalist for "Best Study Guide" in a CertCities reader poll in 2002—and made it even better.

As a "Favorite Study Guide Author" finalist in a 2002 poll of CertCities readers, I know the value of good books. You'll be impressed with Que Certification's stringent review process, which ensures the books are high-quality, relevant, and technically accurate. Rest assured that at least a dozen industry experts—including the panel of certification experts at CramSession—have reviewed this material, helping us deliver an excellent solution to your exam preparation needs.

Best Study Guides

We've also added a preview edition of PrepLogic's powerful, full-featured test engine, which is trusted by certification students throughout the world.

As a 20-year-plus veteran of the computing industry and the original creator and editor of the Exam Cram series, I've brought my IT experience to bear on these books. During my tenure at Novell from 1989 to 1994, I worked with and around its excellent education and certification department. This experience helped push my writing and teaching activities heavily in the certification direction. Since then, I've worked on more than 70 certification-related books, and I write about certification topics for numerous Web sites and for *Certification* magazine.

In 1996, while studying for various MCP exams, I became frustrated with the huge, unwieldy study guides that were the only preparation tools available. As an experienced IT professional and former instructor, I wanted "nothing but the facts" necessary to prepare for the exams. From this impetus, Exam Cram emerged in 1997. It quickly became the best-selling computer book series since "...*For Dummies*," and the best-selling certification book series ever. By maintaining an intense focus on subject matter, tracking errata and updates quickly, and following the certification market closely, Exam Cram was able to establish the dominant position in cert prep books.

You will not be disappointed in your decision to purchase this book. If you are, please contact me at etittel@jump.net. All suggestions, ideas, input, or constructive criticism are welcome!

Ed Tittel

Expand Your Certification Arsenal!

**A+
Exam Cram 2
(Exam 220-301, 220-302)**

James Jones and Craig Landers
ISBN 0-7897-3043-X
$29.99 US/$45.99 CAN/£21.99 Net UK

- Key terms and concepts highlighted at the start of each chapter
- Notes, Tips, and Exam Alerts advise what to watch out for
- End-of-chapter sample Exam Questions with detailed discussions of all answers
- Complete text-based practice test with answer key at the end of each book
- The tear-out Cram Sheet condenses the most important items and information into a two-page reminder
- A CD that includes PrepLogic Practice Tests for complete evaluation of your knowledge
- Our authors are recognized experts in the field. In most cases, they are current or former instructors, trainers, or consultants—they know exactly what you need to know!

www.examcram2.com

About the Author

Max Main is an independent engineering consultant and author with extensive teaching experience in computer networking, wireless networking, and telecommunications engineering at the college level as well as contract teaching and consulting for wireless cellular service providers and network equipment vendors. He has several years' experience with the installation and maintenance of satellite gateway stations for two international teleport operators.

He has designed and tested prototype CDMA location-based technologies. He has conducted formal wireless cellular and microwave technology training for federal government and corporate clients. He is also a consultant to several investment management firms dealing with telecommunications regulatory and market trend issues. He is often invited to speak on federal regulatory trends and investment strategies for the telecommunications industry.

He is the author of two textbooks on wireless network certification and an introductory course textbook on home networking.

He holds CompTIA certifications for A+, Network+, Server+, and HTI+ as well as several Microsoft certifications. He is a certified Microsoft trainer and has experience conducting various instructor-led training courses for Microsoft certified training centers.

About The Technical Editor

Brian Alley has been in the IT field since 1985, first working with NetWare 2.2 and then Windows NT 3.1. Brian has been teaching for Boston University since 1992 as both a staff instructor and consultant. He currently holds CompTIA's A+, Network +, Server +, and I-net + certifications and has sat on various A+ committees. He is an MCSE for Windows NT and Windows 2000 and has been an MCT. Currently, Brian is the owner of Connected Executive and eIT Training, where he provides consulting and training to small- to medium-sized businesses in various IT areas, including networking, security, integration, and automation. He is also co-owner of eIT Prep, an online certification site. Brian has been installing automated systems in client offices and homes since 1990. He is coauthor of several Windows 2000 books and countless training manuals.

Acknowledgements

Writing a book requires the assistance and dedication of a number of people. I want to express my thanks to the editors and staff at Que as well as Charles Brooks and Cathy Boulay at Marcraft International for their assistance and technical consultation.

—*Max Main*

Contents at a Glance

Introduction xxvii

Self-Assessment xxxii

Part I Residential Systems

Chapter 1 Objective 1.0—Computer Networking Fundamentals 3

Chapter 2 Objective 2.0—Audio and Video Fundamentals 81

Chapter 3 Objective 3.0—Home Security and Surveillance Systems 127

Chapter 4 Objective 4.0—Telecommunications Standards 161

Chapter 5 Objective 5.0—Home Lighting Control and Management 201

Chapter 6 Objective 6.0—HVAC Management 233

Chapter 7 Objective 7.0—Home Water Systems Controls and Management 281

Chapter 8 Objective 8.0—Home Access Controls and Management 311

Chapter 9 Objective 9.0—Miscellaneous Automated Home Features 341

Part II Systems Infrastructure and Integration

Chapter 10 Objective 10.0—Structured Wiring—Low-voltage   373

Chapter 11 Objective 11.0—Structured Wiring—High-voltage   415

Chapter 12 Objective 12.0—Systems Integration   449

Part III Practice Exams

Chapter 13 Practice Exam 1: Residential Systems   485

Chapter 14 Answers to Practice Exam 1: Residential Systems   505

Chapter 15 Practice Exam 2: Systems Infrastructure and Integration   525

Chapter 16 Answers to Practice Exam 2   539

Part IV Appendixes

Appendix A What's on the CD-ROM   555

Appendix B Using the PrepLogic Practice Exams, Preview Edition Software   557

Glossary   565

Index   595

Table of Contents

Part I   Residential Systems 1

Chapter 1
1.0—Computer Networking Fundamentals 3

    Network Design Considerations   4
        Wireless Network Protocols and Standards   5
        Remote Access Methods   11
        Home Network Cabling and Transmission   11
        Ethernet Local Area Network Standards   21
        Network Media Considerations   27
    Equipment Location Considerations   28
        Network Equipment: Hardware and Software   28
        Network Equipment Components   32
    Network Physical Devices   35
        Wireless Access Points   36
        Computer Systems   37
        Printers   38
        Residential Gateways   39
        Network-enabled Devices and Appliances   40
    Hardware Configuration and Setting   41
        Operating System Configuration and User Settings   41
        IP Addressing   42
        Using TCP/IP Utility Programs   45
        Dynamic Host Configuration Protocol   48
        The Transmission Control Protocol   48
        User Datagram Protocol   49
        Network Address Translation   51
        Firewall Configuration and Filtering   51
        Network Interface Card Configuration   52
    Device Connectivity   53
        Interfaces with Legacy Devices and Systems   53
        Network Design Access Points   54
        Termination Points   56

Shared In-house Services 59
 Video Surveillance 60
 File and Print Services 61
 Media Services (Streaming Audio and Video) 61
Externally Provided Data Services 62
 Digital Subscriber Line 62
 Integrated Services Digital Network 66
 Cable 67
 Point-to-Point Protocol 68
 Point-to-Point Tunneling Protocol 68
 RAS 69
 Email 70
Standards 71
 The Institute of Electrical and Electronics Engineers 72
 ANSI/TIA/EIA Standards 72
 The ANSI/EIA 600 CEBus Standard 73
Exam Prep Questions 74
Need to Know More? 79

Chapter 2
2.0—Audio and Video Fundamentals 81
Design Considerations for a Connected Audio/Video System 82
 Dedicated Versus Distributed Systems 82
 Whole Home Audio/Video System Design 83
 Remote Access 85
 Zoning Distribution 85
Equipment Location Considerations 86
 Speaker Locations 86
 Television Location 87
 Audio/Video Rack Locations 87
 Touchscreen Locations 88
 Volume Control Locations 88
 Keypad Locations 88
Physical Audio and Video Products 89
 Receivers 89
 Amplifiers 91
 Speakers 92
 Keypads 94
 Video Displays 95
 Source Equipment 99

Configuration and Settings of Audio/Video Components 100
    Volume Settings 100
    Distribution Channels 100
    Equalization 101
    Internal Broadcasting 101
Components Involved in Device Connectivity 102
    Video Signal Formats 102
    Component Video 102
    Composite 103
    S-video 104
    Analog Audio 104
    Digital Audio 105
    Low-voltage Cabling and Connectors 105
    Termination Points 105
In-house Services 106
    Streaming Audio and Video 106
    Personal Video Recorder 107
    Media Server 107
Externally Provided Audio and Video Services 108
    Satellite Service 108
    Cable 108
    External Terrestrial Off-the-Air Broadcast Service 109
    Digital Television Service 109
    Internet Audio and Video Services 110
Audio- and Video-Related Standards 110
    Wiring Standards 110
    Protocol Standards 111
    Resolutions 111
    Video Recording Formats 111
    Surround Sound Formats 113
    Television Transmission Standards 114
Audio and Video Installation Plans and Procedures 117
    Installing a Satellite System 117
    Installing Speakers 119
Audio and Video Maintenance Plans and Procedures 119
    Guidelines for Audio and Video System Maintenance 120
Exam Prep Questions 121
Need to Know More? 126

Chapter 3
3.0—Home Security and Surveillance Systems 127

    Design Considerations   128
        Wireless Security Systems   128
        Hard-wired Security and Surveillance Systems   129
        Remote Access Systems   130
        Fire Detection Systems   131
        Environmental Monitoring   132
        Emergency Response Systems   132
        Temperature Sensors   133
    Location Considerations when Designing a Security or Fire Alarm System   133
        Home Utility Outlet Specifications   133
        Cohesion with Existing Home Systems   134
        Safety and Code Regulations   134
        Existing Home Environments   134
        New Home Construction Environments   135
        Equipment Functionality and Specifications   136
    Physical Devices   138
        Keypads   138
        Sensors   139
        Security Panels   140
        Cameras   141
        Monitors   141
        Switchers   142
    Configuration and Settings   143
        Zone Layout   143
        Passwords   143
        Keypad Locations   144
        Sensor Locations   144
        Camera Locations   146
        Exterior Security Lighting Systems   146
    Device Connectivity   147
        Low-Voltage Wire Connectivity   147
        Wireless System Connectivity   147
        Telephone   147
        Coaxial Cabling   148
        Termination Points   148
    In-house Services   148
        Video Surveillance and Monitoring Services   148
        Alarm Types   149

External Services   149
    External Alarm Monitoring Service   150
    Remote Access   150
Industry Standards for Home Security and Surveillance Systems   150
    The National Electrical Code   150
    The Telecommunications Industry Association and the Electronic Industries Alliance   151
    Underwriters Laboratories   152
Installation Plans   152
    Installing an RJ-31x Telephone Jack   152
    Installing Motion Detectors   153
    Camera and Switching Equipment Installation   153
    Surveillance System Monitor Installation   153
Maintenance Plans and Procedures   153
Exam Prep Questions   154
Need to Know More?   159

Chapter 4
4.0—Telecommunications Standards 161

Telecommunications Design Considerations   162
    Hybrid Systems Design   162
    Analog Communications Systems   162
    Digital Communications Systems   163
    PBX Systems   164
    Key Systems   165
    Voice Over IP   166
    Remote Access Methods, Standards, and Protocols   167
    Home Environmental Factors   168
Telecommunications Equipment Location Considerations   168
    Telecommunications System Characteristics and Restrictions   169
    Industry Standards and Practices   169
    Home Environmental Factors Related to Location   169
Physical Telecommunications Products   170
    Telephone Fundamentals   170
    FAX Machine Communications   171
    PBX Systems   172
    Videoconferencing   172
    Caller Line Identification   172
    66/110 Connection Blocks   173

Standard Configurations and Settings   175
    Phone-line Extensions and Splitters   175
    Voice Mail   177
    Call Restriction   177
    Intercom Systems   178
Methods for Connecting Telecommunications Equipment   179
    Wireless Connectivity   179
    Telephone Connectivity   179
    Physical Wiring Types   180
    RJ-11 Connections   181
    RJ-45 Termination Standards   182
    Termination Points   183
Telecommunications In-house Services   183
    Voice Mail   183
    Intercom Services   183
    Call Conferencing   184
    Extension Dialing   184
Telecommunications External Services   185
    Caller ID   185
    External Voice Mail Services   185
    Call Blocking   186
    Three-Way Calling   187
    Call Waiting   187
    Emergency Response System   187
Telecommunications Industry Standards   189
    Fax Transmission Standards and Protocols   189
    ANSI/TIA/EIA Standards   190
Telecommunications System Installation Plans and Procedures   191
    Telecommunications Wall Outlets   191
    Telecommunications Cable Installation Tips   192
Telecommunications Troubleshooting and Maintenance   193
Exam Prep Questions   194
Need to Know More?   199

Chapter 5
5.0—Home Lighting Control and Management 201
Home Lighting Design Considerations   202
    Load Requirements   202
    Grounding   204
    Wireless/Wire Runs   205

    Home-run   205
    Daisy-chain   206
    Power-line Controls   206
    Conduits   207
    Zoning   208
    Remote Access   208
  Home Lighting Equipment Location Considerations   208
    Planning the Locations for Home Lighting   209
  Physical Lighting Products   209
    Power Outlets   209
    Dimming Modules   211
    Light Switches   213
    Light Fixtures   215
    Automated Window Treatments   217
  Standard Configurations and Settings   217
    Lighting Scenes   218
    Security Lighting   218
    Lighting Zones   219
  Device Connectivity   219
    Nonmetallic Cable   219
    MC Cable   219
    Armored Cable   220
    Low-voltage Wiring   221
    Termination Points   221
  Industry Standards   222
    Wiring Types and the National Electrical Code   222
    Standards and Organizations   222
  Installation Plans and Procedures   224
    Installation Procedures for Single-Pole Switches   224
  Troubleshooting and Maintenance Plans   225
  Exam Prep Questions   226
  Need to Know More?   231

Chapter 6
6.0—HVAC Management 233
  HVAC Design Considerations   234
    Zoned/Non-Zoned Designs   234
    Single or Multiple Pieces of Equipment   236
    Air Handler   238
    Water-based (Radiant) Heating   239
    Remote Access   241

Equipment Location Considerations   242
    Thermostats   242
    Sensors   245
Physical Products and Components   246
    Thermostats   246
    Air Handlers   249
    Damper Controls   250
    Furnaces   252
    Condensers   254
    Distribution Panels   256
Standard Configurations and Settings   256
    Zone Programming   256
    System Programming   257
    Time-of-Day Programming   258
    Seasonal Presets   259
Device Connectivity   259
    Communicating Thermostats   260
    Communications Cable   260
    Termination Points   261
Standards and Organizations   262
    National Electrical Code   262
    TIA/EIA Standards   262
    IEEE Standards   263
    Underwriters Laboratories, Inc.   263
    National Fire Protection Association   264
Installation Plans and Procedures   264
    Thermostat Installations   264
    Duct Installations   266
Troubleshooting and Maintenance Issues   270
    Air Conditioning   270
Exam Prep Questions   273
Need to Know More?   279

Chapter 7
7.0—Home Water Systems Controls and Management 281

Design Considerations   282
    Timed Systems   282
    Zoned Systems   284
    Remote Access   284
Equipment Locations   284
    Irrigation System Control Valve Locations   285
    Backflow Prevention Valve Locations   285

Keypads   286
Relays   286
Heaters   286
Pumps   287
Solenoids   288
Physical Products   288
   Water Alarms   289
   Irrigation Controllers   289
   Control Points   290
   Distribution Panels   291
   Interface Locations   291
   Keypads   292
   Power Supplies   292
   Sensors   293
   Sump Pumps   293
   Solenoids   295
   Sprinklers   296
Standard Configuration and Settings   298
   Zone Programming   298
   Time-of-Day Programming   299
   Seasonal Presets   300
Device Connectivity   300
   Water System Communications Cable   300
   Water System Low-Voltage Wiring   301
   Termination Points   302
Industry Standards   302
   Local Irrigation and Water Conservation Standards   303
Installation Plans and Procedures   303
Water System Troubleshooting and Maintenance Issues   304
Exam Prep Questions   305
Need to Know More?   310

Chapter 8
8.0—Home Access Controls and Management 311

Home Access Design Considerations   312
   Remote Control   312
   Remote Access   313
   Home Access Condition Monitoring   314
Equipment Component Location Considerations   315
   Keypads   315
   Relays   316
   Sensors   317

Physical Products   318
    Keypads   318
    Control Boxes   319
    Access Control Power Supplies   320
    Solenoid-Operated Locks   321
    Distribution Panels   322
    Gates for Home Access Control   323
Standard Configuration and Settings   324
    User Access   324
    Programming   325
    Sensors   325
    Time-of-Day Settings   326
Device Connectivity   327
    Communications Cable   327
    Wireless Communication   329
    Phone Line   330
    Low-voltage Wiring   330
    Termination Points   330
Industry Standards for Home Access Control   331
    The Builders Hardware Manufacturers Association   331
    The National Electrical Code   332
    Underwriters Laboratories, Inc.   332
    The U.S. Consumer Product Safety Commission   332
Installation Plans and Procedures   332
    Reverse Garage Door Sensors   333
Troubleshooting and Maintenance Plans   333
    Wireless Garage Door Openers   334
Exam Prep Questions   335
Need to Know More?   340

Chapter 9
9.0—Miscellaneous Automated Home Features 341
Automated Home Design Features   342
    Automated A/V Cabinetry   342
    Automated Lift Systems   343
    Automated Window Shading Designs   345
    Automated Lift Systems   346
    Stair Lift Designs   346
    Dumbwaiters   347
    Automatic Fireplace Igniters   347
    Automated Home Fans   348

Automated Skylights 348
Location Considerations 348
    Keypads 349
    Relays 349
    Sensors 349
Physical Products 349
    Keypads 349
    Control Boxes 351
    Power Supplies 352
    Solenoids 353
    Distribution Panels 355
    Lift Systems 356
    Portals 357
Standard Configurations and Settings 357
    User Access Programming 357
    Sensors 357
Methods of Device Connectivity 360
    Communications Cable 360
    Low-voltage Wiring 361
    Termination Points 361
Industry Standards 361
    National Electrical Code 362
    TIA/EIA Standards 362
    Underwriters Laboratories 363
Exam Prep Questions 364
Need to Know More? 369

Part II Systems Infrastructure and Integration 371

Chapter 10
10.0—Structured Wiring—Low-voltage 373
Structured-wiring Design Considerations 374
    Wire Types 374
    Conduits 382
    Daisy-chain Wiring Topology 382
    Home-run Wiring Topology 383
Structured-wiring Location Considerations 384
    Remodeling/Existing Wiring Upgrades 384
    New Construction 385
    Equipment Component Placement 387

Physical Structured-wiring Connection Components 388
  Distribution Panels 389
  RJ-45 Connectors 389
  RJ-11 Connectors 390
  RG-6 Coaxial Cable 391
  BNC Connectors 391
  Amplifiers 392
  Filters 392
  Binding Posts 393
  Connectors 394
Configuration and Settings for Structured-wiring Designs 395
  Distribution Panels 395
  Termination Points 395
Device Connectivity for Structured-wiring Design 396
  Coaxial Cable 397
  Category 5 UTP Cable 397
  Fiber-optic Cable 398
  Plenum Cable 399
  Audio Wire 399
  Security Wire 400
  Termination Points 400
Industry Standards for Structured-wiring Design 401
  Standards and Organizations 401
  Federal Communications Commission Rulings 406
Installation Plans and Procedures for Structured Wiring 407
  Installing Structured-wiring Outlets 407
  Installation Tips for UTP and Coaxial Cable 407
Maintenance Plans and Procedures for Structured-wiring Design 408
Exam Prep Questions 409
Need to Know More? 414

Chapter 11
11.0—Structured Wiring—High-voltage 415
High-voltage Wiring Design Considerations 416
  Load Requirements 417
  Grounding 417
  Surge Protection 419
  Power Backup (UPS) 419
  Safety Considerations 422
Audio and Video Equipment Locations 423
  Project Settings 423
  Equipment Component Placement 425

Physical High-voltage Structured-wiring Components   425
    Outlets   426
    Dimming Modules   427
    Light Switches   429
    Fixtures   431
    Source Equipment   433
Configuration and Settings for High-voltage Structured Wiring   434
    Distribution Panels   434
    Circuit Breaker Panels   434
Termination Points   435
    Power Consumption   435
Device Connectivity for High-voltage Structured Wiring   437
    NM Cable   437
    MC Cable   437
    AC Cable   438
    Termination Points   439
Standards and Organizations   439
    National Electrical Code   439
    EIA/TIA Standards   440
    IEEE Standards   440
    Underwriters Laboratories   441
Installation Plans and Procedures   441
    Installing Wall Outlets   441
    Maintenance Plans and Procedures   441
Exam Prep Questions   443
Need to Know More?   448

Chapter 12
12.0—Systems Integration 449

System Integration Design Considerations   450
    Web Pads   450
    Keypads   451
    Touchscreens   451
    Wireless RF/IR Receivers   452
Equipment Location Considerations   452
    Web Pads   453
    Keypads   453
    Touchscreens   453
    RF/IR Receivers   453
Core Components Found in System Integration Designs   454
    Keypads   454
    RF/IR Receivers   454

Distribution Panel   455
Interface Location   455
Control Processor   456
Patch Panel   458
Configuration and Settings   458
  User Interface Programming   459
    Configuring and Programming a Residential Gateway   459
    Configuring and Programming Controller Systems   461
    Configuring and Programming Keypad-based Controllers   462
    Programming Touchscreen User Interfaces   462
    Programming Intelligent Remote Controls   463
Device Connectivity   463
  Communications Cable   464
  Low-voltage Wiring   464
  Wireless   465
  Termination Points   466
Industry Standards   467
  National Electrical Code   467
  ANSI/TIA/EIA Standards   467
  IEEE Standards   468
  Underwriters Laboratories, Inc.   468
  National Electrical Contractors Association   468
Installation Plans and Procedures   469
  Installing Central Home Controllers   469
  Installing User Interfaces   469
Maintenance Plans and Procedures   470
  Testing   470
  Troubleshooting Tools   471
  Maintenance Issues   474
Exam Prep Questions   475
Need to Know More?   481

Part III   Practice Exams 483

Chapter 13
Practice Exam 1: Residential Systems 485

Chapter 14
Answers to Practice Exam 1: Residential Systems 505

Chapter 15
Practice Exam 2: Systems Infrastructure and Integration 525

Chapter 16
Answers to Practice Exam 2 539

Part IV   Appendixes 553

Appendix A
What's on the CD-ROM 555

    The PrepLogic Practice Exams, Preview Edition Software   555
    An Exclusive Electronic Version of the Text   556

Appendix B
Using the PrepLogic Practice Exams, Preview Edition Software 557

    The Exam Simulation   557
    Question Quality   558
    The Interface Design   558
    The Effective Learning Environment   558
    Software Requirements   558
    Installing PrepLogic Practice Exams, Preview Edition   559
    Removing PrepLogic Practice Exams, Preview Edition from Your Computer   559
    How to Use the Software   560
        Starting a Practice Exam Mode Session   560
        Starting a Flash Review Mode Session   561
        Standard PrepLogic Practice Exams, Preview Edition Options   561
        Seeing Time Remaining   562
        Getting Your Examination Score Report   562
        Reviewing Your Exam   562
    Contacting PrepLogic   563
        Customer Service   563
        Product Suggestions and Comments   563
    License Agreement   563

Glossary 565

Index 595

We Want to Hear from You!

As the reader of this book, *you* are our most important critic and commentator. We value your opinion and want to know what we're doing right, what we could do better, what areas you'd like to see us publish in, and any other words of wisdom you're willing to pass our way.

As an executive editor for Que Certification, I welcome your comments. You can email or write me directly to let me know what you did or didn't like about this book—as well as what we can do to make our books better.

Please note that I cannot help you with technical problems related to the *topic* of this book. We do have a User Services group, however, where I will forward specific technical questions related to the book.

When you write, please be sure to include this book's title and author as well as your name, email address, and phone number. I will carefully review your comments and share them with the author and editors who worked on the book.

Email: feedback@quepublishing.com

Mail: Jeff Riley
        Executive Editor
        Que Certification
        800 East 96th Street, Third Floor
        Indianapolis, IN 46240 USA

For more information about this book or another Que title, visit our Web site at www.quepublishing.com. Type the ISBN (excluding hyphens) or the title of a book in the Search field to find the page you're looking for. For information about the Exam Cram 2 Series, visit www.examcram2.com.

Introduction

Welcome to the Home Technology Integrator (HTI+) Exam Cram 2! This book is designed to help you prepare to take—and pass—the Computing Technology Industry Association (CompTIA) certification exams HT0-101, "Residential Systems," and HT0-102, "Systems Infrastructure and Integration." This introduction explains some facts concerning CompTIA's certification programs and how the Exam Cram 2 series can help you prepare for the HTI+ and other CompTIA certification exams.

Exam Cram 2 books help you prepare for CompTIA exams by carefully following the prescribed technology objectives with clear and easy-to-understand coverage of the topics you will need to study in preparation for taking the exams. The Exam Cram 2 series is not structured to teach you everything you need to know about each major home technology topic. Instead, the author has focused on the terminology, physical products, standards, installation problems, and questions you are likely to encounter on each of the exams.

The HTI+ certification exam is different from most of the previous CompTIA exams because it requires you to successfully pass two exams to become certified as a Home Technology Integrator. The exams can be taken on the same day or be scheduled on different days. The Exam Cram 2 staff has worked together as a team to help you prepare to pass both of the HTI+, as well as other CompTIA certification, exams.

Although the HTI+ exam does not require a prerequisite certification, those who currently have an A+ or a Network+ CompTIA certification or equivalent experience will be well prepared to understand the fundamental home networking issues that are an important element of the residential systems exam 1.

If you have limited experience with any of the home automation technologies or home computing systems referenced in the HTI+ exam objectives, you might elect to investigate the availability of instructor-led classroom training or home technology tutorial textbooks that are available from several sources. CompTIA, as well as most testing centers, can advise you on the availability of additional training programs.

This HTI+ Exam Cram 2 book explains many of the current industry practices associated with the installation and testing of the low-voltage and high-voltage structured wiring that is gaining wide acceptance in the residential building trades. This book focuses on understanding many of the National Electrical Code wiring and building trade standards as well as federal government standards associated with residential wiring and automated management systems. Every state and many municipal governments, however, have their own building codes and licensing requirements for installers, electricians, and security system technicians. Local ordinances and building codes are the ultimate governing authority. Obtaining an HTI+ certification offers many advantages for job opportunities in several home technology fields covered in the exam. The HTI+ certification validates your basic knowledge to a potential employer of essential areas associated with automated home management and control systems.

Taking a Certification Exam

After you have prepared for your HTI+ exam, you need to register with a testing center to schedule a specific time to take the exam. Remember, the HTI+ certification includes two exams. Each exam costs $207 for non-CompTIA members and $155 for CompTIA members. Discounts are available for groups larger than 50 people who register at one time to take the exam. You must pass both the residential and infrastructure exams to obtain an HTI+ certification. CompTIA certification testing is administered by Pearson VUE and Prometric. Here's how you can contact them:

- *Virtual University Enterprises (VUE)*—You can register for the exam or obtain phone numbers and locations through the VUE Web site at http://www.vue.com.

- *Prometric*—You can register for an exam through the Prometric Web site at http://www.prometric.com/default.htm.

After you have completed the registration and payment method, you will be informed about the time and location for taking the exam. You should plan your travel to arrive 15 minutes prior to the scheduled time. Be prepared to show two types of identification—one of which must be a photo ID—to be admitted to the exam area.

All exams are closed-book exams, meaning you will not be permitted to take any material into the testing area. However, you will be provided with a blank sheet of paper and a pen or pencil. It is recommended that you immediately write down all the information you have memorized before taking the exam. In Exam Cram 2 books, this information appears on the tear-out Cram Sheet inside the front cover of each book. Review it carefully prior to taking the test.

When you complete a CompTIA certification exam, the test center software informs you whether you have passed or failed the exam. Separate registration is required for each exam, although you can take both exams in sequence. If you fail either exam, you will be permitted to take it again at a scheduled time, but you will be required to pay another $207.

How to Prepare for an Exam

Preparing for any CompTIA exam requires that you obtain study materials designed to provide comprehensive information about the home automation technology topics that appear on the exam. You will need to become familiar with each of the topics covered in this book with emphasis on the exam alerts and practice questions. You can also locate a certified CompTIA training company at http://cla.comptia.org/.

About This Book

The Cram Sheet, located inside the front cover, includes key points and terms you will need to review prior to entering the exam center. Pay particular attention to those topics needing the most review. You can transfer the points you have memorized to paper after entering the exam area.

Each of the Exam Cram 2 chapters follows the hierarchical topical outline of the CompTIA HTI+ objectives for both exams. Chapters 1–9 cover the objectives for the residential systems exam, whereas Chapters 10–12 cover all the objectives contained in the systems infrastructure and integration exam. Each chapter includes the following aids to assist in your preparation for the exam:

➤ *Terms you'll need to understand*—Each chapter begins with a list of the terms and acronyms you will need to know for each topical area of the exam.

➤ *Tasks you'll need to master*—The HTI+ exam has several tasks that are important, such as troubleshooting, installation, and maintenance. These tips are highlighted at the beginning of each chapter. Memorize these tasks when reviewing the material in each chapter.

➤ *Exam alerts*—Each item or topic that is flagged as an exam alert is designed to call your attention to key points you will need to focus your attention on when studying for the HTI+ exam. Although each topic in the book contains information related to an exam objective, the exam alerts help you remember the specific points that are the most important.

This is what the exam alert looks like. An exam alert stresses a home technology standard, a key point, or terms that relate to one or more exam questions. This flag is used to call to your attention a subject you will need to pay extra attention to when reviewing the material in each chapter.

➤ *Tips, cautions, and notes*—In addition to the exam alerts, some areas that might require additional background information or elaboration are highlighted to remind you of a problem with terminology or controversial technical issues.

Examples of how they appear in the book are as follows:

Flat-panel display systems use both plasma and LCD technologies. A CRT is not a flat-panel display. Be aware that some CRT monitors and TVs using a flat-faced CRT design are often listed in vendor specifications as *flat-screen* monitors and TVs.

The Shared Wireless Access Protocol referenced in the HomeRF standard should not be confused with the Wireless Applications Protocol (WAP), which is used primarily for wireless mobile user terminals. The HTI+ objectives include HomeRF and the SWAP protocol. WAP is not a communications protocol applicable to home networking technology.

The 802.11b specification uses the term *distribution system* to refer to the wired network that can be connected to the wireless LAN via an access point. The DS is merely another name for a wired LAN (typically an Ethernet LAN) that can be linked to wireless LAN users.

➤ *Exam prep questions*—A section at the end of each chapter contains questions about key topics that will help you prepare for the HTI+ exams.

➤ *Details and resources*—Each chapter ends with a section titled "Need to Know More?" This section provides additional information on relevant publications, tutorials, and home technology industry Web sites that offer more detailed information on each of the chapter's topics.

Using This Book

The topics in this book follow the 12 main technology areas included in the HTI+ CompTIA objectives for the exam. Each chapter includes the subtopics organized exactly as they appear in the published CompTIA HTI+ objectives outline with emphasis on the terms, standards, automated home designs, installation, and physical descriptions of home technology components.

This book will help you prepare for future opportunities that are predicted by many experts to emerge in the home automated management systems and controls market.

Many home technology industry groups are working with suppliers and standards organizations to make home automation a reality with practical and useful systems and tools. The skills you will need to pass the HTI+ exams are outlined in detail in this book. Please share your thoughts on the book with us. We intend to continue to improve it for future test takers. Your comments and feedback are valuable to us, and all input will be reviewed carefully. Good luck on the exams!

Self-Assessment

The purpose of this section of the Exam Cram series of books is to assist you in the evaluation of your readiness to take the two Home Technology Integrator (HTI+) exams. It will help you understand what you will need to know to not merely pass the exam with a minimum grade, but to be prepared to pass the exam with an exceptional grade.

The HTI+ exam is composed of two exams developed by the Computer Technology Industry Association (CompTIA). The first exam is called HTI+ "Residential Systems" (HT0-101). The second exam is titled "Systems Infrastructure and Integration" (HT0-102). The following section reviews some typical concerns you might have while getting yourself primed for the exam as well as what an ideal home technology integrator candidate might look like.

Home Technology Integrators in the Real World

The purpose of any CompTIA certification exam is to validate your knowledge of a particular technology. The candidate for the HTI+ exam must demonstrate knowledge in not one but several somewhat diverse areas. Although the main thrust of HTI+ technology is to understand the features of automated home management systems, it covers many areas that can initially appear to be unrelated, such as computer networking, irrigation systems, security systems, access systems, and audio/video systems. The common thread for each of these residential management system areas is the perception of how the various systems can be automated, integrated, and managed from a central computer system or programmable device using various protocols that can centralize the features of automated home management. The ideal candidate will be familiar with traditional computer network software and hardware as well as the newer home networking technologies that use existing power wiring and phone lines for connecting and integrating home automation systems and computers.

Many individuals now working in one of the following positions in the building trades or retail home improvement organizations are the ideal candidates for taking the HTI+ examination:

- Licensed electricians or electrician apprentices who are currently working in the home wiring construction
- Heating, ventilating, and air conditioning (HVAC) installers, repair technicians, and HVAC contractor personnel
- Irrigation system installers
- Audio/video/home entertainment retail sales and installation personnel
- Small/office/home office computer network technicians
- Home security system sales and installation personnel
- Home lighting and window control system installers
- Whole home automation contractors

The Ideal Home Technology Integrator Candidate

This section will give you an idea of what the ideal home technology integrator is really like. Do not be concerned if you do not meet all the qualifications. It is unusual, indeed, to find candidates who have work experience and/or formal training in all the HTI+ residential system areas described in the objectives for the exam. If you find areas described here that are outside your expertise, you should concentrate your research and homework in these areas. You should have not less than 1 year (2 years preferred) of work experience *or* academic training *or* a comprehensive understanding in each of the following home technology areas:

- *Computer local area networking installation and troubleshooting experience*—Includes setting up and configuring network interface cards, modems, and high speed Internet access gateways/routers for home network applications.
- *Instructor-led training or e-learning course in computer network theory*—This includes Ethernet and wireless local area network standards and practices.
- *A+ and Network+ CompTIA certifications.*
- *Work experience as an electronic technician, electrician, or electrician apprentice*—Familiarity with home wiring electrical codes and installation

practices is necessary. You should work with or know the basic concept of operation for power-line communications (PLC) and phone-line networking. You should have knowledge of the basic concepts of residential structured wiring and residential wiring standards and understand the features of low-voltage lighting systems and cabling as well as high-voltage home wiring. You should also have a working knowledge of the proper codes for installing wall outlets, ground fault circuit interrupters, light switches, and light fixtures and be able to identify the types of wire, cable, and conduit used in residential wiring.

▶ *Work experience or academic training as an installer for home HVAC systems*—Includes the ability to install and program a digital thermostat and install dampers, air handlers, ductwork, fireplace igniters, heat pumps, condensers, and furnaces.

▶ *Work experience or academic training in the audio/video home entertainment industry*—You should understand the procedures for whole home audio/video distribution system installation; satellite antenna installation and cabling; and high-definition TV setup with DVD and cable interface connectors, receivers, amplifiers, and personal video recorders (PVRs). You should understand the digital video broadcast standards and the principles of operation and installation for plasma TV receivers.

▶ *Work experience or academic training as a home security system installer*—You should be familiar with alarm systems, keypads, motion detectors, glass and window sensors, magnetic switches, video camera/monitor surveillance systems, smoke detectors, remote access systems, and emergency alerting systems.

▶ *Work experience or academic training as an irrigation system or swimming pool/spa installer*—This includes a working knowledge of PVC pipe, valves, programmable controllers, heaters, pumps, sump pumps, sprinkler head types, solenoids, control valves, keypads, and irrigation system low-voltage wiring.

▶ *Work experience or academic training as a telephone installer, cable TV technician/installer, or satellite antenna installer*—You should be familiar with high-speed Internet access technologies such as DSL, satellite, and cable. You should understand the basic features of key telephone systems and PBX phone systems and have a working knowledge or experience with bundled structured wiring, coaxial cable types and ratings, unshielded twisted-pair cable types and ratings, connectors, and the color codes used in structured wiring (including RJ-45 and RJ-11 jacks).

► *Work experience or academic training with residential access systems*—This includes elevators, lifts, dumbwaiters, electrically operated gates, electric/magnetic locks, access system power supplies, solenoid-controlled locks, and remote access locks.

Putting Yourself to the Test

The following series of questions has been included to help you determine any areas where you will need help before taking the exam. Each question is designed to narrow the field on the issues on which you will need to concentrate your efforts. Do not be discouraged if you do not respond to each question with a yes answer. A no answer does not indicate you should give up and not take the test. The questions are designed to help you discover the strong as well as weak areas in your background.

You might be one of those individuals who is dedicated and resourceful enough to train yourself in some of the technology areas. Only you, in the final analysis, can decide the extent of your expertise and knowledge in some areas in lieu of actual on-the-job work experience. You should keep in mind the following tips concerning the background required to pass the HTI+ exam:

► Some experience working with computer systems and knowledge of computer network fundamentals will be helpful.

► Work experience either as an employee or a self-trained homeowner installing electrical appliances, wiring, wall switches, wall outlets, and home electric wiring according to established building codes is also extremely helpful for tackling the HTI+ exam.

Education Background

1. Have you taken any courses in electronic fundamentals or basic electrical circuits? (yes or no)
2. Have you ever taken a class on computer networking? (yes or no)

If you answered yes to both questions, you are on the right path to passing both exams. The HTI+ exam assumes that you have a basic understanding of electrical circuits and a familiarity with fundamental concepts of networking computers in a resource-sharing local area network (LAN). Although the two types of courses mentioned previously will not cover the areas related to HVAC and irrigation systems, they are the core competencies that are the most important areas you will need to pass the exam.

If you answered no to question 1, you might want to consider reading more about electrical wiring fundamentals with emphasis on basic home wiring techniques. Consider any of the following books to help you gain some understanding of basic electrical circuits:

- *Electrical Wiring* by Arthur E. Seale, Jr., Delmar Learning (ASIN: 0790610736)
- *Modern Residential Wiring* by Harvey N. Holzman, Goodheart-Wilcox Co. (ASIN: 1566372763)
- *Electrical Circuit Fundamentals, Fourth Edition* by Thomas L. Floyd, Prentice Hall (ASIN: 013835166X)

If you answered no to question 2, you might need to prepare to pass the exam with some extra study materials on home networking practices. I recommend *Home Networking Survival Guide* by David Strom (McGraw-Hill Companies, ISBN: 0072193115).

Work Experience

If you have current work experience in any of the home technology-related industries, you have the most valuable asset at hand to pass the exam. It is unrealistic to expect that many candidates will have work experience in all the areas of home technologies covered in the exam. If you have any work experience in such areas as electrical residential wiring, security system installation, irrigations system installation, or HVAC installation, you can concentrate your research activities and preparation time on the remaining HTI+ technology areas outside your primary work experience field.

Additional questions to review in your self-assessment exercise on work experience are as follows:

3. Have you ever installed a home theater or stereo system?
4. Are you comfortable with the knowledge you have to install a three-way electrical light switch?
5. Do you know the basic procedures involved in programming a VCR or programmable thermostat?
6. Have you ever installed an underground sprinkler system?

7. Are you comfortable with your knowledge and understanding of how and where to install a smoke detector in a home?

8. Would you have no hesitation installing your own cable modem, network cards, or router in a four-computer home network?

If you answered yes to all six of these questions, you are well prepared as a self-taught home integrator or "do-it yourselfer" to tackle the HTI+ exam. A no answer in any of the topics should alert you to areas where you will need to do some additional research through tutorial articles or publications.

Finally, do not be intimidated by the jargon or plethora of acronyms, terms, and products related to home technology and automation. The Internet has hundreds of Web sites containing helpful articles and technical papers exploring how non-professionally trained homeowners have designed and installed their own home theater, structured wiring, or automated controllers for household lighting systems and landscape irrigation systems.

Testing Your Exam Readiness

As you can see from the preceding discussion, you will need to use all the tools available to ensure your success when you enter the test area. The combined fee for the two HTI+ exams is now more than $400! You do not want to merely take the test as an experiment to see whether you can pass and gamble on luck or retaking the exams later. Prepare yourself by reading any book related to the key home technology areas listed in this self-assessment section and at the end of each chapter, in the section "Need to Know More?" Take the practice exam contained on the CD with this book. Review each area where you missed any correct answers and study the areas you believe need the most additional preparation time.

More Resources for Assessing Your Readiness for the HTI+ Examination

In addition to the sources of information contained in this self-assessment section, you will find it worthwhile to check with the CompTIA Web site often regarding news on the HTI+ exams. Be sure to review the test objectives for both HTI+ exams at http://www.comptia.org/.

On the Internet, you will find many opportunities to meet with other candidates like yourself who are also preparing for the HTI+ exam. An excellent Web site to read what information is available and the experiences others might share with you regarding training materials and related news can be found at the Cram Session Web site at `http://boards.cramsession.com/boards/vbt.asp?b=3550`.

Onward, Through the Fog

After you have assessed your preparation steps and readiness for the exam, conducted research on the latest information on home technology products and services, and obtained the necessary practical experience that will assist you in understanding the basic concepts of residential structured wiring and integration of home automation components, you'll be ready to take some practice exams. Try to take the practice exams as many times as you can until you are familiar with all the correct answers and are comfortable with the terminology. If you follow the assessment steps contained in this section, you will know when you are ready to make an appointment to take both of the HTI+ exams. Good luck!

PART I
Residential Systems

1 1.0—Computer Networking Fundamentals

2 2.0—Audio and Video Fundamentals

3 3.0—Home Security and Surveillance Systems

4 4.0—Telecommunications Standards

5 5.0—Home Lighting Control and Management

6 6.0—HVAC Management

7 7.0—Home Water Systems Controls and Management

8 8.0—Home Access Controls and Management

9 9.0—Miscellaneous Automated Home Features

1.0—Computer Networking Fundamentals

Terms you'll need to understand:

- ✓ HomeRF
- ✓ Bluetooth
- ✓ 802.11a/b/g
- ✓ Shared Wireless Access Protocol (SWAP)
- ✓ Home Phone Network Alliance (HomePNA)
- ✓ HomePlug Powerline Alliance (HPLA)
- ✓ X10
- ✓ Consumer Electronics Bus (CEBus)
- ✓ 10BASE-T
- ✓ Wireless access points
- ✓ TOSLINK
- ✓ FireWire
- ✓ Network interface card (NIC)
- ✓ Network address translation (NAT)
- ✓ RJ-31x
- ✓ Dynamic Host Configuration Protocol (DHCP)
- ✓ The International Corporation of Assigned Names and Numbers (ICANN)
- ✓ Domain name system (DNS)
- ✓ Uniform Resource Locator (URL)
- ✓ Plug and Play (PnP)
- ✓ Universal Plug and Play (UPnP)
- ✓ Firewall
- ✓ ADSL terminal unit (ATU)
- ✓ Digital Subscriber Line (DSL)
- ✓ Data over Cable Interface Specification (DOCSIS)
- ✓ Remote access service (RAS)
- ✓ Point to Point Protocol (PPP)
- ✓ Point to Point Protocol (PPTP)
- ✓ Post Office Protocol 3 (POP3)
- ✓ Simple Mail Transfer Protocol (SMTP)

Techniques you'll need to master:

- ✓ Evaluating design considerations for wireless networks
- ✓ Comparing the features of home networking technologies
- ✓ Selecting the locations for network components
- ✓ Identifying the physical elements of a home network
- ✓ Identifying the configuration and settings for network hardware and software
- ✓ Selecting standard methods for device connectivity
- ✓ Selecting and identifying shared in-house services
- ✓ Describing externally provided data services
- ✓ Describing applicable home networking technology standards

This section of the exam challenges the test taker to identify and describe physical products typically encountered in residential computer networks. Questions from this domain represent 25% of the Residential Subsystems Exam content.

Network Design Considerations

The Residential Systems objective 1.1 states that the test taker should have knowledge of residential network design. Examples of these considerations include

- Wireless protocols and standards
- HomeRF
- Bluetooth
- 802.11a/b/g
- Remote access methods
- Network cabling transmission basics
- Phone line
- Power line
- Ethernet
- Coaxial cable
- Fiber-optic cable
- The extended capability port (ECP)
- Universal serial bus (USB)
- Domain name system (DNS)
- Uniform resource locater (URL)

Wireless Network Protocols and Standards

Wireless networks have been used in business environments for some time. Until recent years, the cost of wireless products exceeded that used for wired networks and as a result, few home office or residential computer networks used wireless products. Wireless standards also were slow to mature, which inhibited the market growth compared to the mature and low-cost wired Ethernet home networking technology. This has changed to the extent that wireless home networking has been standardized and is becoming affordable for the residential computer user.

Three standards have been completed and published for both commercial and residential applications. HomeRF is a standard specifically targeted to the residential user and is a competitive technology with the IEEE 802.11 wireless network standard. Bluetooth has emerged as a third wireless standard but is focused on an embedded technology to make a wide variety of computing and telephone devices free from cords and connecting cables. None of the three wireless networks are interoperable or physically compatible. In this section, you will learn about the terms, capabilities, design considerations, and important performance issues you will need to know when preparing for the HTI+ Residential Systems Examination.

HomeRF

HomeRF is a wireless local area network (LAN) standard supported by the HomeRF Working Group. It is not compatible or interoperable with the 802.11 family of wireless LANs. The founding members of the HomeRF consortium included, among others, Microsoft, Intel, HP, Motorola, and Compaq. The purpose of the HomeRF initiative was to develop a standard for economical wireless home communications including data and voice.

The HomeRF 2.0 a specification supports transmission rates of 10Mbps and operates in the heavily used 2.4GHz radio band using spread spectrum frequency hopping modulation (SSFH). HomeRF also has a built-in capability to avoid interference with other wireless networking devices. It does this by figuring out which "data channel" the interference is on and then telling the radio controller to a not use that channel. The range of HomeRF devices is 50 meters.

Chapter 1

HomeRF uses a protocol a called the Shared Wireless Access Protocol (SWAP), which is derived from the existing Digital Enhanced Cordless Telephone (DECT) standard. The SWAP specification describes wireless transmission devices and protocols for interconnecting computers, peripherals, and electronic appliances in a home network environment.

Corporations supporting the HomeRF standard envisioned a number of home automation features including appliances and home security systems, managed from a central computer.

 The Shared Wireless Access Protocol referenced in the HomeRF standard should not be confused with the Wireless Applications Protocol (WAP), which is used primarily for wireless mobile user terminals. The HTI+ objectives include HomeRF and the SWAP protocol. WAP is not a communications protocol applicable to home networking technology.

Bluetooth Wireless Standard

Bluetooth wireless technology is designed to connect one device to another with a short-range radio link. This standard has evolved from early work and engineering studies performed by Ericsson Mobile Communications in 1994. The Bluetooth project focused on the feasibility of designing a low-power/low-cost radio interface between mobile phones and their accessories. One of the goals of the project was to produce a very small chip-sized radio and signal processor that could be fitted into cell phones and other wireless terminal equipment. In the design, the chip would be required to handle both data and voice communications.

Bluetooth wireless technology eliminates the need for numerous, often proprietary, cable attachments for the connection of practically any kind of communication device. Connections are instant and they are maintained even when devices are not within line of sight. The range of each radio is approximately 10 meters. The supported channels are listed in the Table 1.1.

Table 1.1 Bluetooth Channel Configurations

Configuration	Max. Data Rate Upstream	Max. Data Rate Downstream
Three simultaneous voice channels	64Kbps × 3 channels	64Kbps × 3 channels
Symmetric data	433.9Kbps	433.9Kbps
Asymmetric data	723.2Kbps or 57.6Kbps	57.6Kbps or 723.2Kbps

The Bluetooth wireless technology supports both point-to-point and point-to-multipoint connections. The radio interface operates in the 2.4GHz band and employs frequency hopped spread spectrum (FHSS) modulation. Bluetooth devices can interact with one or more other Bluetooth devices in several ways. The simplest scheme is when only two devices are involved, which is referred to as point-to-point. One of the devices acts as the master and the other as a slave. This ad-hoc network is referred to as a *piconet*. A piconet is defined as any Bluetooth network with one master and one or more slaves; a diagram of a piconet is provided in Figure 1.1. With the current specification, up to seven slave devices can be set to communicate with a master radio in one device. Several piconets can be established and linked together in ad-hoc scatternets to enable communication among continually flexible configurations. Bluetooth is not a network protocol in the same family with the Ethernet, HomeRF, and 802.11 technologies that are tailored to shared network media. Bluetooth is used as a wireless personal area network (PAN) for a single user connecting to one or more peripheral devices.

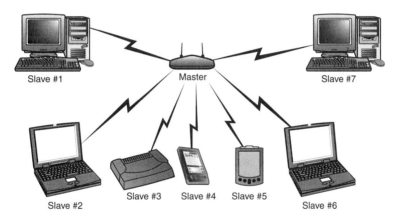

Figure 1.1 Bluetooth piconet.

The 802.11 Wireless Local Area Network Standard

In June 1997, the Institute for Electrical and Electronic Engineers (IEEE) completed work on the initial standard for wireless LANs, defined as 802.11. This standard specified a 2.4GHz operating frequency with data transmission rates of 1Mbps and 2Mbps. Since the ratification of the initial 802.11 standard, the IEEE 802.11 Working Group (WG) has made several revisions through various task groups. The three primary wireless LAN standards important to the home technology market are IEEE 802.11a/b/g.

The 802.11a Standard

802.11a is a Physical Layer (PHY) standard (IEEE Std. 802.11a-1999) that specifies operating in the 5GHz radio band using orthogonal frequency division multiplexing (OFDM). 802.11a supports data rates ranging from 6Mbps to 54Mbps. 802.11a-based products became available in late 2001. 802.11a and b are not compatible and require different hardware configurations (such as antennas).

The 802.11b Standard

The task group for 802.11b was responsible for enhancing the initial 802.11 Direct Sequence Spread Spectrum (DSSS) PHY layer to include 5.5Mbps and 11Mbps data rates in addition to the 1Mbps and 2Mbps data rates of the initial standard. The 802.11 committee finalized this standard (IEEE Std. 802.11b-1999) in late 1999. To provide the higher data rates, 802.11b uses complementary code keying (CCK), a modulation technique that makes efficient use of the radio spectrum. The 802.11b standard provides for operation using radio frequencies in the 2.4GHz band.

Most wireless LAN installations today comply with 802.11b, which is also the basis for Wi-Fi certification from the Wireless Ethernet Compatibility Alliance (WECA). These products have been available since the year 2000.

Many home technology contractors for wireless LANs are selecting products based on the 802.11b standard, which has proven to be very popular; however, the 802.11g standard is gaining in popularity due to its higher transmission speed of 54Mbps and interoperability with 802.11b networks. IEEE 802.11b defines two pieces of equipment. One is a desktop computer or more often a portable laptop with a wireless network interface card (NIC) and access point (AP) that acts as a bridge between the wireless networked computers and a distribution system (DS). *DS* is the term used in the standard for a wired LAN.

The two operational modes in IEEE 802.11b are infrastructure mode and ad hoc mode. The network connectivity for each mode is shown in Figure 1.2.

1.0—Computer Networking Fundamentals 9

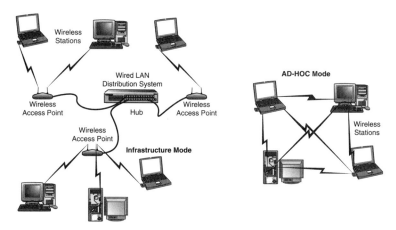

Figure 1.2 Wireless infrastructure and ad hoc modes.

Wireless equipment that meets the 802.11b standard is approved to use the "Wi-Fi" label. The key points to remember for 802.11b devices are that the standard sets a maximum speed of 11Mbps and they are authorized for use in the 2.4GHz unlicensed radio band.

The 802.11b specification uses the term *distribution system* to refer to the wired network that can be connected to the wireless LAN via an access point. The DS is merely another name for a wired LAN (typically an Ethernet LAN) that can be linked to wireless LAN users.

The infrastructure mode of operation consists of at least one AP connected to the DS and can be configured with one of the two following options:

➤ *Basic service set (BSS)*—An AP provides a local bridge function for the BSS. All wireless stations communicate with the AP and no longer communicate directly. All frames are relayed between wireless stations by the AP.

➤ *Extended service set (ESS)*—An ESS is a set of infrastructure BSSs in which the APs communicate among themselves to forward traffic from one BSS to another to facilitate movement of wireless stations between BSSs.

The ad hoc mode operates as an independent basic service set (IBSS) or peer to peer. The wireless stations communicate directly with each other; however, every station might not be capable of communicating with every other station due to the range limitations. No APs exist in an IBSS; therefore, all stations need to be within the range of each other for reliable communications.

The 802.11g Standard

The latest standard in the 802.11 family released by the IEEE is 802.11g. It supports speeds up to 54Mbps and uses the same frequency band as the 802.11b standard. 802.11g is backward compatible with 802.11b, which means the newer 802.11g network devices can operate with legacy 802.11b equipment but, of course, at the reduced throughput rate of 11Mbps.

> The wireless LAN standards and data rates you need to remember include the 802.11a specification that supports data rates ranging from 6Mbps to 54Mbps. 802.11b provides data rates of 5.5Mbps and 11Mbps, and 802.11g supports 54Mbps.

Wireless Infrared LANs

The IrDA infrared transmission specification makes provisions for multiple IrDA devices to be attached to a computer so that it can have multiple, simultaneous links to multiple IrDA devices. IrDA physical specifications require the transmitted signal to be viewed by the receiver up to 1 meter away with the nominal range being 5cm–60cm. Figure 1.3 shows how IrDA links can be used to share computers and devices through a normal Ethernet hub.

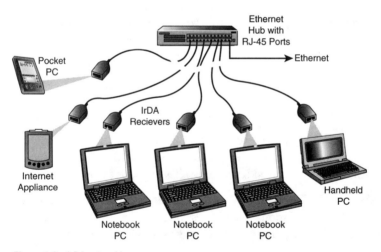

Figure 1.3 IrDA networking.

> IrDA stands for Infrared Data Association, which is an international organization that creates and promotes interoperable infrared data interconnection standards for wireless networks. The name is associated with products that conform to the interconnection standard.

Remote Access Methods

Remote access refers to any type of networking technology that enables you to connect users in geographically dispersed locations. A virtual private network (VPN) is the extension of a private network that encompasses logical links across shared or public networks, such as the Internet, using encryption to keep the data in transit safe and unaltered. This is technically called a *tunnel*. This network technology is called Remote Access Service (RAS).

Remote access methods are also used for *controlling* networks from remote locations. You can create a Web page on the residential server that can be accessed from the Internet using a standard Web browser such as Windows Internet Explorer or the Netscape Navigator. Commercial application packages are available that provide Windows-based X10 Web services. The contents of the Web page can be used to control the operations of the various automated residential subsystems. The remote user needs to have only a Web browser, a modem, and a valid username and password to access and control the resources of the residential network.

Remember that the items that are needed to control automated systems in a home using a Web page are a modem, a Web browser, a valid username, and a password. Several home network control protocols, such as X10, are designed to work with Web browsers for remote access control.

Home Network Cabling and Transmission

Home computer networking design and installation involve connecting computers and peripheral equipment with some type of media, such as copper cable or wireless network interface cards. In the business office environment, buildings are often designed to provide telephone and network cable connections supported with wire ductwork, conduits, and cable trays routed to work locations. Standard residential wiring design, however, has not typically included network cable installation as a standard feature. Recent trends in home automation have introduced new standards for home networking with a "no new wires" strategy that involves borrowing existing telephone or power-line wiring for the network media. This is a solution for existing homes in which installing new wiring is both impractical and expensive.

Structured wiring design for new residential construction with Category 5 cable installed throughout the home offers the optimal solution for home networking but with higher cost. Low-voltage structured wiring standards and practices are covered in Chapter 10, "10.0—Structured Wiring—Low-voltage."

In this section, you will learn about the cabling and transmission standards included in the HTI+ examination. Two types of network hardware have been engineered to use exiting wiring in the home. The power wiring and phone lines are convenient types of media because they already exist in the same general arrangement necessary for computer networking. This eliminates the need to install separate network cabling in an existing home. Specific standards have been developed by industry working groups to *borrow* the phone and power lines to transmit digital information on a home network.

The Phone Line Networking Standard Evolution

The standard adopted for using phone lines for networking is called Home Phone Network Alliance (HomePNA), a name patterned after the standards organization that developed it. The phone-line standard has evolved through three versions. HPNA 1.0, the original version of the standard, operated at 1Mbps. The current specification, HPNA 2.0, operates at 10Mbps. In September 2002, the HomePNA organization announced the approval of the new HomePNA 3.0, next generation specification. The throughput rate for version 3.0 is 128Mbps. The HomePNA new third-generation technology is backward compatible and interoperable with HomePNA version 2.0 network components. Additional information on HomePNA technology can be found at http://www.homepna.org.

You should remember the HPNA acronym and its association with the phone-line networking standard. You should also memorize the transmission speed of 10Mbps for HPNA version 2.0.

The HomePNA standard is based on a process that enables voice and data transmissions to share the available bandwidth of the telephone cabling in a home without mutual interference. It uses frequency division multiplexing (FDM) techniques to divide this bandwidth of frequencies into several communications channels, so HomePNA can be used without interrupting normal voice or fax services. One user can talk on the phone at the same time other users are using the same line to access the Internet or share files on other computers. All HomePNA devices must be connected to a single telephone line. Some homes have more than one telephone line for a fax machine or second voice line. The phone line used does not need to be an "active" telephone line with a telephone number. A spare telephone wire pair can be used, but all HomePNA devices must be connected to the same telephone pair.

Two types of network interface adapters are available for HPNA networking. A network interface card (NIC), shown in Figure 1.4, can be installed in a PCI slot or a laptop card slot. It has two RJ-11 jacks for connecting the adapter to the wall phone outlet jack and a telephone.

1.0—Computer Networking Fundamentals 13

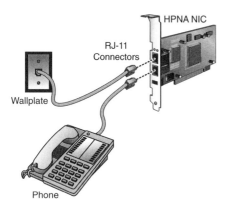

Figure 1.4 HPNA network interface card.

The external HPNA adapter uses the computer USB port for connecting the computer to the adapter (see Figure 1.5). The adapter uses two RJ-11 jacks with cables for connecting the adapter to the wall telephone jack and the telephone set.

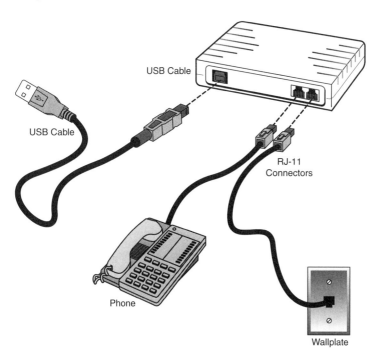

Figure 1.5 HPNA USB adapter.

HomePNA has several distinct advantages for home networking. It is available through existing phone jacks in almost every room of a home, is relatively inexpensive, requires minimal network devices, and is based on a mature industry-wide standard. It is often used as an alternative to power-line networking in situations where homeowners do not want to have any adapters plugged in to electric power outlets. HomePNA 2.0 supports up to 25 devices and is compatible with 10Mbps Ethernet speeds.

Be aware that HomePNA technology is often listed as a network media solution in cases where homeowners do not want adapter modules plugged in to all their power wiring outlets. HomePNA is a competing technology with power-line networking and is discussed in the next section.

Power-Line Networking

Power wiring that carries 60-Hertz (Hz), 120v alternating current in the home can be used with special interface adapters and software to connect computers on a home network. Unfortunately, power lines are a noisy medium and can cause interference with computer systems and networks. Some of the early power-line systems did not work very well. A group of companies joined forces in 2000 to organize the HomePlug Powerline Alliance (HPLA). The alliance has since set standards and reviewed various technical approaches by various vendors. The HomePlug 1.0 specification was released in June 2001, with a theoretical maximum speed of 14Mbps. Several companies now provide components and chipsets that comply with the HPLA standard.

The acronyms for phone-line networking and power-line networking standards organizations can be confusing when you see them both used on the exam. The standard adopted for using the *phone lines* for networking is derived from the standards group that developed it. It is called the *Home Phone Network Alliance (HomePNA)*. The standard is often referenced as *HPNA 2.0*. The *HomePlug Powerline Alliance (HPLA)* is a group of companies that developed the standard for power line networking. The specification is referenced as *HomePlug 1.0*, or sometimes *HPLA 1.0*. Memorize the two standard names so you will be able to recognize the difference between the two technologies.

The home automation products that use power-line technology are also referred to as *power-line carrier (PLC)* devices because they communicate over existing power-line wiring. You might see both PLC and HPLA terms used synonymously when describing power-line products.

Figure 1.6 shows some typical USB and Ethernet-type HPLA devices required to implement power-line networking in a residential environment.

1.0—Computer Networking Fundamentals 15

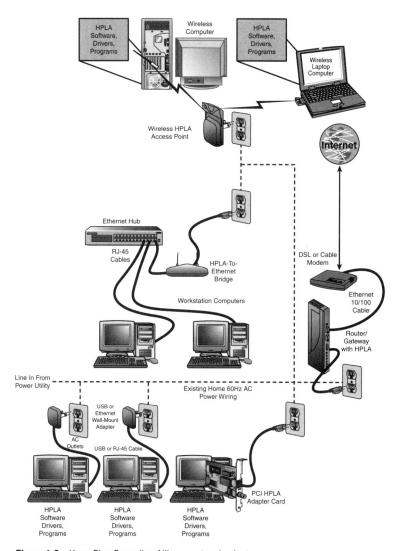

Figure 1.6 HomePlug Powerline Alliance network adapters.

To use the power wiring for a home network, the following components, ports, and resources are needed to complete the design and installation:

➤ *Ethernet or USB power-line adapters carrying the HomePlug 14Mbps Certified logo*—Ethernet or USB adapters plug in to each computer in the home network. Any computer or HPLA device that has a working Ethernet port uses the power-line Ethernet adapter as a bridge device to carry the data to an Ethernet network over the power lines. The USB power-line adapter serves the same purpose but uses the computer USB port instead of an Ethernet port.

- *Spare Ethernet or USB ports in your home computers or other devices.*

- *A HPLA power plug adapter and connecting power cable*—This plugs in to the AC outlet and provides the power-line connection to the HPLA adapters.

- *Open power outlets without surge protection*—Surge-protected power strips must not be used for connecting HPLA adapters because they will block the signals transmitted by the HPLA devices.

- *Software driver programs for the network interface device*—These must be installed in the network computer.

- *A router*—This is an optional component where the home network has access to a broadband Internet access service such as DSL or cable. This enables the broadband access to be shared among all computers connected to the HPLA network.

Be aware that the HomePlug Powerline Alliance digital transmissions appear as high frequency spikes to a surge protector and are essentially blocked from reaching the power-line wiring. Also, remember that surge protectors must not be used between HPLA adapters and the power-line wall outlets. This can appear as a troubleshooting-type question on the exam.

HPLA devices are also available for installation near the device or home appliance that is to be controlled. They include lights, entertainment systems, and security systems. The HTI+ exam will include questions concerning the different protocols used for controlling various HPNA-compatible devices. Information about these protocols is covered in the next section.

Power-Line Control Protocols

The HPNA and HPLA standards support a no-new-wires technology for home networking. As you have learned in the previous section, products developed by numerous companies enable a fast and economical approach for installing home computer networks using phone lines and power lines as the transmission media.

Other types of signaling protocols that use power lines have been developed to allow plug-in devices to talk to each other. They are used to control lights as well as several automated appliances in the home. The protocols in this category you will need to know more about are the X10 and CEBus.

Before proceeding in this section, it is important to gain a clear understanding of the basic differences between HPNA and HPLA home computer networking applications and power line protocols that are limited to control applications. The phone-line and power-line technologies use existing residential utility (phone-line and power-line) wiring to provide Ethernet-type LAN capabilities with no new wires. *Control protocols* are designed to use existing utility wiring to send relatively low-speed control signals to home lighting systems and plug-in appliances for centralized management. Control protocols might or might not use a computer as a system controller. HPNA and HPLA products are focused on high-speed computer-to-computer communications, whereas *control protocols* are focused on low-speed, easy installation; rudimentary functionality; and low cost.

The X10 Power-Line Protocol

X10 is a widely used open standard protocol for home power-line networking. It was developed several years ago by a Scottish firm named Pico Electronics, Ltd. X10 is now an industry standard as well as the name of its manufacturer. Many vendors provide products using the X10 brand name.

X10 controllers send digital signals over existing residential power wiring to receiver modules that are plugged in to wall outlets or permanently installed in a wall switch box. The controllers transmit digital information during the zero crossing time of the AC voltage on the power line. The transmit interval is therefore synchronous with the power-line frequency of 60Hz.

The X10 transmitter is a control box or software-driven menu on a computer display. The protocol employs unique addresses for each receiver, and the receiver plug-in module responds only to commands addressed to its unique location. Transmitters send commands such as All lights On All Lights Off, Dim, Bright, On, and Off. Each command string is preceded by the identification of the receiver unit to be controlled. The X10 protocol transmits 120KHz signal bursts, each 1 millisecond long, at the zero crossing of the voltage on a 60Hz 120v circuit. This broadcast goes out over the electrical wiring in a home (less overhead bits) at a data rate of 18.62bps. As indicated earlier, each receiver is set to a certain X10 address and reacts only to commands addressed to it. Receivers ignore commands not addressed to them. A complete X10 command is transmitted in approximately .78 second.

The X10 specification sets up an address structure supporting a total of 256 addresses. They include 16 unit codes for each of 16 house codes. House codes range from A to P, and unit codes range from 1 to 16. This provides 256 possible unique addresses. In Figure 1.7, the selectors for the house codes and unit codes are shown on the plug-in type X10 receiver module.

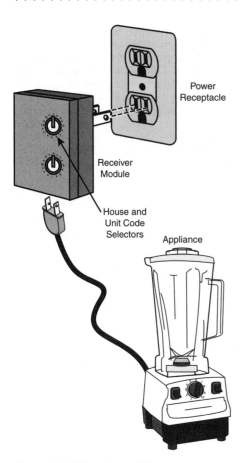

Figure 1.7 X10 receiver module.

Several computer programs are available for controlling X10 devices from a menu-driven display. Wall-mounted X10 control panels and handheld IrDA remote control units can also be employed for managing lights and X10-compatible home appliances.

 Remember the basic performance figures for the X10 protocol. X10 has an address structure that supports a total of 256 receiver locations and transmits at a digital bit rate of 18.62bps. A complete X10 command is transmitted in approximately .78 second. X10 is primarily used to control lights but can be used to control virtually any electrical device in the home that is compatible with the X10 protocol.

Understanding X10 Potential Problems

In a power-line networking installation, a problem can emerge in which the receivers respond to the commands intermittently. This is often caused by the isolation that exists on the 120v residential power wiring connected to the opposite phase of the 240v main service entering the home. In a three-wire system, two 120v distribution systems and a common ground are formed for the home power distribution system from the circuit breaker panel. This is a normal wiring situation in large homes; however, with the X10 protocol, a transmitter connected to one side of the 240v line and the common wire will not transmit reliably to receivers connected to the other side of the service. The solution for this problem is to install a commercial signal bridge (also referred to as a *phase coupler*), which permits signals to pass from one side of the 240v main service to the other phase and thereby bypass the isolation. A powered phase coupler is required to regenerate the signal to compensate for the attenuation of the X10 signal. Figure 1.8 illustrates how an X10 bridge is connected to provide a signal path between the two phases to bypass the isolation created by the pole transformer.

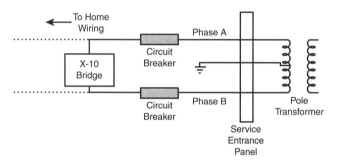

Figure 1.8 X10 bridge.

Another problem can also arise. X10 receiver modules have occasionally turned on by themselves. The solution is to change the house code on the controller and all the receiver modules. A neighbor could have the same system or another compatible system and, if both houses are on the same power pole transformer, signals can couple from one house to another.

 Be aware of the problems that can occur if X10 receiver modules are installed on the opposite phase of a three-wire, 240v hose wiring system. A signal bridge (also known as a phase coupler) is used to provide a signal path across the two sides of the 240v main to remove typical isolation conditions that exist.

The CEBus Standard

The CEBus communications protocol was developed by the Electronic Industries Alliance (EIA) as the ANSI/EIA-600 standard. CEBus is an acronym for *consumer electronics bus*.

The CEBus is designed to control devices on a power line, but it also works on other media. The CEBus standard involves device addresses that are set in hardware at the factory and includes four billion possibilities. The standard also offers a defined language of many object-oriented controls, including commands such as `volume up`, `fast forward`, `rewind`, `pause`, `skip`, and `temperature up 1 degree` or `temperature down 1 degree`. Presently, all the communications hardware, languages, and protocols are available on a chip produced by Intellon Corporation, who sells the chip to other manufacturers for use in their products. Intellon also offers manufacturer private label products and OEM products using the CEBus standard.

The CEBus standard uses spread spectrum modulation on the power line. Spread spectrum involves changing the frequency during transmission: The carrier frequency is swept linearly from 203KHz to 400KHz for 19 cycles, back to 100KHz in 1 cycle, and then back to 203KHz in 5 cycles during a 100-microsecond burst period.

The CEBus standard defines a data communications network that accommodates the following variety of media:

➤ The AC power line

➤ Twisted-pair wires

➤ Coaxial cable

➤ Infrared

➤ Wireless

➤ Fiber optics

All media carry the CEBus control channel and transmit data at the same rate of approximately 8000bps.

With CEBus topology, a device can be located wherever is convenient to the user. CEBus does not use a central controller for managing the delivery of control messages; instead, control is distributed among the CEBus appliances and media routers. The CEBus standard does not specify a particular topology. All the appliance connection points on each medium are treated logically as if they were on a bus. This means that all the appliances on a particular medium sense a data packet at about the same time. All appliances read the destination address contained in the message, and only the appliance with the matching address reads the contents and acts or responds accordingly.

Remember the main performance parameters for the CEBus standard, such as the transmission rate of 8000bps and the fact that addresses are set at the factory. You should also be able to identify EIA-600 as the CEBus standard.

Ethernet Local Area Network Standards

The most popular standard for local area network technology in business and residential networks is the IEEE 802.XX series of Ethernet specifications. The IEEE Ethernet 802.3 standard provides performance specifications for network access control and the underlying physical hardware technology to support it. Ethernet technology has evolved for more than 30 years through progressive upgrades of the standard. These upgrades have improved the specification from its original version that transmitted information at 1Mbps to its newest versions that transmit data at up to 10 gigabits per second (Gbps). Each of the upgrades, however, was dependent on improved performance for the network cable required for each specification revision.

The Ethernet standard can be implemented on any of the following network media by simply installing an appropriate Ethernet NIC in each computer (one that is appropriate for the media type being used) and then connecting the cards to the network media. Almost all modern computer network operating systems support Ethernet network protocols and hardware.

Various Ethernet network adapters called NICs have been designed to operate with the following media types:

➤ UTP cable

➤ Coaxial cable

➤ Fiber-optic cable

➤ Wireless

In each case, the network interface card is equipped with appropriate connectors for each media type. Coaxial cable and associated Ethernet hardware used with the older 10BASE-2 and 10BASE-5 standards have not been used for new installations for several years.

The Ethernet 10BASE-T Specification

In 1990, a major advance in Ethernet standards came with introduction of the IEEE 802.3i 10BASE-T standard. It permitted 10Mbps Ethernet to operate over simple Category 3 unshielded twisted-pair (UTP) cable. 10BASE-T also permitted the network to be wired in a star topology that made it much easier to install, manage, scale, and troubleshoot. Each segment of the star topology has a maximum length of 100 meters.

> The IEEE Ethernet 10BASE-T standard indicates that the maximum recommended length for 10BASE-T Ethernet segments using UTP cable is 100 meters (328 feet). The exam might ask you to answer the question in increments of either feet or meters. You should know the conversion factor is feet = meters × 3.28.

The Ethernet UTP cable uses RJ-45 connectors, and the NIC utilizes a built-in RJ-45 port. The Ethernet cables connect directly into the NIC using this port. These advantages led to a vast expansion in the use of Ethernet. Because of the ease of installation and low cost, 10BASE-T has become the preferred standard for home office networks. An example of the configuration of a 10BASE-T Ethernet LAN is shown in Figure 1.9.

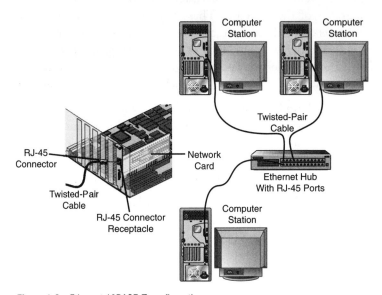

Figure 1.9 Ethernet 10BASE-T configuration.

The Ethernet 100BASE-T Specification

In 1995, the IEEE released a standard that improved the performance of Ethernet technology by a factor of 10 with the introduction of the 100Mbps 802.3u 100BASE-T standard. This version of Ethernet is commonly known as Fast Ethernet and has a transmission speed of 100Mbps. 100BASE-TX has become the preferred standard for implementing Fast Ethernet LANs. 100BASE-T4 is not used for new installations because it is considered an obsolete standard to the universal use of improved Category 5 cable. Three media types are supported by the Fast Ethernet standards:

► *100BASE-TX*—Operates over two pairs of Category 5 twisted-pair cable

► *100BASE-T4*—Operates over four pairs of Category 3 twisted-pair cable

► *100BASE-FX*—Operates over two multimode fibers

Coaxial Cable

Coaxial cable is a special type of cable that gets its name from the common axis shared by a center conductor. It contains an insulating material, a braided metal shield/outer conductor, and an outer insulating cover. The insulating layer, or *dielectric* (nonconductive) material, serves to isolate and maintain a common spacing between the center conductor and the outer metal shield conductor, as shown in Figure 1.10.

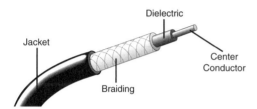

Figure 1.10 Anatomy of coaxial cable.

Coaxial cable was initially used to carry television broadcast programs over long distances and was later adapted for use in computer networking. Coaxial cable was the first media type used with LANs prior to the development and standardization of improved twisted-pair cable. With coaxial cable, the center conductor carries the information and the outer conductor is used as an electrical ground maintained at 0 voltage. Coaxial cable is therefore considered to be an unbalanced media.

Coaxial cable is manufactured in a wide range of electrical properties, sizes, qualities, and cost ranges. Standard, low-cost coaxial cable is used in cable TV distribution systems and backbone circuits for computer networks. The popular type of coaxial cable used for home cabling applications is type RG-6. Some types of coaxial cable used for home video and satellite antenna signal distribution such as RG6 and RG59 have a 75-ohm impendence value, whereas RG-58 type coaxial cable is used for legacy Ethernet (thinnet) networks and is rated at 50 ohms impendence.

Some types of coaxial cable are physically larger than twisted-pair cable. Coaxial cable works well in high-capacity, long-distance applications. However, it also costs more than twisted-pair cable and is less flexible and somewhat more difficult to install in conduit or in-building standard cable runs. It is often used to connect LANs over a circuit called a *backbone connection*. The backbone is where all connections tie together and network resources are most logically placed. In modern LAN environments, twisted-pair cable is the preferred media at the computer port connection point. Coaxial cable is used for video signal distribution in modern residential structured wiring designs.

Coaxial Cable Connectors

The most common type of connector used with coaxial cables is the Bayonet-Neill-Concelman (BNC) connector, shown in Figure 1.11.

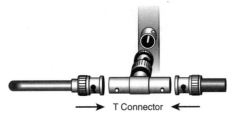

Figure 1.11 BNC connector with T-connector adapter.

Other types of adapters are also available for BNC connectors, including a T-connector, barrel connector, and terminator.

Fiber-optic Cabling

Fiber-optic cable has revolutionized long-distance communications. It works on a simple, basic principle that light waves traveling in a glass medium can carry more information at longer distances than copper wire media can. Fiber-optic cables contain a large number of glass fiber pairs enclosed in a protective sheath. Each fiber strand can be as small as .002". The glass fiber must be free from impurities to pass the light waves with a high degree of efficiency. Each fiber is encased in a layer of glass cladding that has a slightly different refractive index. The cladding serves to direct the light waves back into the central core. Figure 1.12 shows the composition of a fiber-optic cable. The two types of cable shown in the figure are single mode and multimode. The final outer jacket consists of a material to protect the cable from damage. Many fibers can be enclosed in an armored sheath. Fiber-optic cable designed for under the ocean installation is typically equipped with multiple layers of armor for additional strength.

Multimode fiber is the least expensive to produce but also has a lower performance figure because the inner core diameter is larger.

> You should be aware of the purpose of the various layers of fiber-optic cable. Remember that the cladding is used to shape and direct the light waves back into the central core of the cable. Also, you should review the basic differences between single mode and multimode fiber-optic cable.

Single mode cable is used for long-distance applications where the added performance is a major advantage.

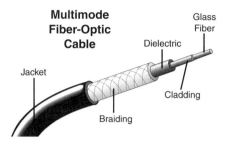

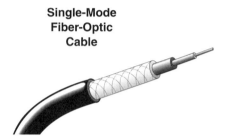

Figure 1.12 Composition of fiber-optic cable.

Fiber-optic systems are a form of hybrid optoelectric transmission systems. They first convert the electrical digital signal into an optical signal. The light source can be one of two types: light emitting diodes (LEDs) and lasers. Light sources are required to generate the light pulses that travel down the fiber-optic cable, and light detectors are used at the receiving end to convert the light waves to electrical signals.

Fiber-optic Connectors

The most common connector used with fiber-optic cable is an ST connector. It is barrel shaped, similar to a BNC connector. A newer connector, the SC, is becoming more popular. As illustrated in Figure 1.13, it has a squared face and is easier to connect in a confined space. Two helpful reminders for remembering these terms are as follows:

➤ *SC*—Try to remember "stick and click." You push the connector gently into the port and it clicks. This is the newer connector in use.

➤ *ST*—Try to remember "stick and twist." You have to twist the connector on to the port to make a connection. This is an outdated method of connectivity, but it's still in use on many networks.

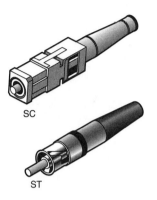

Figure 1.13 ST and SC connectors.

Fiber-optic Patch Cables

Fiber-optic patch cables are short, prefabricated cables with optoelectronic transmitters and receivers integrated into the cable assembly. TOSLINK, developed by Toshiba, is a brand name for a series of fiber-optical interconnects to achieve isolation and noise-free operation. TOSLINK digital optical audio and video applications include satellite receivers, DVDs, CDs, home entertainment systems, digital TVs, and other devices with built-in fiber-optic ports for digital input and output signaling. A TOSLINK cable and connector are shown in Figure 1.14.

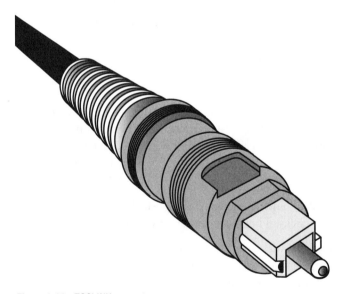

Figure 1.14 TOSLINK connector.

1.0—Computer Networking Fundamentals   27

Optoelectronic is a term used to describe any device that functions as an electrical-to-optical or optical-to-electrical transducer, or an instrument that uses such a device in its operation. This term is applicable to cables such as the Toshiba-developed TOSLINK connecting cables.

Remember that TOSLINK is a type of optoelectronic cable used to connect audio and video components. Also remember that ST and SC connectors are used with fiber-optic cables.

Network Media Considerations

Of all the wired network media types discussed for home networking applications, UTP is the most susceptible to having electronic noise introduced into the wire from environmental sources. This noise is caused by electromagnetic interference (EMI) produced from different electrical sources. Crosstalk is also a type of self-induced interference created by data signals traveling in adjacent wire pairs in a UTP cable. Each pair in a UTP cable is twisted to minimize the effects of crosstalk and EMI.

The shielding in the STP cable significantly increases noise protection. However, coaxial cable offers even greater protection, whereas fiber-optic cable provides extremely high levels of protection from induced noise.

Although UTP cable is the most noise-prone communication media, it also represents the least expensive and easiest media to install. On a cost-per-foot basis, twisted pair is the cheapest medium to install. For residential structured wiring design, UTP cable offers an extensive amount of compatibility with the existing telephone network as well as low-voltage audio distribution systems.

Be aware that UTP cabling is considered to be the most susceptible to electromagnetic interference among the various types of communications cable. It is being improved with progressively higher ratings for speed and EMI immunity with emerging Category 6 and Category 7 cable specifications. The higher categories of cable have more twists per foot, which helps increase the useable bandwidth of the cable. Fiber-optic cable remains the least affected by EMI. The exam might include questions about the relative immunity to EMI for various types of cable. The primary advantages offered by UTP is low cost and ease of installation.

It is important to remember that the specification for the maximum rated frequency for Category 5 cable is expressed as 100MHz. This is the maximum bandwidth rating established for a formal category certification when testing UTP cable. Keep in mind that it is expressed in megahertz (MHz) and not megabits per second (Mbps), which is a measure of data transmission bit rates.

UTP cables are designed to have different twist rates between each pair to minimize crosstalk between adjacent cables. More twists per foot raises the performance and the category rating, as well as increasing the immunity to EMI and crosstalk.

Equipment Location Considerations

The HTI+ Residential Systems objective 1.2 states that the test taker should have knowledge of equipment location considerations associated with residential networking. Examples of these considerations include

- Network equipment, including hardware and software such as the following:
    - Network interface cards
    - Servers (types and uses)
    - Printers
    - Gateways
    - Routers
    - Switches
    - Wireless access points
    - Firewalls
- Network equipment components, including the following:
    - Data ports
    - Jack types

Network Equipment: Hardware and Software

In the following section, the location considerations for home network equipment are discussed. Components include various types of servers, printers, and network interface cards. You will also learn the basics of locating network components included in the exam, such as gateways, routers, switches, wireless access points, and firewalls. The last part of the section deals with the location considerations for various types of ports and jacks.

Network Interface Cards

A NIC is a device that can be installed in a computer workstation or server to provide the interface between the computer and the network media. A

NIC is equipped with a connector that fits in a peripheral component interface (PCI) expansion slot, as shown in Figure 1.15. NICs that can be installed in Personal Computer Memory Card International Association(PCMCIA) expansion slots are also available for laptop computers.

Figure 1.15 shows the configuration and cabling options for installing a NIC in an expansion card slot. A wireless LAN uses a NIC that has a radio transceiver on the card that connects to other wireless NICs.

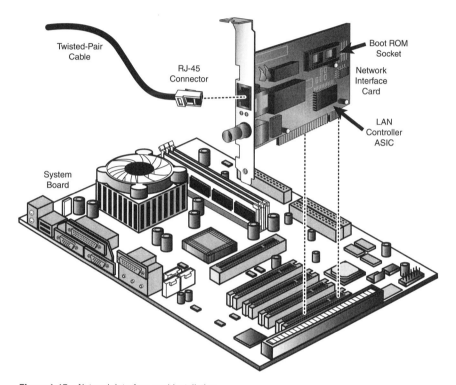

Figure 1.15 Network interface card installation.

For network cards to communicate with each other on a computer network, they must all conform to the same standard. Ethernet is the most widely used LAN standard. Over the past several years, Ethernet has become the preferred LAN standard for corporate and home office networking.

NICs are now available with the capability to operate at either 10Mbps, 100Mbps, or 1000Mbps (10/100/1000) speeds and are designed to fit the peripheral component interface (PCI) expansion slot in desktop PCs. Personal Computer Memory Card International Association (PCMCIA) network cards are designed to be used with laptop computers. NICs are available for both wired Ethernet networks and IEEE 802.11a/b/g wireless networks.

Servers

As discussed earlier, servers are high-performance computers used in client/server networks. This type of network management is employed widely in businesses in which data and resources must be controlled and secured. Normally, client/server networks have been more applicable to a business than a home networking environment. However, they are gaining ground in this environment due to the increasing number of residential subsystems being integrated together. Several guidelines exist for locating servers in a business environment, including

> *Servers are essential to business operations because they hold files and manage databases for a number of departments*—Servers are normally located in a locked, secure area and are accessible only to authorized network administrators.

> *Servers should be connected to a commercial power backup system called an uninterrupted power supply (UPS)*—This ensures continuous service without taking the network down during momentary power outages.

> *Servers should be located in an area where adequate protection is provided from temperature variations*—Incidents of over-temperature are, by far, the most commonly reported cause of computer downtime. Temperature rises during the day can cause server or network problems to appear during the middle of the workday and then disappear when the temperature of the room is lowered. Servers should be located where the temperature can be stabilized over a 24-hour period.

Heat buildup in a location where servers or network workstations are located can cause outages or unreliable operation. If the problem occurs in the middle of the day after a temperature rise in the computer area, it is a heat-related problem and should be corrected to prevent network downtime.

Printers

In a home network, printers are usually connected to the single host computer. Interface connector options include a parallel cable connection and a USB connector, and most printers come with both types of ports. In a home computer network, the printer is connected to one computer and shared with the other hosts on the network to allow printing services to be available from any computer. Printers can be located in any home office location that is convenient to the user.

Routers, Switches, and Gateways

Home computer networks use routers, switches, and gateways to provide Internet access sharing with more than one computer and to provide internal network management and connectivity with other computers in the residential network. A typical residential network might simply use a single gateway that encompasses all three devices, or it might use different combinations of all three devices. For example, you might have a separate switch with a router plugged in to it instead of a router with a switch built in to it. The exact type of connectivity device more or less depends on the type of residential network device you purchase (you might already have a preexisting hub or switch on site, particularly if the network is so big it requires both items to be separate for reasons of security or scalability).

The locations of these devices should be planned in conjunction with the type of Internet access that will be used (dial-up, cable, or DSL). These external services typically enter the home at a central wiring closet or patch panel where the home wiring interface is extended from the cross-connects to the office location for the two or more networked computers.

The routers, switches, or gateways should be located in an area that offers the shortest distance between the cross-connect point and the main host computer (server) that provides the interface to the gateway and the DSL/cable modem.

The terms *gateway* and *router* can be used interchangeably on the HTI+ exam. Remember that *gateway* is an earlier term used for a router and has multiple definitions for enterprise and home networking. *Gateway* has also become the common name for a device that connects external communication services such as an ISP to the internal distribution center inside the home.

Wireless Access Points

Access points need to be installed in locations that are not going to block radio transmissions, and they should be installed as high as possible to maximize the range. The wireless network computers should be within a 150-foot radius of the access point location. The access point should also be accessible for periodic maintenance or troubleshooting. The access point installation software is initially installed on one of the network hosts so that it can be used for configuring the access point with an address and protocol settings.

Firewalls

Firewall is the name applied to a computer program or hardware device that filters the information coming through the Internet connection into your private home network or computer system. Firewalls should be located between the private home network computer (or network computers) and the Internet. Firewalls are often contained as an integrated feature in home network gateways or routers. The home computers are connected to the router, and the router is then connected to the DSL or cable modem.

Network Equipment Components

Network equipment components are used to interconnect the input/output ports on computers and peripheral devices. They include cables, serial ports, parallel ports, and registered jacks. This section describes the standards and connectors used in home networking.

Data Ports and Cable Specifications

In some cases, port and cable specifications dictate how far various types of peripheral equipment can be placed from a computer. When using RS-232C serial ports, for example, cables lengths are limited to 50 feet or less. Therefore, peripheral equipment using this port must be installed accordingly. A description of the locations and uses of various types of home networking components, including ports and jacks, can be found in the next section.

In addition, the type of network being installed limits network cabling between different nodes and hubs. For example, in a network based on an Ethernet topology using CAT5 cabling, the maximum segment length is 100 meters. Standard port and cabling lengths associated with personal computer systems are listed in Table 1.2.

Table 1.2   Standard Port Cabling Lengths and Transmission Speeds

Port/Cable Specification	Maximum Length	Data Transfer Rate
RS-232 serial ports/cables	50 feet	300bps–9600bps
USB 1.1 ports/cables	5 meters	12Mbps
USB 2.0 ports/cables	5 meters	1.5Mbps, 12Mbps, and 480Mbps
Parallel ports/cables	2 meters	10Kbps
IEEE 1284 (EPP) ports/cables	10 meters	2Mbps
IEEE-1394 ports/cables	4.5 meters	100Mbps, 200Mbps, and 400Mbps
10BASE-T Ethernet UTP cables	100 meters	10Mbps
Infrared ports	2 meters	200 Mbps

The IEEE-1394 Serial Port Standard

The IEEE-1394 Port Standard, known as the FireWire port, is a very fast external serial bus standard that supports data transfer rates of up to 400Mbps. It is located on new network-enabled devices. Products supporting the IEEE-1394 standard have different names, depending on the company. Apple, which originally developed the technology, uses the trademarked name FireWire. Other companies use other names, such as i.Link and Lynx, to describe their IEEE-1394 Standard products. A single 1394 port can be used to connect up to 63 external devices.

It is easy to become confused with so many emerging cable standard names and associated brand names. Remember that FireWire, i.Link, and Lynx are brand names for IEEE-1394 Standard serial cables and TOSLINK is a fiber-optoelectronic cable developed by Toshiba.

The USB 2.0 Specification

Also referred to as Hi-Speed USB, USB Version 2.0 is an external serial bus that supports data rates up to 480Mbps. USB 2.0 is an extension of the USB 1.1 specification, is fully compatible with USB 1.1, and uses the same cables and connectors. Some computers come equipped with two or more USB ports. A single USB port can be used to connect up to 127 peripheral devices.

Hewlett-Packard, Intel, Lucent, Microsoft, NEC, and Philips jointly led the initiative to develop a higher data transfer rate than the 1.1 specification speed of 12Mbps to meet the higher bandwidth demands of developing technologies. The USB 2.0 specification was released in April 2000.

The Parallel Port

On PCs, the parallel port uses a 25-pin connector (type DB-25) and is used to connect printers, computers, and removable storage devices such as Zip drives. It is often called a Centronics interface/connector after the company that designed the original standard parallel port.

A newer type of parallel port, which supports the same connectors as the Centronics interface, is the EPP (enhanced parallel port) or ECP (extended capabilities port). Both of these parallel ports support bidirectional communication and transfer rates 10 times as fast as the Centronics port. These modes are defined in the IEEE 1284 "IEEE Std. 1284-1994 Standard Signaling Method for a Bi-directional Parallel Peripheral Interface for Personal Computers."

34 Chapter 1

The EPP is used primarily for CD-ROM drives, tape drives, hard drives, and network interface cards. The ECP, on the other hand, is used by printers and scanners. Most new printers are equipped with both EPP and USB ports to support legacy computer systems and newer computers that employ the improved high-speed USB ports.

Jack Types

Home networks use various types of jacks and connectors to install and configure network components and telecommunications equipment. The most common type of jack in residential network wiring is the familiar RJ type of jack. Common telephone-type jacks are also known as *registered jacks*, using the designation RJ-XX. Registered jacks are a series of telephone-type connection interfaces (receptacle and plug) that are registered with the Federal Communications Commission (FCC). The location of the most common jacks used in home networks are as follows.

RJ-11

RJ-11 is the type of connector used for terminating a telephone cable that connects the telephone to a standard wall outlet. This type of jack is often identified simply as a *phone jack*. The jack is located at the end of a telephone cable and is inserted into a standard telephone wall outlet. The other end of the cable is plugged in to a standard telephone set. RJ-11 jacks are also located on the cable that connects a dial-up modem installed in a PC and the telephone wall jack. The RJ-11 is a 6-4 connector because it actually is large enough to have six contacts on the plug and six wires in the jack, but only the four in the middle are supplied. The cord that is crimped into the plug usually has all four wires in it, but the middle two (red and green) are the only ones used by a single-line phone. The red wire is called the *ring* and the green wire is called the *tip*, and these two wires constitute the twisted-pair wire connection between the telephone and the central office.

Know that the tip (green) and ring (red) are the respective line 1 location color codes for telephone cable on an RJ-11 type jack.

Telephone jacks and plugs are wired to conform to the Uniform Service Ordering Code (USOC) numbers, originally developed by the Bell System and later endorsed by the FCC. One specific type of registered plug and jack hardware can be wired in different ways and have different USOC numbers. USOC has become an acronym, pronounced "you-sock," and jack wiring schemes are generally referred to as USOC codes.

The designation *RJ* means registered jack. The six-position jack can be wired for different RJ physical configurations. For example, the six-position jack can be wired as an RJ-11C (one-pair), RJ-14C (two-pair), or RJ-25C (three-pair).

RJ-45

RJ-45 jacks are four-pair-type connectors attached at both ends of an unshielded twisted-pair cable used to connect components in a home Ethernet network. They are slightly larger than an RJ-11 jack and are the standard termination jacks required for Category 5 UTP cable for Ethernet LANs. This type of connector is also used in homes wired according to structured wiring standards described in ANSI/TIA/EIA-570A. They are used to connect the various components of a home network, such as NICs, hubs, routers, and cable modems.

RJ-14

The RJ-14 is similar to the RJ-11, but the four wires are used for two phone lines. It is also used in residential telephone systems for two-line distribution. Typically, one set of wires (for one line) contains a red wire and a green wire, and the other set contains a yellow wire and a black wire.

RJ-31x

The RJ-31x is a type of telephone jack installed and located near the security system control box. It is installed between all the phones in the home and the outside telephone line. A cable is routed from the RJ-31X jack to the home security alarm control box. When an alarm condition in the security system is triggered, the RJ-31x jack seizes the telephone line and allows the control panel to dial a home security monitoring center. The jack connection avoids the possibility that a burglar might leave a phone off the hook to prevent the system from calling a monitoring service.

The RJ-31x jack is a special type of jack used with home security alarm systems. It is not considered to be a component of a home computer network system. It provides the capability to seize the outside phone line as a priority over any other uses of the phone system in the home. Keep in mind the purpose and location of this type of connector.

Network Physical Devices

HTI+ Residential Systems objective 1.3 states that the test taker should have knowledge of the physical products associated with residential networking. Examples of these products include

➤ Wireless access points

➤ Computers

➤ Printers

➤ Residential gateways

➤ Network-enabled devices and appliances

In this section, you will gain an understanding of the characteristics and physical features of the basic components of a home network system.

Wireless Access Points

The access point is the key component of a wireless network. The name *access point* is referenced in the IEEE 802.11b wireless LAN standard. It acts as a bridging device between a wireless computer network and a wired Ethernet network, as illustrated in Figure 1.16. It has antennas, a processor, software, and a radio receiver/transmitter that is used to communicate with the wireless LAN PCs and laptop computers. It connects with the wired LAN hub using a UTP cable, shown in Figure 1.15. The hub provides connectivity between the access point and the wired Ethernet LAN.

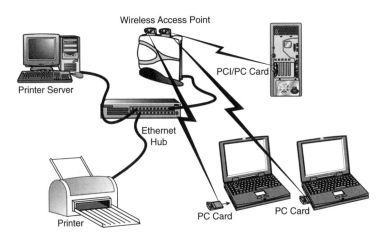

Figure 1.16 A wireless LAN and the access point.

When an access point is used with an 802.11b wireless network, it is defined in the 802.11 standard as operating in what is known as *infrastructure* mode. For the access point to communicate with wireless network adapters installed in PCs, all the components need to be running in the same mode and be set to the same service set identifier (SSID).

Being able to combine both the wired and wireless technologies is useful in a home office network when working with wired desktop PCs and one or more laptop computers. The wireless laptop can be moved around the house while always being connected to the network.

You will need to know that the purpose of an access point within a wireless LAN configuration is to provide a bridge between the wireless LAN computers and the wired LAN, or the DS. *Access point*, *service set identifier*, and *distribution system* are terms used in the IEEE 802.11b Wireless LAN standard.

Computer Systems

A typical personal computer system is depicted in Figure 1.17. A typical PC system includes the following components:

➤ A keyboard

➤ A mouse

➤ The computer system unit

➤ A monitor (or display unit)

➤ Speakers

➤ A printer

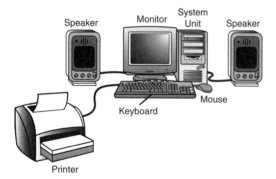

Figure 1.17 A typical computer system.

Modern PC system unit architectures are modular by design, offering components that can be upgraded or replaced during the expected life of the unit. Figure 1.18 shows the basic architecture of a PC. Replaceable units include storage devices such as floppy disk drives and hard disk drives. Expansion slots are designed to allow custom features to be incorporated using plug-in expansion cards, such as modems, sound cards, network interface cards, video adapters, and game port cards.

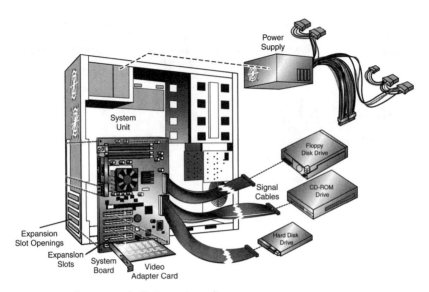

Figure 1.18 Components inside the system unit.

Printers

Printers are used in home office networks to produce hard copy text and high-resolution color graphics. Printers are available with one or more of the following interfaces:

➤ *Parallel port*—This is used with older printers. The IEEE 1284 standard, "Standard Signaling Method for a Bi-directional Parallel Peripheral Interface for Personal Computers," was approved in March 1994. It uses a 25-pin cable with 17 signaling lines and 8 ground lines. It is a relatively slow interface at 10Kbps. Most new printers now come with the improved EPP and USB ports.

➤ *Universal serial bus ports*—Most new PCs and printers are equipped with USB ports. USB 1.1's data rate of 12Mbps is sufficient for PC peripherals such as telephones, digital cameras, keyboards, mouse devices, digital joysticks, tablets, wireless base stations, cartridges, tape drives, floppy drives, digital speakers, scanners, and printers. The higher bandwidth of USB 2.0 will extend the capabilities to 480Mbps.

Ink-jet Printers

Ink-jet printers produce characters and color graphic printing by squirting a precisely controlled stream of ink drops onto the paper. Each ink drop must be controlled very precisely in terms of its aerodynamics, size, and shape; otherwise, the drop placement on the page becomes inexact and the print quality degrades. Ink-jet printers are relatively inexpensive and provide the capability to print excellent color images as well as black print for text printing.

Laser Printers

Laser printers employ a highly focused laser beam to produce images in the form of spots on a photosensitive rotating drum. The spots on the drum are used to attract toner material to the written areas of the drum. Paper is then fed past the rotating drum and the toner image is transferred to the paper. During the final step of the print cycle, a pair of compression rollers and a high-temperature lamp fuse the toner to the paper.

Laser printers tend to be more expensive than ink-jet printers. However, they can be used to produce very high-resolution copies at very rapid page rates.

Residential Gateways

Residential gateways are relatively new home networking devices that integrate several functions into a single device. Combining the functions of a hub or switch, a router, and a firewall, home gateways attempt to combine all the essential features of home networking in one integrated, easy-to-use package. Some newer gateways can also include a DSL or cable modem and typically come with one RJ-45 jack to plug in to a cable or DSL modem. They let you share Internet access by internally managing TCP/IP addresses (covered later in the section "The IP Address Format.")

Wireless gateways combine the functions of an access point and a gateway by allowing all nodes on a home wireless 802.11b network to share a common Internet access modem. Most of the devices now on the market can be configured via an Internet browser on a PC located on the same network. Gateways can also provide a function called *network address translation (NAT)* that enables all the computers on a network to access the Internet using a single IP address (the gateway IP address). The residential gateway sets up a subnetwork using TCP/IP addresses not viewable from the Internet. This enables a home network computer to have a private IP address that is hidden from outside discovery, direct connection, or attack.

Additional information relating to NAT is covered later in this chapter, in the section "Hardware Configuration and Setting."

The residential gateway is a relatively new product in the home networking product line. It differs from the traditional network gateway used to connect dissimilar commercial networks and protocols such as TCP/IP networks and IBM mainframe SNA protocols. Remember that the main purpose of a residential gateway is to integrate several functions, including those of a hub or switch, a router, and a firewall, into a single unit.

Network-enabled Devices and Appliances

A growing variety of home appliances capable of being controlled with a network are becoming available for the home technology integration market. The following are some of the products available with network capability:

➤ Home theater audio/video entertainment systems

➤ Service appliances that prepare food

➤ Lighting systems

➤ Systems and devices that maintain the internal environment, such as heating, ventilating, and air conditioning

➤ Devices that keep the home secure from intrusion or damage from internal and external manmade or natural events

Universal Plug and Play Initiative

Research and development for network-enabled devices is an area of interest for a number of corporations that manufacture computer network equipment. This is the basis for an industry initiative called Universal Plug and Play (UPnP). It is a standard that uses Internet and Web protocols to enable devices such as PCs, peripherals, intelligent network-enabled appliances, and wireless devices to be plugged in to a network and automatically know about each other. With UPnP, when a user plugs a device in to the network, the device configures itself, acquires an IP address, and uses a discovery protocol based on the Internet's Hypertext Transfer Protocol (HTTP) to announce its presence on the network to other devices. For example, if you had a camera and printer connected to the network and needed to print a photograph, you could press a button on the camera and have the camera send a "discover" request asking whether there were any printers on the network. The printer would identify itself and send its location in the form of a uniform resource locator (URL). UPnP networking is also media independent, and UPnP devices can be implemented using any programming language, and on any network operating system.

1.0—Computer Networking Fundamentals 41

Plug and Play (PnP) is a capability developed by Microsoft for its Windows 95 and later operating systems that gives users the ability to plug a device in to a computer and have the computer recognize that the device is there. With Microsoft's participation, Plug and Play has been replaced by the Universal Plug and Play (UPnP) standard. Be aware that PnP and UPnP are different standards.

UPnP is an industry standard for allowing UPnP-compliant network devices to be plugged in to a network and automatically start communicating with other UPnP devices.

Remember that the acronym for Universal Plug and Play is UPnP and keep in mind that UPnP is a later evolution of Microsoft's PnP standard introduced with the Windows 95 Network OS. The exam might include questions concerning the purpose of both the PnP and UPnP network device standards.

Hardware Configuration and Setting

Objective 1.4 of Residential Systems states that the test taker should have knowledge of network configuration. Examples of items that HTI+ technicians should be able to configure include

➤ Operating system configuration and user settings

➤ IP addressing

➤ Dynamic Host Configuration Protocol (DHCP)

➤ Network addressing translation

➤ Firewall configuration and filtering

➤ NIC configuration

Operating System Configuration and User Settings

After the computer's components and peripherals are connected together and their power connectors have been plugged in to a receptacle, the system is ready for operation. However, one thing is still missing—the software. Without network system software to oversee its operation, the most sophisticated computer hardware is worthless.

For the HTI+ technician, software can be divided into two general classes:

➤ Network operating system (NOS) software

➤ Applications software

As with hardware devices, many aspects of software must be configured specifically to match the application the software is being asked to support. Fortunately, many of the operating systems currently in production offer automated (or semi-automated) wizard programs to guide the installation and configuration of software applications and key components. However, as this HTI+ objective points out, some key software components and services exist that residential integration technicians must be able to configure, including

➤ Internet Protocol (IP) addressing

➤ Dynamic Host Configuration Protocol servers

➤ Network address translators

➤ Firewalls

➤ NICs

IP Addressing

The modern Internet we use today began as an experiment funded by the U.S. Department of Defense (DoD) to interconnect DoD-funded research sites in the United States. The 1967 ACM meeting was also where the initial design for the so-called ARPANET—named for the DoD's Advanced Research Projects Agency (ARPA)—was launched. This initiative led to the government-funded development of a suite of protocols called the Transmission Control Protocol/Internet Protocol (TCP/IP) for the first large, packet-switched network known today as the public Internet. All computers, routers, and network-enabled devices that access the public Internet today must be assigned an Internet Protocol (IP) address. A nonprofit organization called the International Corporation of Assigned Names and Numbers (ICANN) issues all IP addresses. In addition, a growing number of LANs communicate using the TCP/IP Protocol with a special set of private IP addresses.

The Internet Corporation for Assigned Names and Numbers is a technical coordination body for the Internet. Created in October 1998, ICANN assumed responsibility for a set of technical functions previously performed under U.S. government contract by Internet Assigned Numbers Authority (IANA). Specifically, ICANN coordinates the assignment of Internet domain names, IP addresses, protocol parameters, and port numbers.

Know that the TCP/IP Protocol must be installed on a computer that has direct access to the Internet. Also, each computer that exchanges information with other computers connected to the Internet must have a valid public IP address. A computer that is connected to a private network can also use IP addressing, but it must use an address that is allocated to private networks.

The IP Address Format

For easier reading, IP addresses are split into four 8-bit fields called *octets*. Each octet is sparated by a dot. An IP address therefore has the format xxx.yyy.zzz.aaa. With 8-bit fields, each octet has a range of 0–255 This format of specifying addresses is referred to as the *dotted-decimal notation*. The decimal numbers are derived from the binary address the hardware understands.

Each IP address consists of two parts: the network address and the host address. The network address identifies the entire network, whereas the host address identifies an intelligent node within the network (a router, server, or workstation).

Three classes of standard IP addresses are supported for LANs: Class A, Class B, and Class C. These addresses occur in four-octet fields, as follows:

- *Class A addresses*—These addresses are reserved for large networks and use the last 24 bits (the last three octets or fields) of the address for the host address. The first octet always begins with a 0, followed by a 7-bit number. Therefore, valid Class A addresses range between 001.x.x.x and 126.x.x.x. This permits a Class A network to support 126 networks with nearly 17 million hosts (nodes) per network. 127.0.0.0 is a Class A address but is reserved for loopback testing (typically, 127.0.0.1).

- *Class B addresses*—These are assigned to medium-size networks. The first two octets can range between 128.x.x.x and 191.254.0.0, and the last two octets contain the host addresses. This enables Class B networks to include up to 16,384 networks with approximately 65,534 hosts per network.

- *Class C addresses*—These are normally used with smaller LANs. In a Class C address, only the last octet is used for host addresses. The first three octets can range between 192.x.x.x and 223.255.255.0. Therefore, the Class C address can support approximately two million networks with 254 hosts each.

Reserved IP Addresses

As noted in the previous addresses, not all possible values are allowed for each octet in the host part. This is because host numbers with octets of all 0s or all 1s are reserved for special purposes. An address in which all host part bits are 0 refers to the network, and one in which all bits of the host part are 1 is called a *broadcast address*. This refers to all hosts on the specified network simultaneously. Thus, 149.76.255.255 is not a valid host address but refers to all hosts on network 149.76.0.0.

Private Network Reserved Addresses

Several regions of IP network addresses have been reserved for private networks. These addresses are used for private networks that do not connect to the Internet. Therefore, these addresses could be used and reused within different corporations because none of the computers in different corporate networks could be reachable from the Internet nor to each other. These private network addresses are as follows:

10.0.0.0	to	10.255.255.255
169.254.0.0	to	169.254.255.255
172.16.0.0	to	172.31.255.255
192.168.0.0	to	192.168.255.255

The address groups reserved for private networks, as well as the 127.x.x.x group reserved for testing networks, are important and should be memorized. The reserved addresses are used for setting up private home networks and are used by gateways and routers for hiding IP addresses from the Internet through the use of NAT.

Subnets

Sections of the network can be grouped together into *subnets* that share a range of IP addresses. Subnets are created by masking off (hiding) the network address portion of the IP address on the units in the subnet. This, in effect, limits the mobility of the data to those nodes in the subnet because they can reconcile only addresses from within their masked range. The three common reasons to create a subnet are as follows:

- ➤ *To isolate one segment of the network from all the others*—Assume, for example, that a large organization has 1,000 computers, all of which are connected to the network. Without segmentation, data from all 1,000 units would run through every other network node. The effect of this would be that everyone else in the network would have access to all the data on the network and the operation of the network would be slowed considerably by the uncontrolled traffic.

- ➤ *To efficiently use IP addresses*—Because the IP addressing scheme is defined as a 32-bit code, only a certain number of possible addresses exists. Although 126 networks with 17 million customers might seem like a lot, in the scheme of a worldwide network system, that's not very many addresses to go around.

- ➤ *To use a single IP address across physically divided locations*—For example, subnetting a Class C address between remotely located areas of a campus would permit half of the 253 possible addresses to be allocated to one location and the other half to be allocated to hosts at the second location. In this manner, both locations could operate using a single Class C address.

Using TCP/IP Utility Programs

The TCP/IP Protocol suite contains a number of small software troubleshooting programs called *utilities*. Some of the utilities are useful for viewing the status of computers connected to a home network; other utilities are used to check the path packets are taking to reach a distant host on a wide area network (WAN).

Most of the programs are run from the command prompt line by entering the name of the program followed by options that are called *switches*. Each program name is referred to as a *command*. Some of the command names are short abbreviations of the functional names of the program to limit them to eight characters. A summary of the various TCP/IP utilities follows:

- ▶ IPCONFIG—Provides information about the TCP/IP settings on the computer you are using. It displays the current IP address and MAC hardware address of your computer. This command works with most operating systems except Windows NT and Windows 2000. To see all the information, you can add the /all switch after ipconfig.

- ▶ WINIPCFG—The Windows utility that displays all the information contained in IPCONFIG. The information is shown on the screen in an easy-to-read format using a Windows-type graphic display. This command does not work with Windows 2000 or Windows XP operating systems.

- ▶ ARP—Stands for Address Resolution Protocol. ARP is used to map logical IP network addresses to the physical addresses of computers on the network.

- ▶ NETSTAT—Shows all the current network connections and the state of the connections. It displays the local computer port numbers as well as the local IP address and the type of protocol being used for the connection.

- ▶ NBTSTAT—NetBIOS over TCP statistics is used to display the NetBIOS names for all the computers on a network, as well as their IP addresses.

- ▶ NET VIEW—Lists all the computers connected to the LAN. Using the command with the name of the computer to which you are connected also shows all the shared resources on that computer.

- TRACERT—A Microsoft-specific tool that enables you to see a path a data packet is taking across the network. When an IP address is specified, TRACERT lists the number of hops to the destination address along with the IP address of each router along the way.

- PING—A TCP/IP utility used to determine whether a remote computer can be reached on a network. The name stands for packet Internet groper. PING sends out an echo request message in the form of a data packet to the host computer you want to reach. It then displays the results of the echo message onscreen. It is in effect saying to the remote computer, "Are you there?" The PING tool sends a packet each second and prints one line for every packet received. When PING is finished, it displays information about the time (in milliseconds) it took for the packet to get to the remote host and a response packet to return. The time can vary due to the amount of traffic on the network.

To run a TCP/IP utility program from the command prompt, select Start, All Programs, Accessories. Then click Command Prompt and type in the name of the utility program. When using the **PING** command, packet filtering policies and settings on routers, firewalls, or other types of security gateways might prevent the forwarding and echoing of test packets. Not all utility programs listed here work with Windows XP.

Internet Domain Names

The IP addresses of all the computers attached to the Internet are tracked using a listing system called the domain name system (DNS). DNS is a database organizational structure whereby higher-level Internet servers keep track of assigned domain names and their corresponding IP addresses for systems on levels under them (that is, it is responsible for resolving host names to particular IP addresses). This system evolved as a way to organize the members of the Internet into a hierarchical management structure.

The DNS structure consists of various levels of computer groups, called *domains*. Each computer on the Internet is assigned a domain name, such as mic-inc.com. The mic-inc portion of the name is the user-friendly domain name assigned to the Marcraft site. The .com notation at the end of the address is a top-level domain that defines the type of organization or country of origin associated with the address. In this case, the .com designation identifies the user as a commercial site.

The following list identifies the Internet's top-level domain codes:

- .com—Commercial businesses
- .edu—Educational institutions

- `.gov`—Government agencies
- `.int`—International organizations
- `.mil`—Military establishments
- `.net`—Networking organizations
- `.org`—Nonprofit organizations

On the Internet, domain names are specified in terms of their fully qualified domain names (FQDNs). An FQDN is a human-readable address that describes the location of the site on the Internet. It contains the hostname, domain name, and top-level domain name. For example, the name www.espn.com is an FQDN.

The letters www represent the hostname, which specifies the name of the computer that provides services and handles requests for specific Internet addresses. In this case, the host is the World Wide Web. Other types of hosts include FTP and HTTP sites.

Be aware that the DNS associates domain names with IP addresses. Also, you should remember that the letters **www** represent the hostname, which specifies the name of the computer that provides services and handles requests for specific Internet addresses. The terms in the FQDN include the hostname, domain name, and top-level domain name. As an example, **www** is the hostname for the World Wide Web.

Uniform Resource Locator

The uniform resource locator is usually abbreviated as simply URL. Think of a URL as an international locator address that can lead you to any file on any computer system anywhere in the world. The URL is also known simply as the Web site address.

Unlike a common postal address, however, a URL is written backward. It starts by naming the protocol that is to be used. Several standard protocols, such FTP, exist on the Web, but the one you'll encounter most frequently is HTTP (the protocol that tells computers the procedure for transporting hypertext documents).

A typical URL might appear as follows:

http://www.mic-inc.com/selfstudy/inet/

This URL indicates that the protocol to be used for transporting the document is HTTP. The next portion tells you that the server on which the document is located is www.mic-inc.com. The address indicates that the folder in which the document is located is selfstudy, which contains the document inet.

Dynamic Host Configuration Protocol

The Dynamic Host Configuration Protocol (DHCP) enables individual computers on an IP network to obtain their IP address configurations from a DHCP server. This reduces the work necessary to administer a large IP network. The DHCP server allocates an IP address to the PC from one of the *scopes* (the pools of addresses) it has available. Each DHCP scope is used for a different TCP/IP network segment. On networks with routers that support DHCP, extra information is added to the request by the router to tell the server from which network the request came. The DHCP server uses this information to pick an address from the correct scope and then replies to the client, allocating it the TCP/IP address and settings required. This is useful for a network that has laptop computers joining the network on a periodic basis because it enables a computer to have a different IP address each time it accesses a network. This simplifies network administration and permits more computers to use a pool of numbers. Most Internet service providers (ISPs) use DHCP to temporarily assign an IP address to each user for the length of time he is connected; then they reuse that number for another customer.

The DHCP is an important resource for setting the configuration of a home office network. The basic purpose of the DHCP is to allocate IP addresses from a pool of addresses. On large networks, the task is usually allocated to a dedicated DHCP server.

The most important configuration parameter carried by DHCP is the IP address. A computer must initially be assigned a specific IP address that is appropriate for the network to which it is attached. It must also be an IP address that is not assigned to any other computer on that network. If a computer moves to a new network, it must be assigned a new IP address for that new network. DHCP can be used to manage these assignments automatically.

The Transmission Control Protocol

The Transmission Control Protocol (TCP) provides for the guaranteed delivery of packets on the Internet. Whereas the IP address identifies the source and destination of the host, the TCP ensures that each packet will be received in the correct order. Every packet is therefore acknowledged and must be error free. If it is in error, a repeat transmission is requested. TCP maintains a flow control mechanism to ensure that the transmission of packets is maintained properly so as not to overflow the buffers at the destination node. TCP is called a *reliable connection-oriented protocol*, and it uses more network resources than UDP, as discussed in the following section.

TCP identifies the port number, which is the final step in addressing between the source and destination computers. Port numbers are needed to connect the two computers to the correct program or process that is running on both computers.

User Datagram Protocol

The User Datagram Protocol (UDP) is a connectionless protocol that is called *unreliable* because it provides no guarantee that packets will arrive at the destination without errors or that they will arrive at all. You might wonder why an application would use UDP with this type of utility. UDP has minimal overhead and is used for routine network management functions that are easy to duplicate for repetitive updates if lost.

Routine messages are often exchanged by network management system software applications that do not need reliable service or want the extra traffic associated with the use of TCP. These are usually short messages that can be repeated periodically. UDP packets do not put a lot of demand on the network because no acknowledgment packets are needed with UDP, resulting in a minimal traffic load on the network. In addition, some types of applications run on host machines that do not require connection-oriented service because they have a way of providing it themselves.

UDP, as is the case with TCP, provides the 16-bit port number in the header bit field. The port number identifies the actual process or application that is running on the source and destination machine. This number is handed off to the upper layers above the Transport layer to match the packet to the correct process that is running in the destination host computer. A further definition of ports and how they are categorized is discussed next.

Well-known Ports

High-layer applications running on network computers are referred to by port identifiers in IP packets. The port identifier and IP address together form a *socket*, and the end-to-end communication between two hosts is uniquely identified on the Internet by the source port, source address, destination port, and destination address.

Port numbers are always specified as 16-bit numbers. Port numbers in the range 0–1023 are called *well-known ports*, and these port numbers are assigned to the server side of an application. Port numbers in the range 1024–49151 are called *registered ports* and are numbers that have been publicly defined as a convenience for the Internet community to avoid vendor conflicts. Server or client applications can use the port numbers in this range. The remaining port numbers, those in the range 49152–65535, are called *dynamic* or *private ports* and can be used freely by any client or server. The Post Office Protocol 3 (POP3) and Simple Mail Transport Protocol (SMTP) are used by mail servers. Many other ports, such as HTTP and FTP, are frequently used port numbers.

Some well-known port numbers are listed in Table 1.3.

Table 1.3 Well-known Port Numbers

Port #	Common Protocol	Service	Port #	Common Protocol	Service
7	TCP	echo	80	TCP	http
9	TCP	discard	110	TCP	pop3
13	TCP	daytime	111	TCP	sunrpc
19	TCP	chargen	119	TCP	nntp
20	TCP	ftp-control	123	UDP	ntp
21	TCP	ftp-data	137	UDP	netbios-ns
23	TCP	telnet	138	UDP	netbios-dgm
25	TCP	smtp	139	TCP	netbios-ssn
37	UDP	time	143	TCP	imap
43	TCP	whois	161	UDP	snmp
53	TCP/UDP	dns	162	UDP	snmp-trap
67	UDP	bootps	179	TCP	bgp
68	UDP	bootpc	443	TCP	https (http/ssl)
69	UDP	tftp	520	UDP	rip
70	TCP	gopher	1080	TCP	socks
79	TCP	finger	33434	UDP	traceroute

When setting up email accounts with an ISP on a network, various protocols such as POP3 and SMTP are used. These protocols enable the user to receive and send email with her favorite email software, such as Microsoft's Outlook Express, Eudora, Netscape Email, and many others.

Memorize the port numbers for the frequently used protocols: HTTP (80), SMTP (25), FTP Control (20), and POP3 (110).

Network Address Translation

Network address translation (NAT) is used to interconnect a private network using unregistered IP addresses with a global IP network using a limited number of registered IP addresses. They provide a method for hiding private IP addresses from the Internet while still permitting the computers on that network to access the Internet.

NAT works by converting a range of IP numbers to another range of IP number(s). The IP numbers to be converted are those assigned to the individual machines on an organization's internal network. Although they can be any IP numbers, they most commonly come from one of the ICANN unassigned ranges (10.x.x.x, 172.16.x.x–172.32.x.x, or 192.168.x.x).

The single IP number to which those numbers are converted, by contrast, is located on the outside network and is a valid, routable IP number. From the inside looking out, the machines can access any host on the Internet directly, but from the outside looking in, it appears that all inbound and outbound TCP/IP traffic is originating from the single valid IP number on the NAT. In essence, the NAT is a special type of IP routing device.

NAT is becoming the preferred method for residential networking consumers to connect several PCs to a gateway using a single IP address for all the computers on the home or small office network. NAT is implemented on most of the popular home office networking products, such as routers, firewalls, and gateways.

Firewall Configuration and Filtering

A *firewall* is a feature that blocks unauthorized, outside users on a public Internet from accessing the intranet site on a private network. A firewall is a combination of hardware and software components that provides a protective barrier between networks with different security levels.

Intranets

Most private corporate intranets require that the internal network segment have a protective gateway to act as an entry and exit point for the segment. In most cases, a router serves as the gateway; however, some networks employ a firewall to act as the gateway to the outside.

Extranets

A relatively new term, *extranet*, is being used to describe intranets that grant limited access to authorized outside users, such as corporate business partners or vendors who might need limited access to a private network. The extranet is that portion of the corporate private network that serves designated users who exist outside the corporate network infrastructure. In this environment, firewalls are used to limit the level of access permitted from the outside.

 A firewall is a computer program or router used for protecting computers on a private network from intrusion by unauthorized users or hackers using the Internet. They are also used to protect corporate intranets and extranets.

Administrators typically configure the software for firewalls so that they pass data to and from only designated IP addresses and TCP/IP ports. In most cases, these ports are the well-known port addresses covered earlier in this chapter. When a private network is unable to communicate with other designated Web sites or URLs, the problems can be caused by the filter settings of the firewall.

Network Interface Card Configuration

Each adapter must have an adapter driver program loaded in its host computer to handle communications between the system and the adapter. These are the Ethernet and Token Ring drivers loaded to control specific types of LAN adapter cards.

In addition to the adapter drivers, the network computer must have a network protocol driver loaded. This program can be referred to as the *transport protocol* or just as the *protocol*. It operates between the adapter and the initial layer of network software to package and unpackage data for the LAN. In many cases, the computer might have several protocol drivers loaded so that the unit can communicate with computers that use other types of protocols.

Typical protocol drivers include the Internetworking Packet Exchange/Sequential Packet Exchange (IPX/SPX) model produced by Novell and the standard TCP/IP developed by the U.S. military for its ARPA network. Figure 1.19 illustrates the various LAN drivers necessary to transmit, or receive, data on a network.

Many newer network cards possess Plug and Play capabilities. The installation and setup of NICs has become a simple procedure with most Microsoft operating systems. Most NICs come with an installation CD that includes the drivers and related setup software for each type of NIC.

For Windows operating systems, using the Network applet in the Control Panel provides access to install the appropriate client, adapter, TCP/IP, and service drivers.

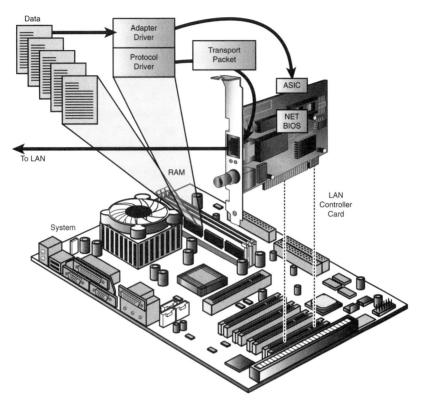

Figure 1.19 Various LAN drivers.

Device Connectivity

Objective 1.5 of Residential Systems states that the test taker should have knowledge of device connectivity concerns. These concerns include such items as the following:

- Interfaces with legacy devices and systems
- Network design access points
- Termination points

Interfaces with Legacy Devices and Systems

In newer personal computer systems, the BIOS and operating system use Plug and Play (PnP) techniques to detect new hardware that has been installed in the system. These components work together with the device to allocate system resources for the device. In some situations, the PnP logic is incapable

54 Chapter 1

of resolving all the system's resource needs and a configuration error occurs. In these cases, the user must manually resolve the configuration problem.

In PnP systems, the adapter cards and peripheral devices can identify themselves to the system during the startup process, along with supplying the system with information about what type of device they are, how they are configured, and which resources they require. In addition, these cards must be capable of being reconfigured by the system software if a conflict is detected between it and another system device.

Most early adapter cards employed hardware jumpers, or configuration switches, that enabled them to be configured specifically for the system in which they were being used. The user had to set up the card for operation and solve any interrupt or memory addressing conflicts that occurred. Such cards are referred to as *legacy cards*.

Network Design Access Points

Residential and home office network design involves planning how the components of a particular network will be configured and connected together.

Figure 1.20 shows a basic residential network connectivity design showing all the network access points.

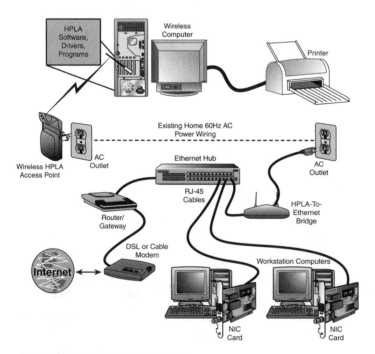

Figure 1.20 Home network connectivity.

1.0—Computer Networking Fundamentals 55

This design illustrates the connections required for a three-computer home office Ethernet network with shared Internet access. The connection access points for this home office network example are as follows:

➤ *Each networked computer has an Ethernet 10/100 NIC installed and configured in a PCI expansion slot*—Each computer is connected to a port on the Ethernet hub or router using Category 5 cable and RJ-45 connectors. This configuration enables a 100BASE-T LAN to serve the home office as well as the Internet access port.

➤ *The hub or router serves as the gateway to the DSL or cable modem*—It is connected to the router port using Category 5 cable and RJ-45 connectors. This enables the high-speed Internet access port to be shared by all the networked computers. Internet access is available through NAT included in the router, so the Internet "sees" only the IP address of the router and the IP addresses of the computers are hidden from the Internet. Normal file and printer sharing is provided by the installed network operating system on each computer.

➤ *The cable or DSL modem provides the high-speed connection to the ISP*—The ISP address for the router is provided during setup by the ISP. The modem is connected to the hub using a standard Category 5 cable and RJ-45 connectors. The external access connection is either a phone line with RJ-11 connectors or F-type cable connectors using RG-6-type coaxial cable.

➤ *The printer is connected to any computer on the LAN by a parallel or USB cable*—The printer is shared by the selected host with the other two computers.

The example shown for home network design access points is typical, although it might have several variations. The use of a gateway can include the router, hub, and modem into a single integrated unit. The DSL modem also uses an internal splitter to provide a standard telephone jack and the high-speed data port for the Internet connection. The advantage offered by the NAT-type router is that no Internet server is used to control access by the other two computers. An Ethernet cable/DSL router (also known as a *gateway* or *switch*) provides access to all computers at the same time without any other computer needing to be turned on.

A modem is required to connect a computer or a home network to the Internet. Modem types vary according to the type of connection used, such as dial-up analog modems, cable modems, or DSL modems.

Termination Points

Termination points for a computer home office network include the following areas:

➤ The common interface points between all low-voltage cables entering and exiting the home that are connected to the home network

➤ All cabling, connectors, patch panels, and termination blocks that provide an interconnection with computers and computer network components

The following paragraphs summarize the key points you will need to know concerning home network cable termination points.

Crossover Network Patch Cable Termination

The simplest type of network in a home consists of two computers connected together. You can connect the commuters without a hub, using a crossover cable. The crossover cable pin connections are shown in Figure 1.21.

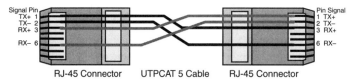

Figure 1.21 Crossover cable connection.

The crossover cable connects the TX (transmit) and RX (receive) functions for each NIC connection in a two-computer direct connection. A straight-through cable does not perform the crossover because it is designed to connect to a hub when more than two computers are used in an Ethernet star topology. The crossover connection is usually performed in the hub; therefore, a straight-through cable must be used.

Star Topology Termination Points

A star topology has been defined as the standard for network cabling in accordance with ANSI/TIA/EIA-568-A. In addition, the IEEE 802.3 10BASE-T standard requires a star topology using UTP cable. A star topology calls for networked computers and telephone outlets to be wired directly to a central equipment hub that establishes a central termination point. When a star topology is used, locating and isolating wiring problems is much easier than it is with daisy-chain wiring. The concept of a star topology is shown in Figure 1.22, where any device (including computers, telecommunications equipment, and video and audio cables) is terminated at a common location.

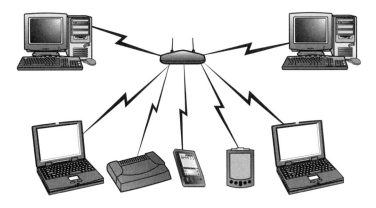

Figure 1.22 Star topology.

Insulation Displacement Connection Type Termination Blocks

Insulation displacement connection termination is the recommended method of copper termination recognized by ANSI/TIA/EIA-568-A for UTP cable terminations. These termination devices are the popular choice for home network star topology connection points. Commonly called *punch-down* connections, these connections require the use of a small punch-down tool to properly secure the cable to the terminal block. Punch-down connections remove or displace the conductor's insulation as it is seated in the connector. During termination, you press the cable between two edges of a metal clip, which displaces the insulation and exposes the copper conductor. This ensures a solid connection between the copper conductor and terminating clip. Screw-type terminal faceplates commonly used in voice applications are not recommended by ANSI/TIA/EIA-568-A for UTP terminations.

Type 66 Punch-down Termination Blocks

The Type 66 IDC termination block is the choice for connecting voice applications such as PBX, Key Telephone Systems (KTS), and some LANs. Several manufacturers of the Type 66 termination block designs have updated their termination blocks to handle high-speed data applications and comply with ANSI/TIA/EIA-568-A Category 5 specifications. They typically contain a plastic block 3" wide by about 10" high with 50 rows of four small, forked clamps sticking out. The four clamps in each row form two paired terminals; the installer can punch a wire into the forked clamp on the left, punch another into the forked clamp on the right, and then install a bridge clip across the two middle clips to cross-connect the two connections. You can disconnect the bridge clip to separate the circuits for testing and measuring.

Type 110 Punch-down Blocks

Type 110 punch-down blocks allow higher density wiring than Type 66 punch-down blocks do. 110 type blocks are preferred for home network use. A typical 110 module includes a standoff, through which the home network wiring is routed and fanned out to the appropriate location. Cross-connect wire is punched down to the upper terminals of the block, similar to the 66 type block.

Category 5 Twisted-pair Cable Termination Points

UTP cables are used for connecting home network components. The ANSI/TIA/EIA-570 standard also lists Category 5 UTP cable as the minimum requirement for the design of residential, low-voltage, structured wiring. UTP cable is terminated using RJ-45 jacks and plugs. The ANSI/TIA/EIA-568-A and B color codes and pinouts are shown in Figure 1.23. The two standards have the same color code for each pair—the difference is the switched pin designation for pairs 2 and 3. The ANSI/TIA/EIA-568-A standard is the preferred wiring scheme for Ethernet connector terminations. Residential structured wiring uses pairs 2 and 3 for the Ethernet LAN data communications, whereas pairs 1 and 4 are reserved for telephone lines 1 and 2, respectively.

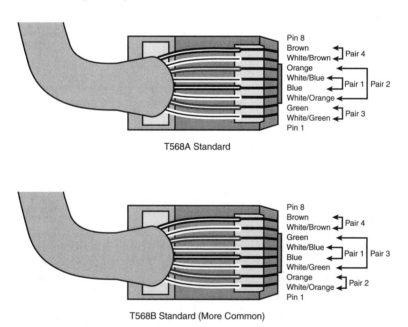

Figure 1.23 RJ-45 wiring standards.

Tip and Ring Designations

The colors for RJ-45 jacks are also referred to as *tip* and *ring*. These are terms that originated with the old plug-and-cord–based switchboards. The plug was a phono jack type with a tip element, an insulating disk, and the shaft (or ring) of the plug. The conductors of the pair were terminated in their respective elements of the plug.

The tip and ring pin and color designations for the RJ-45 plugs and jacks are shown in Table 1.4.

Table 1.4 Tip and Ring Color Codes

Tip/Ring	Color Code	Wire Pair
Pin 1 tip	White/Orange	Pair 2
Pin 2 ring	Orange	Pair 2
Pin 3 tip	White/Green	Pair 3
Pin 4 ring	Blue	Pair 1
Pin 5 tip	White/Blue	Pair 1
Pin 6 ring	Green	Pair 3
Pin 7 tip	White/Brown	Pair 4
Pin 8 ring	Brown	Pair 4

NOTE: Notice that the tip wire terminations are striped and the ring terminations are solid colors. Also, the tip connections are always the odd-numbered pins.

Shared In-house Services

HTI+ Residential Systems objective 1.6 states that the test taker should have knowledge of in-house services commonly used with residential networks. These services include items such as

➤ Video surveillance

➤ Print services

➤ File services

➤ Media services, such as streaming video and audio

Video Surveillance

Video surveillance systems employ closed-circuit TV equipment and are used for viewing the outside area and some inside areas, such as a nursery. The components of a video surveillance system can vary according to the number of areas to be viewed, indoor versus outdoor locations, night versus daylight viewing, black-and-white versus color monitors, and cameras. An option also can include a microphone for monitoring sounds in the vicinity of the camera location. Video surveillance cameras are usually integrated with a home security system.

A basic video surveillance system consists of a camera, connecting cables, and a video cassette recorder (VCR), shown here in Figure 1.24. The monitor accepts the raw video signal from the camera and displays the image in a location inside the home. The VCR provides a time-lapse history of the viewing area for later review.

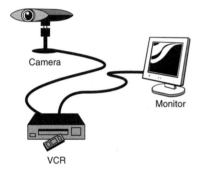

Figure 1.24 Video surveillance system components.

Additionally, numerous control system options are available, including wireless, X10-based controllers and computer control from the home network. A video system can also be used with a standard TV receiver through the use of radio frequency (RF) modulators, which take the video signal from the camera and retransmit the modulated signal on an unused standard TV channel. This configuration can be used with multiple cameras, each on a different channel. This allows switching to different channels for monitoring multiple locations inside and outside the home. Additional information on security and video surveillance systems is covered in detail in Chapter 3, "3.0—Home Security and Surveillance Systems."

File and Print Services

In business environments, it is not uncommon to have special computers reserved to perform key functions for the entire network. For example, certain types of servers are designed just to handle and process Web operations for the organization. Other machines might host a printer used by the other members of the network (the company therefore needs to buy only one printer instead of one for each user).

In these situations, the remote computer to which the physical printer is attached is referred to as a *print server* (even though the machine might be working in a peer-to-peer network and might not be a server in the client/server network sense). The machine simply provides a specific service for the other members of the network.

Using this definition, some typical network functions routinely are assigned to certain computers, thereby making them servers. The most basic server type is a *file server*, which is used to store data that is shared with network users. When a server is used as a print server, it is capable of sharing the local print device that is attached to it or a print device that has its own network adapter.

Residential networks are capable of providing all the same services available to a business. In these types of networks, servers are usually dedicated to managing files, printers, and all media services in the home. In most server applications, the computer is designed and equipped in certain ways to optimize its operations for the services it is delivering.

For example, servers designed to handle audio and video media must include fast microprocessor speeds, large amounts of installed RAM, and high-capacity disk drives. File and printer servers for home networks can be lower-end devices with smaller memory and disk drive requirements.

Media Services (Streaming Audio and Video)

Streaming audio and video technology provides the capability to view full-motion video clips and audio information on a computer. The information is processed in real-time at high bit rates on-the-fly (streaming). This is an emerging technology made possible by dramatically improved computer microprocessor clock speeds, high-speed Internet access techniques, and efficient video and audio compression standards. New application programs are available to process streaming video files, such as Windows Media Player, Apple's QuickTime, RealOne Player by Real Networks, and several other proprietary media player software applications. Streaming audio and video is made possible by the development of very efficient compression standards that reduce the number of bits required to produce quality video and audio at modest transmission bit rates. The most important video coder/decoder (codec) standards for streaming video are H.261, H.263, MJPEG, MPEG1, MPEG2, and MPEG4.

Externally Provided Data Services

HTI+ Residential Systems objective 1.7 states that the test taker should have knowledge of externally provided data services. These services include

- DSL
- ISDN
- Cable
- PPP
- PPTP
- RAS
- Email

Digital Subscriber Line

Digital Subscriber Line (DSL) is a high-speed Internet access service that uses the existing telephone lines between the subscriber location and the local office to deliver Internet access transmission rates considerably faster than analog dial-up 56Kbps modems. Many types of xDSL service exist and are based on numerous standards, proprietary versions, different bandwidths, and various upstream and downstream transmission speeds. The number of DSL variations and service brand names has expanded because of the standards published by both the ITU and ANSI organizations.

The most recent set of standards for DSL that you will need to know about has been released by the International Telecommunications Union (ITU) standards body as the G.920 series of recommendations. They are described in more detail at the end of this chapter in the "Standards" section.

The xDSL family tree includes two main branches—symmetric and asymmetric. *Symmetric* DSL services provide identical data rates upstream and downstream; *asymmetric* DSL provides relatively lower rates upstream but higher rates downstream. Ideally, DSL service remains on all of the time. With an always-on connection, subscribers no longer need to physically dial up the number for the ISP to log in to the Internet. The most popular DSL service offerings are described in the following paragraphs.

Asymmetric Digital Subscriber Line

The maximum benefit for a high-speed Internet access service is the capability to download content as rapidly as possible. Asymmetric DSL meets this need with a higher download speed but a slower upstream speed. The actual network bandwidth a customer receives from a DSL service in the home depends on the total length of his telephone local loop distance to the local office. The longer the line, the less bandwidth DSL can support. Likewise, its thickness (wire gauge) can affect the performance. To be eligible for DSL service, the local loop phone line must meet specific requirements. First, the subscriber must lie within the distance limitations of DSL. Generally, this means no more than approximately 18,000 feet of phone wire length to the public exchange.

Table 1.5 illustrates the downstream performance of ADSL as a function of the distance from the subscriber to the local office.

Table 1.5   ADSL Performance Versus Local Loop Cable Length

Cable Length (Feet)	Bandwidth Availability (Kbps)
18,000	1544
16,000	2048
12,000	6312
9,000	8448

One disadvantage of ADSL service is that when the ADSL local loop length increases, the available bandwidth decreases for both upstream (not shown) and downstream traffic. These numbers assume 24-gauge wire; performance decreases significantly if 26-gauge wire exists on the local loop.

ADSL works by splitting the phone line into two frequency ranges. The frequencies below 4KHz are reserved for voice, and the range above that is used for data. This means you can use the line for phone calls and data network access at the same time. It is called *asymmetric* because more bandwidth is reserved for receiving data than for sending data.

Figure 1.25 illustrates the allocation of bandwidth of an ADSL line for the voice, upstream, and downstream portions of the total available bandwidth.

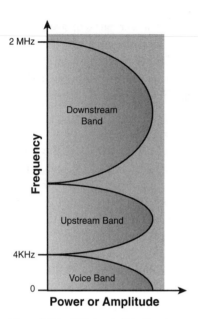

Figure 1.25 ADSL bandwidth allocation.

DSL Modems and Splitters

A DSL modem (also known as an *ADSL terminal unit [ATU]*) provides the interface between the home computer or computer network and the phone line. DSL modems are available in both internal and external configurations. *Internal* modems are installed in a computer expansion slot similar to an analog dial-up modem. The *external* version connects to the computer using a USB port or an Ethernet NIC jack. Most home networks use an Ethernet router between the modem and the home network computers, which enables all the home network computers to have equal access to the DSL Internet access port. With this type of connection, the DSL modem connects to a Category 5 cable port on the Ethernet router.

A Plain Old Telephone Service (POTS) splitter is required to separate the telephone voice band (0KHz–4KHz) and the DSL band used to transmit digital information. It acts as a low-pass filter to keep the high-frequency signals from causing interference to the lower-frequency voice channel. The splitter works in one of the three following configurations:

- *Splitter-based DSL*—Any DSL service that requires that a signal splitter be manually installed at the subscriber location
- *Splitterless DSL*—Any DSL service in which the signal splitting is provided remotely from the telephone exchange carrier local office

➤ *Distributed splitter DSL service*—Lowers the complexity at the subscriber location but is more complex to implement at the local office

The technical term used to describe a DSL modem has many confusing variations due to the wide interpretations of acronyms used by international and U.S. standardization groups and modem suppliers. Most consumers refer to the modem as a *DSL modem*. The engineering term is *ATU* or *ADSL terminal unit*, but the ADSL terminal unit (ATU) is also known as the ADSL transceiver unit, ADSL transmission unit, and ADSL termination unit. The ATU-R (remote) unit is installed in the home, and the ATU-C (central office) unit is installed in the local telephone company's central office. Exam questions might include references to any of these terms, which all mean the same thing. Regardless of what it's called, it's the point at which data from the user's computer or network is connected to the DSL line.

In addition to the DSL modem, the splitter is important to proper operation of the DSL data transmission and analog voice service, which both share the same phone line. If the splitter or ATU-R is not installed properly or malfunctions, a high-frequency noise will cause interference as well as considerable distortion to voice signals. Be aware of this problem and the symptoms resulting from a bad or missing splitter on a DSL line. Also remember that the splitter can be installed at the subscriber location or the telephone exchange central office.

G.Lite and G.DMT

Most service providers offer only two types of DSL service for home subscribers; they are called G.Lite and G.DMT. All the other DSL types of service are generally available only to business customers.

Available since 1999, G.Lite (ITU Recommendation G.992.2) is a form of ADSL that does not require a splitter installation at the subscriber location, but it does so at the expense of lower data rates. G.Lite can be used with a distributed splitter configuration but with added complexity and limited benefit.

G.Lite represents the most consumer-friendly version of DSL. Its equipment and service cost less than other varieties, and it reportedly has a do-it-yourself installation. G.Lite supports a maximum of 1544Kbps downstream and 384Kbps upstream, whereas regular full rate ADSL (as shown earlier) can support more than 8000Kbps.

G.DMT (G.Discrete Multitone) ADSL denotes the other standard for home DSL service and is defined in ITU Recommendation 992.1. The main difference between G.Lite and G.DMT is bandwidth. Sometimes called *full-rate ADSL*, the G.DMT variety can download data at up to 8Mbps and send data upstream at up to 1.5Mbps. However, to obtain such high speeds, the subscriber's modem must be located within 10,000–12,000 feet of the local telephone company central office. Outside the 12,000-foot range and ranging up to 18,000 feet away from the central office, G.DMT ADSL can reach up to 1.5Mbps downstream. Even though G.DMT ADSL offers higher speeds, it requires a more complicated installation. G.Lite allows a do-it-yourself installation with no local splitter, but G.DMT DSL requires the telephone company to install a splitter on the phone line. Figure 1.26 shows the system configuration for a G.DMT installation and a G.Lite installation, with the two POTS splittered and splitterless variations and ATU locations.

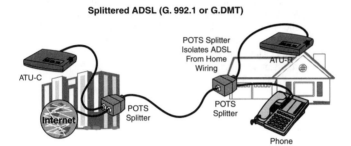

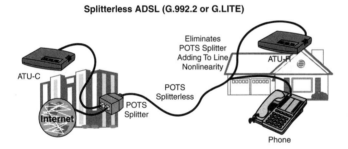

Figure 1.26 ADSL system components.

Discrete multitone (DMT) is a method of separating a DSL signal so that the usable frequency range is separated into 256 frequency bands (or channels) of 4.3125KHz each. Quadrature amplitude modulation (QAM) is used within each channel. By varying the number of bits per symbol within a channel, the modem can be rate-adaptive. Both G.Lite and G.DMT use DMT.

Integrated Services Digital Network

Integrated Service Digital Network (ISDN) is a high-speed service offered by some local exchange carriers to home and business subscribers. The introduction of faster DSL service in recent years has diminished the interest and market for ISDN service to residential subscribers. Similar to DSL, ISDN service uses existing voice telephone lines from the local telephone office to the subscriber equipment to send and receive data and voice information.

ISDN service requires a digital modem called a terminal adapter (TA) connected to a home office computer or gateway.

Three levels of ISDN service are available: basic rate interface (BRI) services, primary rate interface (PRI) services, and broadband ISDN (B-ISDN) services.

The ISDN Basic Rate Interface Service

ISDN BRI service offers two B channels and one D channel (2B+D). BRI B-channel service operates at 64Kbps and is meant to carry user data; BRI D-channel service operates at 16Kbps and is meant to carry control and signaling information, although it can support user data transmission under certain circumstances. BRI also provides for framing control and other overhead, bringing its total bit rate to 192Kbps.

ISDN Primary Rate Interface Service

ISDN PRI service offers 23 B channels and 1 D channel for a total bit rate of 1.544Mbps (the PRI D channel runs at 64Kbps). ISDN PRI in Europe, Australia, and other parts of the world provides 30 B channels plus 1 64Kbps D channel, with a total interface rate of 2.048Mbps.

ISDN Broadband Service

B-ISDN is a proposed service that might include interfaces operating at data rates of 150Mbps–600Mbps. B-ISDN uses asynchronous transfer mode (ATM) to carry all services over a single, integrated, high-speed packet-switched network. Additional work is expected to be accomplished on the B-ISDN standard before it is commercially successful.

Cable

High-speed Internet access service is also provided by TV cable service providers. Originally, cable network video feeds were gathered at a central headend and propagated down through a hierarchy of splits and branches until they reached the home. The modern version of these networks has replaced the copper branches in the upper levels with fiber-optic links directly to fiber nodes, which serve neighborhood-size areas. Such networks are known as *hybrid fiber/coax (HFC)* networks.

High-speed data service to a residential subscriber requires a cable modem. The cable modem typically has two main connections—one to the computer USB port or Ethernet 10/100 NIC and another connection to the cable coaxial cable outlet on the wall. When the residential subscriber has multiple computers configured as an Ethernet LAN, the connection to the Ethernet computer network can include an Ethernet gateway/router with other computers on the residential network connected to a port on the router. The router enables each computer on the residential LAN to have equal access to the modem.

With the cable modem connection, a user can upload and download information 1,000 times faster than the fastest analog telephone modem using dial-up service. The cable modem is similar to an ASDL modem because it establishes two different transmission (asymmetric) rates. The cable modem enables the uploading of data (information leaving the subscriber PC to the server) to have a slightly slower speed when compared to downloading information (the server sending information to the subscriber).

The North America cable modem standard is called Data Over Cable Service Interface Specification (DOCSIS). Cable modems compliant with this standard have been available since 1998. To deliver DOCSIS data services over a cable television network, one 6MHz radio frequency channel in the 50MHz–750MHz spectrum range is typically allocated for downstream traffic to residential subscribers and another channel in the 5MHz–42MHz band is used to carry upstream signals. A headend cable modem termination system (CMTS) communicates through these channels with cable modems located at the subscriber location.

The cable modem specification called DOCSIS was developed by a consortium of cable companies who grew impatient waiting for the IEEE standard to be completed. You will need to know that the full name of this cable modem standard is Data Over Cable Service Interface Specification (DOCSIS).

Point-to-Point Protocol

The most widely used protocol for accessing ISPs through dial-up connections is the Point-to-Point Protocol (PPP). PPP is a very versatile RAS protocol that can assign IP addresses to remote computers over telephone connections).

The PPP protocol runs in a point-to-point manner across typical RS-232, RS-422, or RS-423 serial connections, as well as through standard telephone connections. The type of physical interface being used with the protocol determines the actual data transmission rate that can be achieved.

Point-to-Point Tunneling Protocol

The Point-to-Point Tunneling Protocol (PPTP) is an extension of the PPP protocol. It is used to send PPP packets over a nonsecure network such as the Internet. *Tunneling* is a technique in which a datagram is contained within the envelope of another higher-level protocol—in this case, IP—during transfer across the Internet. This embedding technique is called *encapsulation*.

With the use of PPTP, a company no longer needs to lease its own dedicated private lines for wide-area communication but can securely use the public Internet. This type of interconnection is also known as a *virtual private network (VPN)*. The use of PPTP is very useful for business travelers who need to dial in to the private corporate network from a remote location, such as a hotel room. The use of the Internet and an ISP through a local access number is less expensive than using long-distance telephone connections. Both the corporate remote access server and the remote access client need PPTP software to complete the proper handshaking and connection sequence.

When the packet arrives at the designated computer, the PPP packet is removed from the IP packet and delivered to the destination computer. PPTP uses only IP, IPX, or NetBEUI datagrams contained inside an IP packet.

RAS

Remote access is the capability of a user to access a network from a remote location using the public switched telephone network (PSTN) or using the Internet through dial-up access, DSL, or cable. All modern network operating systems come with built-in support for RAS.

A *remote access server* is a computer on a LAN that is dedicated to handling users who are not on the LAN but need to have access to it. When the remote user is authenticated, she is permitted to access the resources of the LAN as if she were physically located in the local facility.

Network connections are usually supported using PPP and PPTP, discussed earlier.

RADIUS

Remote authentication dial-in user service (RADIUS) is a client/server protocol that enables remote access servers to communicate with a central server on a LAN to authenticate dial-in users and authorize their access to the requested system or service. RADIUS enables a company to maintain user profiles in a central database that all remote servers can share. It provides better security, allowing a company to set up a policy that can be applied at a single administered network point.

Figure 1.27 illustrates the connectivity and network components required for remote users to connect to a private LAN.

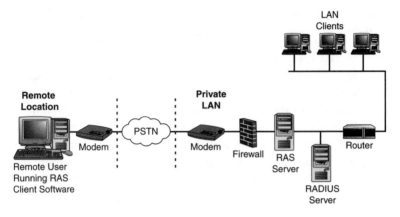

Figure 1.27 Remote access service.

As Figure 1.27 illustrates, several important components are involved in a RAS connection, including the following:

➤ A RAS server running RAS protocols

➤ A firewall to protect the LAN from attacks

➤ A router for sending requests to other hosts on the LAN

➤ A RADIUS server to perform the authentication and authorization function

➤ A local client running RAS client software

➤ A physical communication link that connects the client with the remote server

 NOTE: The RAS server might perform the authentication on medium-size networks. As the network load grows, a dedicated server such as a RADIUS server is often needed to perform the authentication for remote clients.

Email

One of the most widely used services available for use over computer networks is electronic mail (email). This feature enables users to send and receive electronic messages over the Internet. As with the regular postal service, email is sent to an address, but the email address is not the address of the recipient's computer. It is the address of the server on which the client has an email account, either on a private network or an ISP. When the recipient opens his email client program, the email server notifies him that it has email in the server mailbox. Access to the mailbox is achieved using a password and user account ID. Some well-known email clients are Microsoft Outlook, Outlook Express, Eudora, and Pegasus. The default email reader supported by the Windows Outlook Express applet is the POP3 standard. Likewise, it includes a standard SMTP email utility for outgoing email.

When setting up an email account, you must supply the following configuration information:

➤ Account name

➤ Password

➤ POP3 server address

➤ SMTP server address

The two protocols for setting up an email account are the POP3 server address and the SMTP server address. You should also memorize the well-known port addresses for SMTP (25) and POP3 (110).

Standards

HTI+ Residential Subsystems objective 1.8 states that the test taker should have knowledge of the standards associated with residential subsystems and networking. These standards include topics such as

➤ Wiring standards

➤ Protocol standards

➤ Standards and organizations, including

  ➤ National Electrical Code

  ➤ EIA/TIA standards

  ➤ IEEE standards

  ➤ Underwriters Laboratories, Inc. (UL)

The HTI+ objectives for each domain chapter include the same standards organizations, although they all do not apply in each case. The standards applicable to each chapter are described where appropriate.

The Institute of Electrical and Electronics Engineers

The Institute of Electrical and Electronics Engineers (IEEE) is a standards organization that develops a large number of standards for computer networking. One of the major areas of responsibility discussed in this chapter includes the IEEE 802.3 series Ethernet LAN standards and the IEEE 802.11 Wireless LAN standards.

ANSI/TIA/EIA Standards

The most widely referenced standards included in this chapter are the ANSI/TIA/EIA standards group. These terms refer to the following organizations:

- *ANSI*—The American National Standards Institute
- *EIA*—The Electronic Industries Alliance, formed to address the electronic properties of cable
- *TIA*—The Telecommunications Industry Association

The ANSI/TIA/EIA 568 Standard

The ANSI/TIA/EIA 568 standard title is "Commercial Building Telecommunications Cabling Standard." The 568-A and 568-B specification describes the Category 5 UTP cabling terminations standards, including pinouts and color codes. In April 2001, all the addenda and previous TSB publications were revised and folded into the new ANSI/TIA/EIA-568-B, which is now published in three parts, as follows:

- ANSI/TIA/EIA 568-B.1, "Commercial Building Telecommunications Standard—General Requirements"
- ANSI/TIA/EIA 568-B.2, "Commercial Building Telecommunications Standard—Balanced Twisted Pair Cabling Components"
- ANSI/TIA/EIA 568-B.3, "Optical Fiber Cabling Components Standard"

ANSI/TIA/EIA 570-A

The 570-A standard was approved on September 1, 1999. It was developed specifically to address the previously neglected residential wiring standard area and applies to telecommunications premises wiring systems installed within an individual building with residential end users.

TIA/EIA-570-A establishes minimum grades of residential cabling for residential construction. The two grades are described as follows:

➤ *Grade 1*—Provides a minimum of one Category 3 UTP and one 75-ohm coaxial cable at each location

➤ *Grade 2*—Requires two Category 5 UTP cables, two 75-ohm coaxial cable at each location, and (as an option) optical fiber-optic cabling

Both ANSI/TIA/EIA-568A and 570-compliant wiring systems are based on a star wiring topology. This type of wiring is also called *structured* wiring. In a star topology, all wiring is routed directly to a central location. Home cabling routing for grades 1 and 2 is also known as *home run* wiring.

You will need to memorize the required number and cable types to meet the requirements for grade 1 and grade 2 cables for structured wiring in the ANSI/TIA/EIA 570A standard. Although Category 3 cable is included in class 1 minimum requirements, Category 5 or higher is typically used in residential wiring. But remember the question relates to the language in the standard and not to best practice for installing a higher-grade cable. Grade 2 requirements state the minimum requirements as two Category 5 outlets and two 75-ohm coaxial cable outlets in each room.

A star topology is required for IEEE 10/100BASE-T Ethernet LANs as well as residential structured wiring requirements contained in ANSI/TIA/EIA 568-A and 570-A standards. It is considered to be the most reliable topology as opposed to daisy-chain of wiring. Wiring all coaxial cable and UTP cables to a common location is called home run wiring.

The ANSI/EIA 600 CEBus Standard

The CEBus standard adopted by the EIA as EIA 600 is a protocol specification to support the interconnection and interoperation of consumer products in a home. The CEBus standard's power-line carrier technology uses the home's 120v, 60Hz, electrical wiring to transport messages between household devices. CEBus is an acronym for consumer electronics bus.

Exam Prep Questions

Question 1

Which type of cable is the most susceptible to electromagnetic interference?
- ○ A. Coaxial cable
- ○ B. Fiber-optic cable
- ○ C. Unshielded twisted-pair
- ○ D. Shielded twisted-pair

Answer C is correct. Unshielded twisted-pair cable is the correct answer. Although it is the most popular type of cable used for local area networking, it is subject to interference from adjacent cables and electrical equipment such as radio transmitting sources, motors, and power lines. Fiber-optic cable offers the highest protection from EMI sources, so answer B is incorrect. Coaxial cable and shielded twisted-pair cables offer greater immunity to interference than UTP but less than fiber-optic cable, so answers A and D are incorrect.

Question 2

Which of the following is a type of fiber-optoelectronic cable?
- ○ A. TOSLINK
- ○ B. Lynx
- ○ C. i.Link
- ○ D. FireWire

Answer A is the correct answer. It is the trademarked name of a fiber-optic (optoelectronic) cable that is used for connecting components such as DVDs, HDTVs, and PVRs as well as other high-bandwidth applications. The other three cable types listed in answers B, C, and D are brand names often used synonymously for IEEE-1394 connecting cables. They are not associated with optoelectronic type cables and are therefore incorrect.

The HTI+ exam rarely includes questions concerning trademarked names or vendor-specific products. However, the use of brand names associated with industry standards for cable (or de facto standards) such as FireWire, Monster cable, TOSLINK, i.Link, and others is widespread in the jargon of home automation products. It is important to know the origin of these names and the standards each represents.

Question 3

> Which of the following IP address groups can be used for assignment to computers using the public Internet?
>
> ○ A. 10.0.0.0–10.255.255.255
> ○ B. 12.222.0.0–12.222.255.255
> ○ C. 172.16.0.0–172.31.255.255
> ○ D. 192.168.0.0–192.168.255.255

Answer B is the correct answer. It is the only IP address shown in the four options that is a routable IP address. The address groups listed in answers A, C, and D are reserved for private networks and are not routed by the Internet, so those answers are incorrect. You will need to memorize the address groups reserved for private IP networks.

Question 4

> Which of the following is not a high-speed Internet assess service?
>
> ○ A. ADSL
> ○ B. G.Lite
> ○ C. PnP
> ○ D. G.DMT

Answer C is the correct answer. PnP is the Plug and Play standard developed by Microsoft for the automatic recognition and configuration of PnP-compatible installed devices. Answers A, B, and D are all variations of DSL types of services and are therefore valid types of high-speed Internet access service offerings, making those answers incorrect.

Question 5

> Which standard supports home networking using existing phone lines as the network media?
>
> ○ A. HPLA
> ○ B. HomePNA
> ○ C. X10
> ○ D. DOCSIS

Answer B is the correct answer. The standard adopted for using phone lines for networking is called Home Phone Network Alliance (HomePNA). The HPLA and X10 protocols are used on the existing home power wiring; therefore, answers A and C are incorrect. DOCSIS is the cable modem standard, so answer D is incorrect.

Question 6

Which organization coordinates the assignment of Internet domain names, IP addresses, protocol parameters, and port numbers?
- A. The FCC
- B. The IEEE
- C. The IETF
- D. The ICANN

Answer D is the correct answer. The Internet Corporation for Assigned Names and Numbers (ICANN) is a technical coordination body for the Internet. ICANN coordinates the assignment of Internet domain names, IP addresses, protocol parameters, and port numbers. The FCC (Federal Communications Commission) is a regulatory agency of the U.S. government, so answer A is incorrect. The IEEE is another standards body that develops standards for computer networking and many other telecommunications areas, so answer B is incorrect. The Internet Engineering Task Force (IETF) is a large, open international community of network engineers, researchers, and vendors that provide technical guidance for the Internet community, so answer C is also incorrect.

Question 7

How many addresses are supported by the X10 protocol?
- A. 128
- B. 256
- C. 512
- D. 1024

Answer B is the correct answer. The X10 specification sets up an address structure supporting a total of 256 different addresses. Answers A, C, and D are each incorrect for the same reason—they each contain the wrong values for the total number of addresses available with the X10 protocol.

Question 8

Which protocol uses tunneling to send PPP packets over a nonsecure network such as the Internet?
- A. TCP
- B. DHCP
- C. HTML
- D. PPTP

Answer D is the correct answer. The Point-to-Point Tunneling Protocol (PPTP) encapsulates IP packets for transport over a public network. PPTP is used by remote access servers to support communications between remote users and a private home or corporate network. Answer A is incorrect because TCP is a Transport layer protocol and has no role in tunneling. Answer B, the Domain Host Configuration Protocol, is used to assign and track IP address allocations on an IP network and is not related to tunneling, so it's incorrect. Answer C is also incorrect because it is a language used on the Internet's World Wide Web. HTML is basically ASCII text surrounded by HTML commands.

Question 9

Which transmission rate is utilized for home networks using the HPNA 3.0 specification?
- A. 128Mbps
- B. 10Mbps
- C. 100Mbps
- D. 1Mbps

The correct answer is A. In September 2002, the HomePNA organization announced the approval of the new HomePNA 3.0, next-generation specification. The throughput rate for the latest version 3.0 is 128Mbps. Answer B is incorrect because 10Mbps is the transmission rate for HPNA 2.0. Answer C is incorrect because no existing HPNA specification is rated at a transmission rate of 100Mbps. Answer D is incorrect because 1Mbps was the original transmission rate of HPNA version 1.0 and not HPNA 3.0.

Question 10

> Which standards document describes the performance parameters for CEBus control devices?
>
> ○ A. HPNA 1.0
> ○ B. ANSI/TIA/EIA 570-A
> ○ C. EIA 600
> ○ D. ANSI/TIA/EIA 568-B

Answer C is correct. The CEBus standard adopted by the EIA as EIA 600 is a protocol specification to support the interconnection and interoperation of consumer products in a home. The CEBus standard's power-line carrier technology uses the home's 120v, 60Hz electrical wiring to transport messages between household devices. CEBus is an acronym for consumer electronics bus. Answer A is incorrect because HPNA 1.0 is the earlier specification for a home phone-line networking protocol adopted by the Home Phone Network Alliance. Answer B is incorrect because the ANSI/TIA/EIA 570-A is the Residential Telecommunications Cabling Standard and is not associated with the CEBus EIA 600 standard. Answer D is incorrect because the ANSI/TIA/EIA 568-A is the Commercial Building Telecommunications Cabling Standard and is not associated with the CEBus standard.

Need to Know More?

 Main, Max, and Sarka, Caleb. *The Complete Introductory Networking Course*. Kennewick, WA: Marcraft International Corporation, 2002. (ISBN: 1-58122-044-8. This textbook covers the basics of network operating systems, network media, wireless LANs, network operations, and troubleshooting common network problems.)

 O'Driscoll, Gerard. *The Essential Guide to Home Networking Technologies*. Upper Saddle River, NJ: Prentice Hall PTR, 2001. (ISBN: 0-13-019846-3. This book provides detailed information on the new home networking technologies, power-line and phone-line network standards, and related home technology initiatives.)

2.0—Audio and Video Fundamentals

Terms you'll need to understand:

- ✓ Whole home
- ✓ Remote access
- ✓ Surround sound
- ✓ Speaker impedance
- ✓ Cathode ray tube
- ✓ Flat-panel display
- ✓ Plasma display
- ✓ Modulator
- ✓ Bridged amplifier
- ✓ Personal video recorder
- ✓ Composite video
- ✓ Component video
- ✓ S-video
- ✓ Internal broadcasting
- ✓ Streaming audio and video
- ✓ HAVi
- ✓ Digital TV
- ✓ HDTV
- ✓ NTSC
- ✓ ATSC

Techniques you'll need to master:

- ✓ Configuring an audio/video distribution system
- ✓ Configuring a surround sound speaker system
- ✓ Connecting speakers to a bridged amplifier
- ✓ Connecting a DVD player to an HDTV
- ✓ Explaining the advantages of a flat-panel display
- ✓ Planning the installation of a satellite antenna
- ✓ Listing the minimum viewing distance for a large-screen TV
- ✓ Describing interlaced and progressive scanning formats
- ✓ Making a simple drawing of a dynamic speaker
- ✓ Planning the component hookups for a home theater system
- ✓ Listing the types and advantages of projection TV systems

Design Considerations for a Connected Audio/Video System

The Residential Systems objective 2.1 states that the test taker should have knowledge of residential audio/video system designs. Examples of considerations for these designs include

- Dedicated versus distributed system
- Whole home systems
- Remote access
- Zoning distribution

Dedicated Versus Distributed Systems

A key design consideration for home audio/video systems is the choice between a dedicated versus distributed system. Features of each of these designs are summarized in the following paragraphs.

Dedicated Audio/Video System Design

A dedicated system design places all the audio/video equipment in one location. This is the typical design choice for a home theater system where a location such as a den or a custom-designed entertainment center is the single dedicated location for home entertainment. This is also the preferred design for home theater surround sound systems where the entire room can be wired for the optimum location of surround sound speakers and home theater source equipment such as large-screen high-definition television (HDTV) systems or video projection systems. This design approach offers the best overall performance and sound quality but requires a room with good acoustics, an adequate viewing area, and a separation of speakers for the best performance.

Distributed Audio/Video System Design

A distributed audio/video system design is favored by those who desire TV and audio entertainment in different areas, or *zones*, of the home rather than the dedicated home theater design. Depending on the type of cabling installed, various levels of control and automation can be designed into a distributed system. The minimum audio/video distributed system would include separate TV and source equipment (tuner, VCR, DVD, and CD) in one or more locations but without multiple zone options. The top of the line distributed design would include a single A/V rack for all source equipment feeding multiple zones for both audio and video entertainment; this is also known as a *whole home* system design.

Whole Home Audio/Video System Design

A *whole home* audio/video system design enables you to see and hear all the shared video and audio sources in multiple locations throughout the home. Audio equipment in a central location (family room, den, and so on) distributes sound to speakers located in other rooms and areas. Video distribution uses a central video distribution panel, amplifiers, and modulators to distribute high-quality video information to multiple areas in the home. The design of a whole home audio/video distribution system depends on a prewired structured cabling plan for dual coaxial cabling (RG6 quad-shield coax) and dual Category 5E cabling to each room.

 Audio/video system design alternatives include both *dedicated* and *distributed* architectures. This terminology is generally used by contractors to describe the dedicated home theater design option versus a distributed design with audio/video components in multiple areas of the home. A distributed architecture should not be confused with the audio/video *distribution system* associated with a whole home design that *distributes* shared audio and video sources to multiple locations.

Video Distribution System

A model for the design of a home video distribution system is shown in Figure 2.1. Variations on this model are possible depending on the distance of each cable run, home size, cost, and number of video sources. The figure represents all the basic features of an average video distribution system.

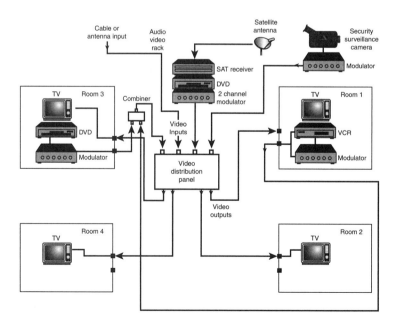

Figure 2.1 Video distribution system.

Features offered by the whole home video distribution system shown in Figure 2.1 are as follows:

- *Video distribution panel*—The central hub for all video signals. It contains amplifiers, splitters, and modulators and provides the capability to have all home video sources distributed to any television receiver in the home.

- *RG6 coaxial cable*—Provides the input and output connections to the video distribution panel. Each room has dual coaxial cable wall outlets. The bottom connector on each outlet takes signals generated in that room from a VCR or video camera and feeds them back to an input on the distribution panel so that the video signals become available to all other TVs in the house.

- *Outputs from the video distribution panel*—Provide amplified modulated video inputs to each of the TV receivers in the separate rooms. The amplifiers are adjusted to provide the same video signal quality to each room.

- *Modulators*—Convert the video source signal (surveillance camera, satellite receiver, VCR, and DVD player) to an unused television channel for internal broadcasting on the video distribution system. This enables the home users to choose which video sources they want to view by tuning the television receiver to the desired channel.

- *Combiner*—Consolidates video inputs from source equipment and connects them to the video input jack on the distribution panel.

Audio Distribution System

The design of an audio distribution system shares some of the features of a video distribution system. Audio distribution requires audio source equipment located in a central location to be shared by different users in various zones. A zone includes a group of speakers in one area of the home. Audio sources can be designed with either a single zone or multizone distribution plan. A *single zone* is a basic arrangement in which all zones hear the same audio source at any interval of time. *Multizone* design permits different users in different areas of the home to listen to different sources independently. Volume controls and source equipment keypads are usually mounted in wall outlets. They're wired into the system so they can control the sound in a room or the entire home. Keypads can be used to select the source and control the volume either manually or with handheld infrared remote units.

Audio distribution systems use a variety of wire and cable types for connecting components, including Category 5 UTP cable, speaker cable, and shielded audio patch cables. Source audio equipment is usually located in a headend location where all equipment for the audio and video system components is installed in a shared audio/video equipment rack.

Audio source equipment and controls selection depends on the size of the home, number of rooms, and number zones; however, a basic design includes the following types of audio source equipment:

➤ *Speakers*—These are mounted in each zone with either wall, ceiling, outdoor, or bookshelf-designed mounts.

➤ *Preamplifier*—This is used for multisource input and output switching and source component selection.

➤ *AM/FM radio tuner and amplifier*—The tuner can be either an integrated amplifier/tuner design or separate components. The tuner/amplifier can also provide input and output switching for other sources such as CD players and tape.

➤ *CD player with either single tray or multiple-CD selection.*

➤ *Wall-mounted volume controls and component selection keypads.*

➤ *Tape cassette players/recorders*—These have all but disappeared as an audio source; however, they are often included in some home designs to provide an easy and economical means for recording audio sources from FM radio broadcasts. Digital audio tape (DAT) offers superior quality when compared to analog tape players/recorders.

Remote Access

Many remote access systems are available for use with home audio/video. Most components have their own wireless infrared remote controllers. Universal handheld infrared controllers can be programmed to work with several components.

Zoning Distribution

As discussed earlier, zoning distribution is an option for the home entertainment system in which different sources can be selected for different areas of the home. A multiroom, multisource system can deliver audio and video to several areas throughout the home with the capability to play different sources (tuner, CD player, or cassette player) in different areas simultaneously. Any of these systems can be controlled in any area of the home by a simple wall-mounted keypad or handheld remote. A down-sized system is called a single-zone system, in which, at any given time, all speaker pairs will play the same audio. This type of system is the least expensive and simplest to install.

Equipment Location Considerations

The HTI+ Residential Subsystems objective 2.2 states that the test taker should have knowledge of the equipment location considerations associated with residential audio and video components. Examples of these considerations include

- Speakers
- Televisions
- A/V rack
- Touch screens
- Volume controls
- Keypads

Speaker Locations

Speaker systems are designed for installation in a variety of locations in a home, including the outdoors. Small bookshelf speakers, flush mount in-wall speakers, and weather-resistant speakers can be used around the home. Some speakers are designed to be partially buried in the ground or disguised as rocks for pool or patio use. In modern home theater installations, surround sound is the preferred system configuration for speaker locations.

Surround Sound Speaker Location Terminology

Locating the speakers is an important step in planning a home theater system. A typical home theater layout with surround sound speakers is shown in Figure 2.2.

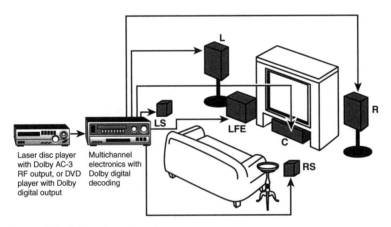

Figure 2.2 Surround sound speakers.

As shown in Figure 2.2, the Dolby Digital Surround Sound "5.1" format provides up to five discrete (independent) channels (center, left, right, surround left, and surround right) of *full frequency effects* (from 20Hz to 20,000Hz). Thus, the "5" designation. The plan places two speakers in the front of the room located to the left and right of the listener. The third speaker should be located in the center in front of the listening position. A pair of surround sound speakers—one on each side and slightly above and behind the audience—completes the surround sound effect. A sixth channel dedicated for low-frequency effects (LFE) is reserved for the subwoofer speaker. The LFE channel gives the Dolby Digital format the ".1" designation. The ".1" signifies that the sixth channel is not a full-frequency channel because it contains only deep bass frequencies (3Hz–120Hz). Subwoofer placement is not critical because the low-frequency sounds are not sensitive to direction. However, it should be facing out into the room without obstructions from furniture or other objects to avoid absorption of the low-frequency propagation.

The Dolby Digital Surround Sound format has specific names and locations for the five surround speakers and the sixth LFE channel. The locations are center (C), left (L), right(R), right surround (RS), and left surround (LS). The LS and RS are ideally placed to the side and slightly behind the listener's seating position. *There is no designated rear surround sound speaker location.* The subwoofer (LFE) can be located anywhere in the room. The subwoofer does not contribute to stereo imaging; therefore, adding a second subwoofer does not enhance the surround quality.

Television Location

The placement of the television is dependent on several issues. It should be in a location where the light from exterior sources can be controlled. The size of the room will determine the maximum TV screen size as well as the optimum viewing distance. The minimum recommended viewing distance for a 40" television screen is 10 feet. Optimum distances are illustrated in Figure 2.3.

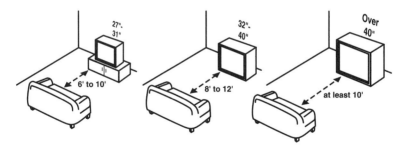

Figure 2.3 TV viewing distance.

Audio/Video Rack Locations

The best location for the audio/video rack in a home theater room is in the front of the viewing area. This simplifies the layout for wiring between the source equipment, TV, and speakers. For a whole home design, the audio/video equipment rack is typically located near the main audio/video distribution panel.

Touchscreen Locations

A *touchscreen* is an input device that enables users to operate audio and video equipment by simply touching icons on a display screen. Touchscreens are available as standalone, large-screen panels similar to a computer monitor, wall-mounted units, or handheld remote controllers using wireless infrared technology. A typical touchscreen remote controller is shown in Figure 2.4.

Figure 2.4 Touchscreen remote controller.

Touchscreen controllers can be programmed to meet the requirements of most custom home audio and video systems. They can be used to control amplifiers, DVD players, and CD players; set volume levels; or switch source equipment located in another location.

Volume Control Locations

Volume controls are often located away from the source equipment. In a multizone system, you might want to have the volume controls wall mounted in each room. This type of control is implemented at the speaker location instead of the source amplifier.

Where control of several speakers from a single volume control is required, an impedance matching volume control is necessary. This type of control eliminates the need for separate switching between speaker volume controls and maintains the same impedance as seen by the amplifier for two, four, or eight speakers. Impedance matching volume controls can also be wall mounted.

Keypad Locations

Keypads need to be located in every room where access to the audio and video systems is required. Keypads can also be accessed with remote handheld infrared units similar to TV remote controllers.

In whole house systems, keypads must be located throughout the home to manage source equipment, control the volume, and turn power on and off on selected audio/video systems.

Physical Audio and Video Products

Residential Systems objective 2.3 states that the test taker should have knowledge of the physical products associated with residential audio/video technology. Examples of these products include the following:

- Receivers
- Amplifiers
- Speakers
- Keypads
- Video displays
- Source equipment

Receivers

A *receiver* is one of the main components of an audio/video system. It is used to process audio, video, satellite, and AM/FM broadcast signals received from external sources and output amplified or decoded signals to television video input jacks, speakers, or recording devices. Most receivers used in home audio/video systems combine the functions of a radio tuner, preamplifier, and amplifier in a single chassis. Receiver systems are also available in which the preamplifier, tuner, and amplifier are separate units. Although the separate units require more shelf space, they usually provide more features and options that can affect or enhance sound quality. A block diagram of a receiver with the three embedded component systems is shown in Figure 2.5.

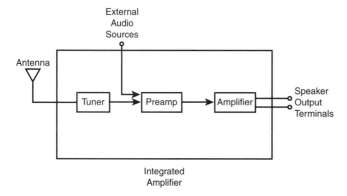

Figure 2.5 Audio/video receiver.

Most receivers are sold under various names with varying capabilities and configurations. The common types of receivers are described as follows:

- *Audio/video (A/V)*—These receivers have an integrated AM/FM tuner and process from four to six channels with amplifiers capable of driving multiple speaker channels. They have built-in surround sound decoders that separate the sound signal from the video source (such as a DVD player) into several channels and then route it to several speakers. Numerous surround sound formats are available and are covered later in the section titled "Surround Sound Formats." The most common type of surround sound decoders are Dolby Pro-Logic, which processes four channels of sound, and Dolby Digital, which processes six channels and is known as the 5.1 format, discussed earlier. A/V receivers are also available in Dolby Pro-Logic/Digital Ready and Dolby Digital 5.1/DTS (Dolby Theater System) modes. A/V receivers usually have both audio and video input jacks for DVD, S-video, and component and composite video.

- *Stereo receivers*—Less sophisticated than A/V receivers, they also have an AM/FM integrated tuner for receiving commercial broadcast stations. Stereo receivers split the signal from an audio source, such as a CD player, into two channels—left (L) and right (R)—and then route it to two speakers. Many stereo receivers have outputs for an extra set of speakers but are limited to two channels of sound. Stereo receivers are the least expensive type receivers and lack the capability for full surround sound channel processing found in later model A/V receivers.

- *Satellite receivers*—Used to receive and process high-frequency digitally encoded signals from a satellite dish-type antenna. They are used to demodulate and decode digital television programs received from Direct Broadcast Satellite (DBS) service.

Tuners

The tuner portion of a stereo or A/V receiver selects radio stations in the AM and FM broadcast bands. The received signal is sent to the preamplifier and ultimately through the amplifier to the speakers. The FM band provides the best quality for music due to the added bandwidth allocated to each FM broadcast channel. FM broadcasting uses the wider channel to capture frequencies in the higher portion of the audio spectrum presenting a superior quality of sound compared to AM broadcasting stations.

Modern tuners use a digital display that shows the status of the input sources, including the frequency of the station selected, the received signal strength, and a stereo indicator to show when the FM station has sufficient signal strength for the tuner to lock on to the stereo signal. Tuner specifications include the sensitivity of the tuner with respect to weak signals for both the AM and FM bands. A wide and narrow selectivity control is often included on a tuner to provide the best reception for weaker distant stations and local stations.

Amplifiers

Amplifiers are available as an integrated component in A/V receivers or as a separate unit. The amplifier's job is to take audio inputs from a number of sources that have been selected and processed in the preamplifier and amplify them to a level suitable for driving multiple speakers. In multiple-channel surround sound amplifiers, a separate channel is included for each of the speaker outputs.

Amplifier Power Ratings

All power amplifiers are designed to deliver an output signal rated in watts. The power rating is usually rated for various load impedances. The most common is 8 ohms, with 4 ohms and 2 ohms also sometimes listed. These impedances are designed to match the load impedance of speakers connected to the output of the amplifier. Ohms are the unit of measurement for what is normally listed as the "nominal impedance."

It is important to remember that impedance ratings for speakers are not the same as resistance, although both are measured in ohms. The value of impedance varies with the frequency of the signal out of the amplifier. The impedance of a speaker is its opposition to an alternating current (AC). The lower the impedance, the higher the current; therefore, the speaker impedance varies above and below the nominal impedance when audio information such as music is amplified and passed to the speaker for reproduction. Keep in mind that resistance is the direct current (DC) resistance and impedance is a value that changes with frequency.

Amplifier Bridging

Bridging an amplifier is a technique used to configure a two-channel stereo amplifier to drive a single load (speaker) with more power than the sum of the two original channels. An amplifier running in bridged mode has a single output channel to which a load (speaker) can be connected. It is no longer a two-channel (stereo) amp as far as the input signals and loads are concerned. If you have two speakers and want to use bridged amplifiers, you will need two stereo amplifiers.

With bridged stereo amplifiers, a single input signal is applied to the amplifier. Inside the amplifier the signal is split into two signals: One is identical to the original, and the second is also identical, except it is inverted. The original signal is sent to one channel of the amplifier, and the inverted signal is applied to the second channel. The output produces two channels, which are identical, except one channel is the inverse of the other.

As shown in Figure 2.6, the speaker is connected between the left channel positive output terminals and the right channel negative output terminal. In other words, one channel pulls one way while the second channel pulls in the opposite direction. This allows considerably more power to be delivered to a single load than is the case with a single stereo amplifier. Amplifiers running in bridged mode are typically limited to speakers with impedance ratings of no less than 8 ohms.

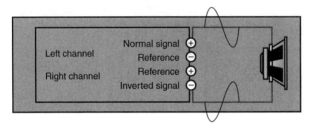

Figure 2.6 Bridged amplifier.

 A bridged amplifier can be used to increase the power available to a single speaker because the output voltage is effectively doubled, which theoretically increases the power output by a factor of 4.

Speakers

Speakers produce sound from the electrical audio signal developed by the amplifier. Speakers are available in various design categories and location options. Dynamic cone-type loudspeakers are the most common type of speaker used in home audio systems.

Dynamic Speakers

A *dynamic* speaker, also referred to as a *cone* speaker, is a type of speaker in which the audio signal is applied to a voice coil that is part of the moving system. As indicated in Figure 2.7, the voice coil is mounted in a way that enables it to move freely within the strong field of a permanent magnet. The cone assembly is attached to the voice coil, and the cone is attached to the outer suspension ring. The speaker cone moves air back and forth when the audio signal is applied to the voice coil, which in turn produces an audible sound reproduction of the electrical audio signal.

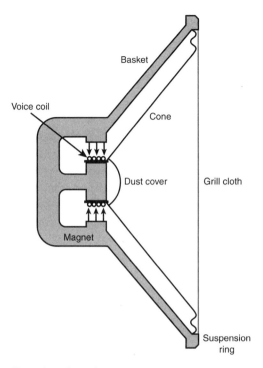

Figure 2.7 Dynamic loudspeaker.

Surround Sound Speakers

Surround sound systems use several speakers called *satellite* speakers and a subwoofer. *Subwoofers* are speakers that are designed to produce only the lower frequencies in the region of 20Hz–80Hz and usually include their own amplifiers.

Speaker Impedance

The impedance specification for a speaker refers to the resistance an amplifier will encounter when trying to drive a given speaker. Today, most loudspeakers are rated at 8 ohms, but this is an often misunderstood specification because, in reality, the impedance of a loudspeaker varies with its input audio frequency.

Speaker Power Rating

The power rating is a speaker specification that indicates how much power in watts the speakers can handle without damage.

Speakers are more likely to be damaged by an underpowered amplifier than a high-powered amplifier. This is because clipping with an overdriven low-powered amplifier causes distortion of the sound and can cause the amplifier or speaker to fail.

| When an amplifier is too small for the speaker system, high volume levels can cause the amplifier to automatically shut down due to clipping of the output audio signal. This is a problem that can be cured by using a higher-powered amplifier consistent with the power rating of the speakers.

Keypads

Keypads are used to control the components in a home audio/video system. They can be mounted in wall outlet boxes and be used to select the source for music and video components, turn power on and off, set the volume, and adjust the treble and bass settings for optimum room and area acoustics. Some keypads have IR sensors to enable remote control of functions with remote portable units. A typical wall-mounted keypad is shown in Figure 2.8.

Figure 2.8 Wall-mounted keypad.

Portable handheld remote keypads, shown in Figure 2.9, are also available for controlling audio/video equipment.

Remote units use IR wireless signals to control wall-mounted keypads equipped with IR receivers or to control the components at the source location. Many so-called "universal" remote units are programmable and can learn all the IR codes from other handheld remotes. This avoids the problem of having to deal with several handheld remotes from different equipment manufacturers.

Figure 2.9 Remote-control keypad.

Video Displays

Video displays accept and process video information from various sources and provide the output data as visual images. Examples of video sources found in home audio/video systems include desktop computers, laptop computers, TVs, closed-circuit TV cameras, camcorders, video projection systems, and high-definition television (HDTV) receivers.

Video displays are manufactured using one of the following three basic display technologies:

➤ Cathode-Ray Tube (CRT) Display Technology

➤ Flat-Panel Display Technology

➤ Video Projection Technology

Cathode-Ray Tube Displays

Cathode-ray tubes (CRTs) have been used as the main display component of television receivers and desktop computer monitors for several years. Although popular and relatively inexpensive, they are limited by economical

Chapter 2

and technology constraints to display areas of approximately 36". CRT monitors and TV displays are specified in size by the diagonal measurement of the tube front area.

The design features of a CRT are illustrated in Figure 2.10.

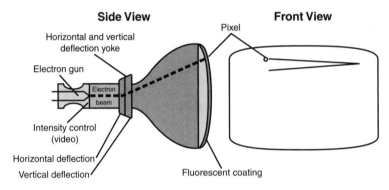

Figure 2.10 A CRT display.

Flat-Panel Display Technologies

Flat-panel displays get their name from the type of physical shape and appearance that distinguishes them from CRT displays. They are suitable for lightweight portable laptop computers and new large-screen HDTV receivers. The leading technologies for flat-panel displays are liquid crystal displays (LCDs) and plasma displays.

 Flat-panel and *flat-screen* displays are terms often used to describe both liquid crystal and plasma-type thin-panel displays. New CRT technologies such as the patented Sony Trinitron CRT as well as other newer computer CRT-type monitors also have a flatter surface than conventional CRT displays. They are often advertised as flat-screen displays. Be aware that the term *flat-screen* can be confusing because it can refer to either type of video display in some vendor specifications.

Liquid Crystal Displays

Liquid crystal displays are the most common type of flat-panel display. They are used for a wide range of small, portable electronic and computing devices, including laptop computers, cell phones, games, and personal digital assistants (PDAs). LCD displays use two sheets of polarizing material with a liquid crystal solution between them. An electric current passed through the liquid causes the crystals to align so that light cannot pass through them. Each crystal, therefore, is like a shutter, either allowing light to pass through or blocking the light.

Color LCD displays are created by adding a three-color filter to the panel. Each pixel in the display corresponds to a red, blue, or green dot on the filter. Activating a pixel behind a blue dot on the filter produces a blue dot onscreen. Similar to color CRT displays, the dot color on the screen of the color LCD panel is established by controlling the relative intensities of a three-dot (RGB) pixel cluster.

A display screen made with thin-film transistor (TFT) technology is an advanced liquid crystal display design that is common in top-of-the-line laptop computers and has a transistor for each pixel. TFT is also known as *active matrix* display technology and contrasts with *passive matrix* LCD technology, which does not have a transistor at each pixel. An active matrix display can be switched on and off very rapidly, presenting a fast, high-resolution appearance. Active matrix is the superior technology for LCDs.

Plasma Display Panels

Plasma display panels are a further improvement in flat-panel display technology. A plasma display is a type of advanced large-screen TV display, as shown Figure 2.11.

Figure 2.11 Large-screen plasma display.

Plasma display panels contain hundreds of thousands of tiny cells (called *pixels*) containing minute amounts of an inert gas sandwiched between two sheets of glass. Electrodes are placed in pairs on the inner side of the front plate. When electrically charged, they produce an ultraviolet beam, which activates the phosphorous coating of the cell transmitting light through the glass surface. Plasma displays are available for large-screen TV displays offering high resolution and a 16:9 aspect ratio compatible with the HDTV format. The flat-panel design allows them to be wall mounted or in component cabinets. Plasma displays are expensive and range in screen sizes from 42" to 50" with a thickness of only 4".

Keep in mind that flat-panel display systems use both plasma and LCD technologies. A CRT is not a flat-panel display. As mentioned earlier, be aware that some CRT monitors and TVs using a flat-faced CRT design are often listed in vendor specifications as *flat-screen* monitors and TVs.

Video Projection Systems

Video projection is a type of display that uses a high-intensity light source to project video images on a screen. They are used for large-screen TV displays as well as multimedia meeting presentations. Unlike plasma, LCD, and CRT displays, which are direct view technologies, projection systems are indirect view video displays in which the image size is enhanced by projecting the video image on a large screen. Video projection is accomplished using either front projection or rear projection.

Front Video Projection

Front projection is a method of viewing that uses a video projector and a separate pull-down screen to show the projected image similar to a movie screen. The video projectors are often ceiling-mounted for home theater installations, as shown in the Figure 2.12. Front-projection high-end systems can achieve an image size ranging from 3 feet to 25 feet. Front-projection systems are also used for computer-based presentations as well as projection TV systems for large rooms or auditoriums.

Figure 2.12 Front-projection TV.

Rear Video Projection

Rear video projection is a method of projection that combines a projector and viewing screen into one television unit. A typical rear projection unit is shown in Figure 2.13.

Figure 2.13 Rear-projection TV.

A rear-projection TV display contains an integrated front projector that is aimed at a mirror. The image is then reflected onto the rear of a display screen. The mirror enables the image to travel a sufficient distance between the projector and the screen to provide an image of 40"–80". Rear-projection systems are popular choices for home theater systems in which room size prohibits the use of a front projection system.

Touchscreens

Touchscreens are input devices that present graphical images for users who can then select options by touching the icons, symbols, or displayed keypads on the screen. Touchscreens use both CRT and LCD video technology. In addition, they are often used in home entertainment systems but are more applicable to commercial kiosks, library searches, point-of-sale cash registers, and computer-aided design applications. A touchscreen display is illustrated in Figure 2.14.

Figure 2.14 Touchscreen.

Source Equipment

Source equipment includes all components in a home audio/video system that play back, decode, capture, or process information and output the analog signals for processing by speakers or video display systems.

Audio/video distribution systems are used to select various source components for playing music or routing different video channels from source equipment, such as satellite receivers and DVD and VCR playback equipment, to different rooms. Source equipment for a home audio/video system includes the following components:

- ➤ DVD player/decoder
- ➤ Laser disk player
- ➤ Cable TV set-top box/decoder
- ➤ Off-the-air TV antenna
- ➤ Compact disc (CD) player
- ➤ Audio/video receiver

- Tuner
- Amplifier
- Personal video recorder (PVR)
- Video cassette recorder (VCR)
- Audio/video distribution system
- Satellite receiver/decoder
- Modulator

Configuration and Settings of Audio/Video Components

Residential Systems objective 2.4 states that the test taker should have knowledge of residential audio and video systems configuration. Examples of A/V items that HTI+ technicians should be able to configure include

- Volume settings
- Distribution channels
- Equalization
- Internal broadcasting

Volume Settings

Volume settings for audio programs in a whole-home system can be configured for speakers located in different zones. Wall-mounted volume controls and remote keypads can be used for each zone to set the volume for personal listening choices in each area. Home theater volume settings can be set for the surround sound speakers to fit the room acoustics.

Distribution Channels

Distribution channels are configured in the audio/video distribution panel mentioned earlier. Video and audio sources are connected by cables to the distribution panel. From the panel, they are routed using Category 5 UTP and coaxial cable to TV receivers and speakers throughout the home. Modulators are used for each video source, which allows separate video content to be viewed on television receivers by simply tuning to the correct channel set by each modulator. Audio channels are also switched in the distribution panel to enable separate content to be selected for different zones.

Equalization

Equalization is a sound processing function that is used to selectively boost or decrease bands of audio frequencies of a sound system. The most common equalization system used in home audio systems is the tone controls. They provide a quick and easy way to adjust the sound and partially compensate for the room acoustics. These controls are labeled "bass" and "treble."

Graphic equalizers, as shown in Figure 2.15, are a step up from tone controls in terms of capability and control.

Figure 2.15 Graphic equalizer.

A *graphic equalizer* is simply a set of filters, each with a fixed center frequency that cannot be changed. The only control you have is the amount of boost or in each frequency band. This boost or cut is most often controlled with sliders. This interface is pretty intuitive because the frequency response of the equalizer resembles the positions of the sliders themselves. The sliders are a graphic representation of the frequency response, hence the name "graphic" equalizer.

Internal Broadcasting

Internal broadcasting is a method that enables different video sources to be viewed at different locations in the home by tuning a cable-ready television receiver to an unused channel. *Unused channel* refers to selecting an unused UHF TV channel for each installed modulator. This unused channel must be in the range of channel 14–78 or CATV channels 65–135 and not be used in the local over-the-air or cable broadcast domain. Video sources can then be privately "broadcast" internally. Channels can be selected independently on each television receiver by each user to allow cable channels, a VCR, or a DVD to be viewed on each TV by selecting the correct configured modulator channel.

 Modulators are a key piece of hardware in a video distribution system. They provide the capability to create your own internal TV stations from local video sources such as a VCR, laser disc, DVD, or CCTV cameras. You can combine these new TV stations with your existing cable or off-air stations and easily distribute all signals throughout your home on a single coaxial cable.

Components Involved in Device Connectivity

Residential Systems objective 2.3 states that the test taker should have knowledge of A/V device connectivity concerns. These concerns include such items as follows:

- Component
- Composite
- S-video
- Analog audio
- Digital audio
- Low-voltage power lines
- Termination points

Video Signal Formats

Video connections on the rear panel of video consumer electronic devices are configured with different jacks to provide options for connecting television receivers, DVDs, VCRs, PC monitors, laser disc players, and camcorders. This section describes the various types of formats for both audio and video connections.

Video formats include

- Component video
- Composite video
- S-video
- RGB video

Audio formats include

- Analog audio
- Digital audio

Component Video

Component video is a format that provides the best color reproduction when connecting home theater products. Component video divides the signal into three components called the Y signal and the intermediate color difference signals: R-Y and B-Y. The luminance signal has a bandwidth greater than 6MHz, and the color difference signals have bandwidths greater than 3MHz. It is the format used by DVDs to store analog video information in a digital format and is also the preferred method for connecting DVD players to HDTV receivers. Component video connections require three bundled 75-ohm coax cables, which carry the Y, R-Y, and B-Y signals separately. Each of the three coax cables is terminated at both ends with RCA connectors. The component video input jacks to home theater products, shown in Figure 2.16, are color coded with the designations Y (green), Cr (red), and Cb (blue).

Figure 2.16 Component video jacks.

 Analog component video refers to a standard consisting of analog video signals transmitted on three cables—one for luminance (Y), one for the red component from which is subtracted the total luminance (R-Y), and one for the blue component from which is subtracted the total luminance (B-Y). The green component is derived by combining these three components. The designations Y/Pb/Pr and Y/Cb/Cr are also used loosely to represent analog component video and can appear on input/output jacks for home theater products.

 When connecting a DVD player to a TV receiver that uses component video inputs, three cables are required. Two additional cables are required for analog stereo sound (left and right) connections. Component video should not be confused with composite video, which is transmitted using a single cable plugged in to jacks usually colored yellow and labeled "video in" or "video out."

Composite

Composite video is a video format in which color (chrominance), luminance (brightness), and synchronization signals are mixed into a single radio carrier wave. It is the video standard used for television broadcasting. Inside the television receiver, the composite signal is separated into the red, green, and blue (RGB) color information. Composite video is the most widely used analog standard for connecting TVs with VCRs, laser disc players, and camcorders. As illustrated in Figure 2.17, a single coaxial cable with an RCA jack is required to connect audio/video using the (yellow color-coded) composite video jack.

Figure 2.17 Composite video connection.

S-video

S-video (separated video) is a compromise between component analog video and composite video because it separates the luminance (brightness) and chrominance (color information). It uses twin coaxial cables integrated into a singe cable and terminated with a DIN connector. The S-video DIN connector configuration is shown in Figure 2.18. Pins 1 and 3 in Figure 2.18 carry the luminance (brightness) signal and pins 2 and 4 carry the chrominance (color) information.

Figure 2.18 S-video connector.

> **NOTE** DIN is a common type of connector used for video and computer interfaces. DIN is an acronym for Deutsches Insitut für Normung eV, a standards setting organization for Germany.

Analog Audio

Analog audio connections are available on most TVs, VCRs, and CD players. They require a RCA-type jack with coaxial cable. The analog outputs include left and right stereo or mono, as illustrated in Figure 2.19. They are used with audio/video components that are not equipped with the preferred digital audio connectors.

Figure 2.19 Analog audio connectors.

Analog outputs available on amplifiers with digital surround sound decoders can be routed to each speaker for surround sound effects in a home theater setting.

Digital Audio

Digital audio refers to the transmission or reproduction of sound that is stored in a digital format. CDs, DVDs, digital audio tape (DAT), streaming audio, and sound files stored on a computer are examples of media that contain digital audio and video information. Digital sound files must be processed by a decoder to convert them to analog audio files. Analog audio information can be processed by an amplifier and coupled to a speaker system for the reproduction of sound.

Digital audio output jacks on home theater equipment use either coaxial cable RCA-type connectors or optical (TOSLINK) connectors. Figure 2.20 shows the optional coaxial cable or optical cable connectors available on most home theater sound system components.

Figure 2.20 Digital audio connectors.

TOSLINK is the name of a brand of optical cable developed by the Toshiba Corporation. Toshiba invented the concept of transferring digital audio signals inexpensively and efficiently using clear plastic cable and patented TOSLINK connectors. CD players, DVD players, mini-disc players/recorders, CD recorders, some MP3 equipment, and some Dolby Digital receivers have TOSLINK input and output jacks.

Low-voltage Cabling and Connectors

Low-voltage wiring includes all cabling in a home audio/video distribution system. Low-voltage cable also includes audio and video patch cables and speaker wire used to connect individual system components. Homes wired with modern structured cabling standards include dual Category 5 and dual RG6 cabling outlets in each room. All cables are run from each room to a central distribution box. This is a *home run*-type wiring topography that supports all the distribution system connectivity discussed earlier for both video and audio distribution for whole home designs.

Termination Points

Termination points include all connectors, jacks, and cable terminations used in home audio/video systems. Table 2.1 lists the various types of cable terminations used for specific audio and video applications.

Table 2.1 Audio and Video Cable Terminations and Jacks

Jack Type	Type of Cable or Wire	Application
RCA	Coaxial cable	Analog audio
RCA	Coaxial cable	Digital audio
Optical (TOSLINK)	Optical cable	Digital audio
F-type	Coaxial cable	Satellite/cable TV
S-video	4-pin DIN coaxial cable	Analog video
RJ-45	Cat 5 UTP	Audio/digital data
Binding posts	Speaker wire	Audio for speakers

In-house Services

Residential Systems objective 2.6 states that the test taker should have knowledge of in-house services commonly used with residential audio and video systems. These services include items such as

- Streaming audio
- Streaming video
- PVR
- Media server

Streaming Audio and Video

Streaming is a popular technology for capturing audio and video files and playing them on a computer while they are being downloaded from an Internet Web server. This avoids time-consuming downloads of large files. Viewing or listening to streaming files requires a player application on the user's client computer. Most players are free and are available for installation on Windows and Macintosh computers. Home audio/video distribution systems can be integrated with home computer networks to play streaming audio and video files throughout the home.

Some typical streaming audio and video players and file extension names are as follows:

- RealNetwork's RealOne Player (.rm file extension)
- Microsoft's Media Player (.wmf file extension)
- Apple Computer's QuickTime (.mov file extension)

When audio or video is being streamed from the Internet, a small buffer space is created on the user's home computer and data starts downloading into it. As soon as the buffer is full, the file starts to play. As the file plays, it uses up information in the buffer, but while it is playing, more data is being downloaded. As long as the data can be downloaded as fast as it is used in playback, the file plays smoothly.

Personal Video Recorder

The personal video recorder (PVR) is a new type of consumer electronics equipment that records television programs digitally encoded in the compressed MPEG format on a nonremovable hard disk. Similar to a VCR, a PVR has the capability to pause, rewind, stop, or fast forward a recorded program. Because the PVR can record a program and replay it almost immediately with a slight time lag, real-time programming can be paused and resumed by the user without the loss of any portion of a TV broadcast. It also can be programmed to record programs at selected times similar to a VCR but without the use of tape cassettes, which have limited playing/record capacity.

PVRs should not be confused with VCRs. PVRs use nonremovable hard drives and record television programs using digital compression technology. VCRs use tape cassettes and record composite video signals in an analog format. Also remember that PVRs can also be loosely referred to as digital video recorders (DVRs).

Media Server

A *media server* is a high-performance computer system designed to provide downloads for audio and video streaming files stored on large-capacity hard drives. Most media servers are used by commercial service providers for pay per view (PPV) and video on demand (VOD) multimedia services. Products for the home audio/video media server market have recently been introduced that are derived from the technology used for personal video recorders mentioned earlier.

A media server can also be used with a home network media management system. A media server for the home has a high-capacity hard drive and an operating system. The audio and video outputs from the server go to the home theater system, and the server is connected to the home Internet gateway for downloading MP3 files and other streaming audio and video files. Music and video libraries can be developed by storing hundreds of CDs and DVDs on a home media server hard drive. Play lists can be retrieved on a display screen and played back on demand to any location by the server software.

Externally Provided Audio and Video Services

Residential Systems objective 2.7 states that the test taker should have knowledge of externally provided audio and video services. These services include

- Satellite service
- Cable service
- Terrestrial/off-air service
- Internet service

Satellite Service

Direct broadcast satellite (DBS) service is a class of service defined by the Federal Communications Commission whereby home subscribers receive television programs by direct paid subscription access to a satellite rather than through an intermediate cable service provider.

DBS service is provided by satellites operating over the United States with assigned Ku band frequencies using MPEG-2 compression, which supports a technology for the use of small (18") earth station antennas. DBS service offers the advantage of hundreds of channels and reception anywhere in the 48 mainland states. Currently, subscription DBS service is available from two service providers: DirecTV (Hughes Electronics Corporation) and DISH Network (EchoStar Communications Corporation).

DBS decoder boxes are required for receiving scrambled TV signals from the DBS satellite. The decoder/descrambler decoder boxes are usually installed with other audio/video components in a home theater area.

Cable

Cable TV service provides television programming service to homes over coaxial cable networks. Cable TV service providers design the network in a way that permits regular TV sets to receive cable channels similar to the way they receive over-the-air channels. Cable systems usually provide approximately 60 analog channels. The modern cable networks that have been upgraded to hybrid fiber/cable (HFC) can carry more than 110 channels. Many service providers offer digital service using MPEG-2 compression that enables hundreds of channels over a single cable. Premium digital channel service requires a decoder set-top box at the subscriber home location to receive the digital/MPEG-2 encoded channels.

Cable service was originally called community antenna television (CATV). It is often referred to (incorrectly) as *ca*ble TV (CATV). CATV was developed originally in 1949 in Pennsylvania and Oregon where entrepreneurs built antennas on top of mountains or buildings to improve television reception in fringe reception areas. The received signals from distant broadcast TV stations were amplified and distributed for a fee to subscribers in the local community. This was the genesis of the cable TV industry that provides most of the cable-delivered TV programming services today. So-called *cable channels* such as CNN are initially distributed worldwide by satellite and downloaded to cable companies for local cable network distribution.

External Terrestrial Off-the-Air Broadcast Service

Terrestrial television and AM/FM radio reception is the mode used with an outside antenna pointed toward commercial broadcast stations. Signals are received by the antenna and routed by a cable to the TV receiver and AM/FM tuner. Terrestrial/off-the-air reception requires an antenna and receiver that tunes to the TV channel frequencies. AM and FM reception can often be accomplished with indoor antennas in urban areas. All radio and TV broadcasting stations are assigned TV channels and radio frequencies by the FCC. The two main TV frequency bands are

- *VHF (very high frequency)*—Provides channels 2–13 in the frequency range 54MHz–216MHz.

- *UHF (ultra high frequency)*—Provides channels 14–69, in the range of 470MHz–806MHz. Previously, they went up to channel 83 (890MHz), but channels 70–83 were reallocated in 1974 to make spectrum available for cellular telephone service.

The AM radio frequency band extends from 530KHz to 1700KHz, whereas the FM broadcast band extends from 88MHz to 108MHz.

Digital Television Service

DTV is a new over-the-air digital television system that will be used by the nearly 1,600 local broadcast television stations in the United States. The DTV standard is based on the Advanced Television System Committee (ATSC) standard A/53 covered later in the section titled "Television Transmission Standards." Television stations will operate two channels during the transition, which includes an existing analog channel as well as a new DTV channel. The analog channel will allow consumers to continue to use their current TV sets to receive traditional analog programming during the transition. The DTV channel, on the other hand, will allow consumers to receive new and improved services with new DTV sets or with special converter boxes that will enable some DTV programs and services to be viewed on existing analog TV sets.

High-definition TV

Technically, HDTV is a type of DTV as defined by the rules and regulations of the Federal Communications Commission. The emerging DTV system is really a multifaceted standard. One particular facet that is receiving most of the publicity is known as HDTV.

External radio and television broadcasts that are received over the air are free. They are not encoded or encrypted and can be received by installing the proper antenna, AM/FM receiver, and television receiver. Other television programming services discussed in this section such as cable TV and DBS are paid subscription services. Keep in mind that HDTV is one type of DTV service that is also being inaugurated by broadcasters as a free over-the-air service on new channels similar to existing standard definition (SDTV) broadcasts.

Internet Audio and Video Services

As discussed earlier, Internet streaming audio and video is a type of externally provided service that can be integrated into the design of a home audio/video system. Streaming technology enables movies, animation, distance learning, teleconferencing, Internet seminars, MP3, Internet radio, and many new Internet applications.

Audio- and Video-Related Standards

Residential Systems objective 2.8 states that the test taker should have knowledge of standards associated with residential audio and video subsystems. These standards include topics such as

- ➤ Wiring standards
- ➤ Protocol standards
- ➤ Resolutions
- ➤ Video formats
- ➤ Organizations
- ➤ IEEE standards

Wiring Standards

Homes that are wired for A/V systems use a structured home run wiring scheme that conforms to the EIA-600 Standard. This wiring standard is applicable to all types of home wiring and is explained in greater detail in Chapter 12, "12.0—Systems Integration."

Protocol Standards

Protocols are sets of rules usually implemented in software to standardize a communications or control process in a computer and telecommunications system. A recent protocol development is used to enable audio and video components to exchange control information under a single central controller. This major new protocol adopted for the home entertainment industry is called the Home Audio Video Interoperability Protocol.

Home Audio Video Interoperability

Home Audio Video Interoperability (HAVi) is a standard protocol used in audio/video equipment and was developed by several leading electronics and computer manufacturers. It is a digital networking initiative that provides a home networking software specification for seamless interoperability among home entertainment products.

The specification uses the IEEE-1394 external bus standard as the interconnection medium.

Resolutions

The *resolution* of an image describes how fine the dots are that make up that image. The more dots there are, the higher the resolution. A 300dpi (dots per inch) printer is capable of printing 300 dots in a line 1" long. This means it can print 90,000 dots per square inch. When displayed on a monitor, the dots are called *pixels*. A 640 × 480-pixel screen is capable of displaying 640 distinct dots on each of its 480 lines, or about 300,000 pixels.

Later in this section, we will investigate further the resolution of the analog and digital television transmission standards as well as audio and video recording formats.

Video Recording Formats

Video formats used for consumer electronic products are classified according to the number of horizontal lines of resolution contained in each frame. A higher number of lines produces a progressively higher-quality picture.

The three basic video formats are called VHS, 8mm, and digital and are described in the following paragraphs.

VHS

VHS (Video Home System) is a recording and playback method used by standard VCRs. It has a horizontal resolution of 250 lines. VHS-C

(VHS-compact) format was introduced in 1982 as a derivative of the VHS format for small camcorders.

Super VHS

Super VHS (S-VHS) uses a higher bandwidth to process signals with superior quality compared to the standard VHS format. Used in both the large and VHS-C cassette types, S-VHS achieves a higher picture quality to VHS due to the improved signal processing. It has a higher horizontal resolution of 410 lines.

8mm

The 8mm tape cartridge is the same size as the VHS-C cartridge (it's much thinner, though), yet contains up to 120 minutes of video at SP. Although it supports the same lines of resolution as VHS (less than broadcast quality), it has vastly improved stereo audio capabilities and is a much higher-quality medium, making duplication of 8mm videotapes practical. It has a resolution of approximately 260 lines.

HI-8

HI-8 is also an 8mm tape format but has a much higher oxide quality, enabling a higher density of signal information (more lines of resolution) to be recorded on the same length of tape. HI-8 systems provide more than 400 lines of resolution, significantly better than NTSC broadcast quality. HI-8 tape is used by many brands of camcorders.

Mini DV

Digital video (DV) is a digital format that uses a miniature cassette that is only 1/12 the size of a standard VHS format tape. The standard play recording time for these mini DV cassettes is one hour. DV offers the best video quality available, in excess of 500 lines of resolution, compared to the 410 lines for HI-8 and SVHS. Audio is recorded at CD quality.

Digital8

The Digital8 format was released in February 1999 by Sony in support of its line of consumer camcorders. This format uses the 8mm or HI-8 tape cartridge to record the digital information. With a HI-8 cartridge, resolution in excess of 500 horizontal lines can be achieved with this digital format. These camcorders have the added advantage of also being able to play back standard HI-8 and 8mm tapes.

Betacam

Betacam is a high-quality professional video format based on the component video standard, and it was developed by Sony. The Betacam format records all three signal components (Y, U, and V) independently, so minimal signal loss occurs during the record/playback process. The signal carrying the color information—U and V—is time compressed and recorded onto one video track, whereas the luminance—or Y—signal is recorded onto a second track. The use of two separate tracks eliminates the cross-color and cross-luminance effects inherent in composite recording. Both tracks are recorded using high-frequency FM carriers.

Betacam SP

Betacam SP, also developed by Sony, uses cassettes and transports similar to the old Betamax home video format, but the performance is dramatically improved over Betamax. Tape speed is six times higher and luminance and chrominance are recorded on two separate tracks. Betacam SP uses metal particle tape and a wider bandwidth recording system (which provides better picture quality than Betamax).

Surround Sound Formats

Surround sound has become the preferred audio format for home theater installations. Dolby Laboratories licenses many of the audio technologies described in this section that provide multichannel sound systems for audio/video home theater systems. Dolby has introduced several variations of the Dolby Digital surround format. The formats you will need to remember are included in the following list of surround sound options:

- ➤ Dolby Surround
- ➤ Dolby Pro Logic
- ➤ Dolby Pro Logic II
- ➤ Dolby Digital (AC-3)
- ➤ Dolby Digital
- ➤ THX Surround EX

Dolby Digital AC-3 has six discrete channels (5.1) of digital sound. 5.1 indicates five channels—five with full bandwidth and one dedicated to bass enhancement called low frequency effects (LFE).

The THX Surround EX format was jointly developed by Lucasfilm THX and Dolby Laboratories and is the home theater version of Dolby Digital Surround EX, an extended surround sound format used by state-of-the-art movie theaters. Lucasfilm THX licenses the THX Surround EX format for use in receivers and preamplifiers. In November 2001, Dolby Laboratories started to license THX Surround EX under its own name, Dolby Digital EX, for consumer audio/video home theater equipment.

Digital Theater System Formats
Digital Theater System (DTS) is another digital surround format. Similar to AC3 or Dolby digital, DTS also has a 5.1 channel format.

Television Transmission Standards
The four existing analog world television transmission standards in use today are

- National Television System Committee (NTSC)
- Advanced Television Systems Committee (ATSC) Digital Television Standard (DTS) A/53
- Phase Alternating Line (PAL) (used in Europe)
- Sequential Couleur Avec Memoire (SECAM) (used in France)

The National Television Systems Committee
The National Television System Committee is an organization that developed the standard as well as the name for transmission of television signals in the United States. It prescribes a vertical resolution of 525 lines on the TV tube (the cathode-ray tube). The lines are stacked on top of each other from the top to the bottom, and the field rate is set at 60 fields displayed per second.

A *field* is defined as a set of even lines or odd lines. The odd and even fields are therefore transmitted to the TV set sequentially. All odd and even sets of lines are transmitted and interlaced on the screen producing a full-frame period of interlaced lines in approximately 1/30 of a second. The NTSC system is therefore called a 30 frames-per-second system. NTSC is also used to describe the standard for recording television signals on tape such as VHS. NTSC is used in the United States and several other countries.

The analog commercial television transmission standard for analog TV in the United States is the NTSC standard. It is a 525-line 30-frames-per-second interlaced standard. PAL and SECAM are transmission standards used in Europe and France, respectively.

Advanced Television Systems Committee and DTV
DTV is a new over-the-air digital television standard that will be used by the nearly 1,600 local broadcast television stations in the United States. The DTV standard was incorporated into the FCC Rules, by reference to ATSC Document A/53 ("ATSC Digital Television Standard, 16 Sep 95"). The DTV standard will enable commercial broadcasters to provide

higher-quality services. First, DTV will permit transmission of television programming in new wide-screen, high-resolution formats known as HDTV. In addition, the new DTV television system will enable transmissions in SDTV formats that provide picture resolution similar to existing television service. Both the HDTV and SDTV formats will have significantly better color rendition than the existing analog television system. None of the DTV transmissions will be capable of being received on existing analog TV receivers.

High-definition TV Standards

All DTV formats are transmitted using MPEG-2 compression. Sixteen DTV formats with different frame rates, interlaced and progressive scanning, and aspect ratios exist. HDTV uses 6 of the 18 formats adopted for the DTV standard by the ATSC. The remaining 12 standards are collectively identified as SDTV. The 6 HDTV standards with the corresponding screen resolution, aspect ratio, frames per second, and scan mode are shown in Table 2.2.

Table 2.2 HDTV Standards

Horizontal Lines	Vertical Lines	Aspect Ratio	Frames per Second	Scan Mode
1,920	1,080	16:9	30	Interlaced
1,920	1,080	16:9	30	Progressive
1,920	1,080	16:9	24	Progressive
1,280	720	16:9	60	Progressive
1,280	720	16:9	30	Progressive
1,280	720	16:9	24	Progressive

The DTV standard is a transmission standard, not a display standard. Broadcasters are not forced to use an interlaced display just because it exists in the standard. Most broadcasters, however, have elected to use the 1080i interlaced format. It is the only one out of the six standards that uses interlaced scanning. A few broadcasters are planning tests in the future with the 720p (progressive scan) format. These are the only two HDTV formats adopted at this time by commercial broadcasters. The two formats are described as follows:

> *The 1080i interlaced transmission standard*—Provides 1,080 interlaced lines of resolution (more than double the current 525 lines) using interlaced scanning. It first transmits all the odd lines on the TV screen and then transmits all the even lines.

► *The 720p progressive format*—Offers 720 lines of resolution. This offers slightly less resolution but constructs, or *paints*, the screen image in a single pass. Progressive scan transmission eliminates some irregularities inherent to the interlaced display that engineers refer to as *artifacts*. Due to the higher horizontal scan rate, 720p requires more transmission bandwidth than the 1080i standard and is not compatible with most existing TV broadcast transmission equipment.

Unlike the current NTSC TV standard, the HDTV screen has a 16:9 aspect ratio, which means the screen is not square but similar to the letter box mode occasionally used for movie telecasting.

The FCC has mandated that all stations be capable of broadcasting HDTV by 2006.

It is important to remember that the FCC policy requires commercial TV stations to broadcast all their transmissions in HDTV format by the year 2006. The HDTV transmission mode includes 6 of the standard transmission rates, which are a subset of the 18 DTV standards adopted by the Advanced Television Systems Committee. You also need to keep in mind that the majority of broadcasters have adopted the 1080i interlaced DTV standard for HDTV transmissions.

Institute for Electrical and Electronic Engineers Standards

The IEEE-1394 standard is a high-speed external bus interface standard adopted by the Institute of Electrical and Electronics Engineers (IEEE) for very fast (400 megabits per second) digital data transfer, especially applicable for streaming video, camcorders, data storage products, and many other audio/video components. All camcorders use a 4-pin connector for their 1394 interfaces, and nearly all computers use a 6-pin connector for their 1394 interfaces. A 6-pin-to-4-pin configuration is required to connect a computer to a camcorder. Cables can be up to 15 feet in length and can connect up to 63 computer peripheral devices. Products supporting the IEEE-1394 standard are sold under various brand names, depending on the company. Apple, which originally developed the technology, uses the trademarked name FireWire. Sony has a trademarked 1394 product called i.Link, and the Texas Instruments product name is Lynx.

The IEEE-1394 external bus standard might appear in the exam under the product names of FireWire, i.Link, and Lynx. They are all product names that support the IEEE-1394 bus standard. The bus speed specification is 400Mbps, and it is capable of supporting up to 63 peripheral devices.

Audio and Video Installation Plans and Procedures

The Residential Systems objective 2.9 states that the test taker should have knowledge of the installation plans and procedures associated with residential audio and video systems.

Installing a Satellite System

DBS outdoor antennas must be installed using the instructions included with the hardware kit. Most DBS antennas include provisions for making the final adjustments using the signal strength meter display on the TV receiver screen.

NOTE The term *LNB* used in the text is an acronym for low noise block-down converter (so called because it converts a whole band, or block, of frequencies to a lower band). A triple LNB with an elliptical dish can receive signals for all DBS and EchoStar satellites located in the orbital slots at 101°, 110°, and 119° west longitude with moving the antenna.

DBS dish-type antennas are equipped with an LNB that converts the received satellite signal to a lower frequency. The LNB requires a connection to a coaxial cable for sending the received signal to the indoor decoder box.

Guidelines for installing a satellite antenna system for receiving DBS service are as follows:

➤ The antenna location must have a clear view of the southern sky. The antenna distance above the ground is not important.

➤ You can determine the correct look angle to the selected DBS satellite from the installation geographical coordinates using a program that is usually included with the installation kit. Figure 2.21 illustrates the two coordinates of azimuth and elevation that must be calculated.

➤ The antenna must be grounded according to the National Electrical Code (NEC) and local building codes to protect the system from lightning damage. It is very important that both the antenna and the coaxial cable be grounded to a central common point building ground.

➤ Type RG6 quad-shield coaxial cable should be used to connect the feed assembly on the antenna to the indoor decoder box.

➤ Dual LNB antennas designed for feeding to decoder boxes requires two RG6 coaxial cable feeds, as illustrated in Figure 2.22.

118 Chapter 2

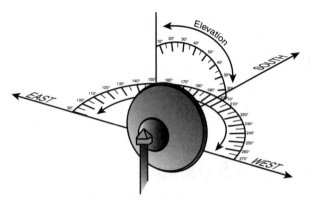

Figure 2.21 Antenna azimuth and elevation.

Figure 2.22 Dual LNB.

➤ The multisatellite triple LNB antenna, as shown in Figure 2.23, is capable of receiving signals from all three DirecTV satellite locations at 101°, 110°, and 119° west longitude simultaneously. The dish comes with an integrated 4 × 4 multiswitch for easy connection to four receivers.

Figure 2.23 Multiple satellite antenna.

Remember the basic requirements for selecting a sight and installing a satellite antenna. The height above ground for the antenna is not important; however, the site must have a clear view of the southern sky with no trees or buildings in the receiving path of the antenna. The antenna must also be grounded to protect against lightning damage.

Installing Speakers

Speakers can be installed in walls or in the front and side areas from the audience in a home theater room.

The speaker distance from the amplifier must be considered when selecting the proper gauge speaker wire. Speaker wire is available in a variety of lengths and thicknesses; the thickness is measured the same as power wiring using the American Wire Gauge number. The lower the gauge number, the thicker the wire. Speaker wire is generally available in the range from 12-gauge to 16-gauge. The thicker wire is capable of passing the audio signal from the amplifier output to the speaker with less resistance.

When wiring a multiroom audio system with long runs from the amplifier, use the guidelines recommended in Table 2.3 for selecting the minimum wire gauge.

Table 2.3 Minimum Wire Gauge Guidelines	
Distance from Speaker to Amplifier	Gauge
Less than 80 feet	16
80–200 feet	14
More than 200 feet	12

The general rule for installing speakers is to first limit the distance between the amplifier and speaker to a maximum of 50 feet. If this is not possible, a minimum 14-gauge should be used for runs of 100 feet. 16-gauge is the minimum gauge speaker wire recommended for any location.

Audio and Video Maintenance Plans and Procedures

The Residential Systems objective 2.10 states that the test taker should have knowledge of maintenance issues associated with typical audio and video systems found in residential settings.

Guidelines for Audio and Video System Maintenance

The following are recommended maintenance and operating guidelines that will aid in extending the life of components in a home audio/video system:

- *Maintain good speaker connections and avoid shorting out the wire pairs.*

- *Turn down the volume before you turn off a receiver*—The volume will remain at the level you turned it off at.

- *Keep the equipment dry*—If it does get wet, wipe it dry immediately. Liquids might contain minerals that can corrode the electronic circuits.

- *Use and store the audio/video components only in normal temperature environments*—High temperatures can shorten the life of an electronic device, damage batteries, and distort or melt plastic parts.

- *Handle the system components gently and carefully*—Dropping them can damage the circuit boards and cases.

- *Install the components away from dust and dirt, which can cause premature wear*—Wipe the audio/video components with a dampened cloth occasionally to keep them looking new. Do not use harsh chemicals, cleaning solvents, or strong detergents to clean the system.

Exam Prep Questions

Question 1

> When connecting a DVD player to a new HDTV, which type of connection would you use to obtain the best video quality?
> - A. S-video
> - B. Composite video
> - C. Component video
> - D. VHS

The correct answer is c. S-video and composite connections are also used for connecting video components but at a reduced level of quality; therefore, answers A and B are incorrect. VHS is an older standard used for analog video tape recording, so answer D is incorrect.

Question 2

> Which type of connector is used with an S-video cable?
> - A. DIN
> - B. RCA
> - C. RJ-45
> - D. Optical

The correct answer is A. S-video coaxial cable uses twin coaxial cables integrated into a singe cable and terminated with a DIN connector. Pins 1 and 3 of a DIN connector carry the luminance (brightness) signal, and pins 2 and 4 carry the chrominance (color) information. RCA jacks are used with single coaxial cable, so answer B is incorrect. RJ-45 cables are used for terminating UTP cable, so answer C is incorrect. Optical cables are normally used with digital audio and video cables, so answer D is incorrect.

Question 3

Which of the following is not a name for IEEE-1394 external bus products?
- ○ A. FireWire
- ○ B. i.Link
- ○ C. Lynx
- ○ D. TOSLINK

The correct answer is D. TOSLINK is a product name for optical connectors and cables developed by Toshiba. All the other three names are brand names for IEEE-1394 external bus products, so answers A, B, and C are incorrect.

Question 4

Which one of the following statements describes the function performed by a modulator?
- ○ A. It converts the video source signal (surveillance camera, satellite receiver, VCR, and DVD) player to an unused television channel for internal broadcasting on the video distribution system.
- ○ B. It performs a sound processing function that is used to selectively boost or decrease bands of audio frequencies.
- ○ C. It converts the received satellite signal to a lower frequency for use by the indoor decoder box.
- ○ D. It consolidates video inputs from source equipment and connects them to the video input jack on the distribution panel.

The correct answer is A. Answer B describes the functions performed by an equalizer, so it is incorrect. Answer C describes the purpose of an LNB, so it is incorrect. Answer D is the task performed by combiner, making it incorrect.

Question 5

Which one of the following camcorder tape formats provides the highest resolution?
- ○ A. HI8
- ○ B. Digital8
- ○ C. Super VHS
- ○ D. 8mm

The correct answer is B. Digital8 camcorders use a digital recording format with a resolution in excess of 500 lines. The other formats listed in answers A, C, and D are analog tape formats with less resolution, making them incorrect.

Question 6

Select the one statement that is not true concerning the location or performance of a subwoofer in a home theater installation.

- A. A subwoofer should always be placed behind the listener for the best stereo and low-frequency effect.
- B. The subwoofer does not contribute to stereo imaging; therefore, adding a second subwoofer does not enhance the surround quality.
- C. Subwoofer placement is not critical because the low-frequency sounds are not sensitive to direction.
- D. The ".1" in the Dolby 5.1 designation signifies that the sixth (subwoofer) channel is not a full-frequency channel because it contains only deep bass frequencies in the range of 3Hz–120Hz.

The correct answer is A. Because the location of the subwoofer is not critical and does not contribute to stereo imaging, answer A is the untrue answer. Answers B, C, and D are all true. As indicated in answer D, the ".1" designation in 5.1 surround sound systems is not a full-frequency channel and is added for LFE.

Question 7

Which analog television transmission standard is used for broadcasting television programs in the United States?

- A. SECAM
- B. PAL
- C. NTSC
- D. VGA

The correct answer is C. The National Television System Committee developed the standard as well as the name for the transmission of television signals in the United States. SECAM and PAL are television standards used in France and Europe, respectively; therefore, answers A and B are incorrect. VGA is an older graphics video mode used with computer displays, so answer D is incorrect.

Question 8

Which type of video display technology is most often used in the manufacturing process of portable electronic and computing devices, including laptop computers, cell phones, games, and PDAs?

- ○ A. Plasma
- ○ B. CRT
- ○ C. Front projection
- ○ D. LCD

The correct answer is D. LCDs are the choice for a large number of portable devices. Plasma displays are more expensive and are used for large flat-panel TV displays, so answer A is incorrect. CRT and front projection systems are not portable technologies and are used for medium and large screen television and computer display systems, so answers B and C are incorrect.

Question 9

Which of the following statements is not true regarding speakers? (Select all that apply.)

- ○ A. Impedance ratings and resistance for speakers are not the same, although both are measured in watts.
- ○ B. A cone speaker is a type of speaker in which the audio signal is applied to a voice coil that is part of the moving system.
- ○ C. The power rating of a speaker is measured in ohms.
- ○ D. The general rule for installing speakers is to first limit the distance between the amplifier and speaker to a maximum of 50 feet if possible.

The correct answers are A and C. Answer A is untrue because impedance and resistance are measured in ohms and not watts. Answer C is also an incorrect statement because power ratings for speakers are measured in watts, not ohms. The other two answers are true statements and therefore are incorrect answers. Answer B is not the correct answer because a cone speaker has a component called a voice coil that moves. Answer D does not qualify as a correct answer because the general rule for installing audio equipment requires a practical limitation of 50 feet between the amplifier and the speaker.

Question 10

> Which class of service is defined by the FCC whereby home subscribers can receive television programs by direct paid subscription access to a satellite rather than through an intermediate cable service provider?
>
> ○ A. VHS
> ○ B. DTS
> ○ C. PVR
> ○ D. DBS

Answer D is correct. The FCC term for the type of service licensed for broadcast direct from the satellite to subscribers is the direct broadcast service (DBS). Answer A is incorrect because VHS is an acronym for video home system, a recording and playback method used by standard VCRs. Answer B is incorrect because DTS is an acronym for digital theater system, a surround sound format that is similar to AC3 or Dolby digital. Answer C is incorrect because PVR is an acronym for personal video recorder. PVRs are designed to record television programs digitally encoded in the compressed MPEG format on a nonremovable hard disk.

Need to Know More?

 Fleischmann, Mark. *Practical Home Theater: A Guide to Video and Audio Systems*. 1st Books Library, 2002. (This book answers most of the questions regarding the purchase and installation of home theater products.)

 Lampen, Stephan H. and Steve Chapman. *Audio/Video Cable Installer's Pocket Guide, First Edition*. Berkeley, CA: McGraw-Hill Professional, 2002. (This book is intended for those who are planning to wire or remodel a home for audio/video systems. It includes a description of cables, connectors, and audio/video components.)

 Harley, Robert and Tomlinson Holman. *Home Theater for Everyone: A Practical Guide to Today's Home Entertainment Systems, Second Edition*. Acapella Publishing, 2002. (This book serves as a buyer's guide for today's home entertainment systems. It also includes instructions for connecting components in a home theater system.)

3.0—Home Security and Surveillance Systems

Terms you'll need to understand:

- ✓ Remote access
- ✓ Bypass mode
- ✓ Quad
- ✓ ANSI/TIA/EIA-570
- ✓ Video switchers
- ✓ Passive infrared sensor
- ✓ Charge coupled device (CCD)
- ✓ Zone layout
- ✓ Hard-wired
- ✓ RJ-31x
- ✓ Glass break sensor
- ✓ Camera resolution
- ✓ Lux rating
- ✓ CCTV

Techniques you'll need to master:

- ✓ Connecting a quad switcher to a VCR and cameras
- ✓ Planning the installation of a smoke detector
- ✓ Identifying the location of a glass break detector
- ✓ Planning a security system zone layout
- ✓ Selecting locations for keypads
- ✓ Evaluating a wireless and hard-wired security system
- ✓ Selecting a proper password for a monitoring station
- ✓ Comparing the light sensitivity levels of various brands of surveillance cameras
- ✓ Identifying the components contained in a security control panel

A home security and surveillance system is an essential part of any modern automated home. The basic design of a security system begins with analyzing the needs of the inhabitants, surveying existing technology and hardware, reviewing system costs, considering monitoring choices, and finally planning the installation.

In addition to perimeter and interior protection offered by a security system, surveillance monitoring includes features that enable the inhabitants to observe environmental conditions inside and outside the home when at home or away.

In a home surveillance system, video cameras and display systems are considered by most contractors to be optional items. Subject to the homeowner's choice, surveillance equipment can function independently from the basic features included in home security monitoring.

This chapter provides the information you will need to know as a home integrator for designing, integrating, and installing a home security and surveillance system.

Design Considerations

The design of a security and surveillance system should provide for the protection of the entire perimeter of a home as well as visual- and audio-based surveillance monitoring. Security system sensors are available that are designed to detect sound, window and door intrusion, air movement, body heat, motion, and other conditions that indicate an intruder is present. A good security system design should consider the best plan for existing homes as well as new construction. It should also consider the lifestyle of all the inhabitants, the location of valuables or any items to be protected, how the system is to be controlled, adequate smoke and fire alerting sensors, and the type of emergency response required. The design choices are numerous and varied due to advances in home security technology and the wide availability of compact, low-cost video surveillance systems.

Wireless Security Systems

Wireless home security systems use battery-powered radio transmitters and receivers to connect the various components such as cameras, sensors, area motion detectors, sirens, central controllers, smoke/fire detectors, keypads, and video displays. These types of security systems are usually available at a local hardware store or on the Internet and are often designed for do-it-yourself installation. The basic advantages of wireless security systems are

➤ *Wireless systems are easy to install*—They avoid the expensive and time-consuming task of installing new wires in the walls of existing homes.

➤ *Wireless systems enable you to take the components with you when moving to a new location.*

➤ *Wireless sensors are designed to transmit a unique identification code to a controller*—The controller learns the identity of each sensor and links it to an appropriate zone. Each sensor also transmits status information such as battery voltage, condition of the sensor switch, and other diagnostic messages.

➤ *Some homeowner-installed wireless security systems can be set up to record a voice message and to call programmed numbers in the event of an alarm*—This saves the cost of a professional monitoring service.

➤ *Wireless sensors, motion detectors, and video cameras can often be installed in locations that are not accessible for wired equipment.*

The basic disadvantages of wireless security systems are

➤ Wireless system design specifications can limit the distance between sensors, cameras, and the central controller.

➤ Wireless systems can be vulnerable to electromagnetic interference (EMI) in some locations.

➤ They require periodic replacement of batteries.

➤ Most professional builders recommend wireless systems as a last choice.

Hard-wired Security and Surveillance Systems

Hard-wired security and surveillance systems use wires installed inside the walls, attics, crawl spaces, and underground to connect the sensors to a central controller. Surveillance cameras or microphones are also wired to speakers, video switchers, and video display monitors. A hard-wired system design normally uses power from the home AC power wiring as the primary source. A rechargeable battery pack is used by the controller for backup during power outages. The main components of a hard-wired system are similar to a wireless system but without the radio receiver and transmitter components. They include a central control panel, sensors, one or more keypads, motion detectors, smoke and fire sensors, cameras, camera switchers, video displays, and sirens. The advantages for a hard-wired security system are

- Hard-wired security systems are considered by most contractors to be more reliable than wireless systems.
- Hard-wired systems are usually installed by a professional security system contractor with warranties and maintenance support.
- Hard-wired systems avoid the problem of EMI and radio range limitations inherent in some wireless security systems.
- The hard-wired components are usually less visible and more aesthetically pleasing than wireless components.
- Hard-wired systems do not depend on batteries except for power failure backup protection.

The disadvantages of a hard-wired security system are

- Hard-wired systems are more expensive than wireless systems.
- Hard-wired systems are usually leased from the company that installs the system. Unlike a wireless system, the hard-wired system remains an integral part of the home. The components are not capable of being moved to another home when the owner relocates.
- Problems can arise in the installation of sensors in existing homes where some areas are not accessible for pulling wires inside the walls.

Remote Access Systems

A remote access system provides the capability to monitor and control a home security system from a location away from the home. A telephone call to the home followed by a key number code allows the caller to obtain status information concerning environmental and alarm system condition. Remote systems can also be programmed to call a specific phone number when certain environmental conditions exceed an established threshold. A special synthesized voice response system provides the caller with an audible report. The caller, with proper coded inputs, can also perform all the same control functions from a distant location that are available on the keypad in the home.

Features of remote access systems vary among vendors, but most systems have features similar to those listed here:

- Monitoring and reporting temperature inside and outside the home
- Reporting on any sensors that have exceeded preset thresholds

➤ Reporting on the date and time of any alarm conditions that existed

➤ Monitoring loud noises that exceed a set time interval using a built-in microphone in the home security system

➤ Reporting the status of smoke alarms or heat sensors

Fire Detection Systems

Fire detection sensors are available in two categories called heat detectors and smoke detectors. They operate on a principle of detecting heat rise or smoke in the home and can be either hard-wired with voltage supplied by the AC power wiring or battery operated. Most fire detectors currently available are powered by 9-volt DC transistor radio batteries, 120-volt AC power wiring, or 120-volt AC power with battery backup. Wired-in smoke detectors connected to a fire or security system are usually powered by DC from the security panel. This gives the detectors a natural battery backup in the event of a power failure. Common voltages are 6v, 12v, and 24v DC with 12v DC being the most common.

Some 120v AC units have the capability to interconnect so that when one unit activates, it causes the audible alarm in the other units to sound. Units can also be purchased that have a relay output for connection to any security system control panel or wireless transmitter. The most common types available in hardware stores operate as independent sensors that are battery powered, are not connected to the security system, and have their own audible alarm sounder. Heat sensors operate using a different technology from smoke detectors. The basic design features of each type are summarized in the following paragraph.

Heat Sensors and Smoke Detectors

Heat sensors are designed to detect a rapid rise in temperature. They also have a feature that sets off an alarm when a fixed temperature is reached. *Smoke* detectors do not react to heat but use one of two common sensor designs to detect smoke. An *ionization* type of detector forms an electrical path inside a small chamber with a very small amount of radioactive material. When smoke enters the chamber, the particles attach themselves to the ions and change the electrical current flow. A *photoelectric* type of detector works by using a photoelectric cell and a light source. The light does not usually reach the photoelectric cell, but when smoke is present the light is dispersed and reaches the photoelectric cell, triggering the alarm. The main difference between the two types is photoelectric types are more sensitive to large particles and ionization types are more sensitive to small particles.

Modern home design should include at least one of each type. Smoke and heat detectors should be located in each sleeping area and on each story of the home and placed on the ceiling or on the wall 6"–12" from the ceiling.

 Remember that heat and smoke detectors operate on different types of technology. Heat detectors react to abrupt changes and go into an alarm condition when a temperature changes rapidly or reaches a fixed value. Smoke alarms do not react to heat but go into an alarm condition when smoke enters the sensor area. The two main types of smoke detectors are ionization and photoelectric.

Environmental Monitoring

Environmental monitoring can be incorporated into the design of a home security and surveillance system to monitor the status of specific conditions. This might be applicable to rural homes, vineyard owners, farmers, fruit growers, and ranchers. Environment monitoring includes the tracking and measurement of external as well as internal parameters. The following are examples of the types of environmental data that can be monitored with existing sensor systems:

➤ Inside and outside temperatures

➤ Barometric pressure

➤ Frost alarm

➤ Minimum and maximum temperature memory

➤ Temperature trend indicator

➤ Humidity

Emergency Response Systems

An emergency response system is a valuable optional enhancement to a home security system. A large number of designs are available from which to choose, depending on the personal needs of members of the household.

A personal emergency response system is usually designed for elderly or disabled individuals living alone who need to contact a doctor or relatives in an emergency with a minimal number of actions. Systems are available incorporating some or all of the following physical and functional features:

➤ *Emergency response requests*—These are initiated by the user using a miniature wireless radio transmitter with a built-in help button carried or worn as a pendant by the user. Transmitters are lightweight, battery-powered devices activated by pressing one or two buttons.

➤ *A receiver console connected to the telephone*—This receives radio signals from the user's transmitter and dials an emergency response center number.

➤ *An emergency response center*—This is similar to an alarm monitoring service except the personnel in an emergency response center are experienced medical response personnel.

Temperature Sensors

Temperature sensors are used as one component of an environmental monitoring and security system mentioned earlier. They are normally used to monitor high and low temperature values in vacation homes, water pipes, furnace and heating vents, outside farm buildings, computer equipment, utility rooms, or areas that might sustain damage with extreme temperature swings.

Temperature sensors are often integrated with a central controller that includes a display, a keypad, and an automatic telephone dialer that can alert the homeowner or monitoring center when temperature limits are exceeded. Temperature sensors use a variety of electronic designs and components to measure temperature, including both analog and digital signal outputs that change value with a change in temperature and operate remote switches or active alarms.

Location Considerations when Designing a Security or Fire Alarm System

Each component of a home security and surveillance system is designed to be used in a specific location. Window sensors have requirements that are different from door sensor requirements, and smoke and heat detectors each have special location requirements. Computer systems can be integrated with control protocols such as X10 and CEBus to operate and manage security systems. In the following paragraphs, you will discover how smoke detectors work and where they are located. Also covered are the functional design requirements for various types of home security system components.

Home Utility Outlet Specifications

Home security systems require a number of utility outlets to support the installation of sensors, sirens, controllers, surveillance cameras, security lights, and motion detectors throughout the home. New construction provides the opportunity to install structured wiring outlets in all rooms of the home. This includes 120v AC power outlets, Category 5 UTP cable outlets, #22-gauge two-pair wiring for sensors, and RG-6 coaxial cable outlets.

Cohesion with Existing Home Systems

The security system should have sensors located in areas where components of existing home systems require monitoring or protection from theft.

Heat sensors are the most common type of sensor for monitoring components of the heating, ventilation, and air conditioning (HVAC) system. They are mounted near the furnace to monitor rapid changes in temperature that could warn of a possible fire.

Security systems area often designed to be integrated with the home computer system. Computer interfaces are used with X10 and CEBus protocols to monitor and control interior and exterior security system components.

Safety and Code Regulations

The safety and code regulations for installing and locating fire alarms, sensors, and smoke detectors are usually governed by local building codes. The local codes are enforced by a fire marshal who conducts inspections of public places and sets the standard for local residential inspections.

Smoke Detector Installation and Location Requirements

On the national level, codes and standards have been published that establish the requirement for fire detection equipment installation. As an example, a home must have at least one smoke detector installed to meet National Fire Protection Association (NFPA) Rule 72 and Underwriters Laboratories, Inc. (UL) 985 standards. The UL 985 standard covers household fire warning system control units intended to be installed in accordance with the National Fire Alarm Code (ANSI/NFPA 72) and the National Electrical Code (ANSI/NFPA 70).

ANSI/NFPA 72 is the standard document that defines the spacing for smoke detectors, which is typically 30 feet when installed on a smooth ceiling.

Existing Home Environments

Existing home environments require new wire and cabling to support security system components where hard-wired systems are installed. Wireless systems are popular for existing home environments because of the convenience of locating sensors without the need to install additional wiring in the walls.

Smoke Alarm Requirements

Smoke alarm requirements are not the same for every home. They vary according to local municipal standards and the age of a specific dwelling. For homes built prior to 1979, battery-powered smoke alarms are permissible. As to smoke alarm placement, requirements also vary according to the age of the dwelling.

New Home Construction Environments

New home construction facilitates the location of hard-wired security system wiring and components during the early phase of the construction. Structured wiring in new homes supports most of the needs of a well-planned multizone security system. Phone lines, coaxial cable, and Category 5 cable are components of a structured wiring installation. They provide the basis for integrating the security system with the computer network and other home automation systems. The exception is the wiring for door, window, and motion detector sensors. Inexpensive two-conductor wire must be installed in the interiors of walls between the controller location and each sensor. In addition, two-pair #22-gauge wire must be installed for all low-voltage powered sensors such as motion detectors. One pair is used for power and the other pair is used for signaling.

Smoke Alarms in New Home Construction

For all new home construction, fire alarm sensors must be powered by the home AC power electrical wiring. Although this overcomes the problem of neglecting to replace batteries on a periodic basis, there remains the problem of power outages that would also disable a fire warning sensor that uses the home wiring as a power source. Sensors can be purchased with integrated battery powered backups.

For homes built prior to 1979, battery-powered smoke alarms are permissible. In newer dwellings, alarms must be powered by the electrical wiring. The problem with battery units is that people often neglect battery replacement. On the other hand, what good are wired-in smoke alarms if you have an electrical fire accompanied by a power outage? The safest arrangement, therefore, is to install wired-in alarms equipped with battery backup. This type of integrated alarm can be obtained at most hardware stores and is required for homes built as of 1993.

As to smoke alarm placement, requirements also vary according to the age of the dwelling. In older homes, most municipalities require alarms in the following locations: within close proximity to all bedroom entrances, on each story of a multilevel home, and in basements. The latest standards, enacted in 1993, require that there be an additional alarm in each bedroom. Another practical location, although not required, is the garage.

Equipment Functionality and Specifications

Each component of a home security and surveillance system has a specific function to perform. The basic security system includes a control panel connected by cables to sensors at various locations throughout the home. A perimeter protection system must include sensors at every opening, including doors, windows, garage doors and windows, and doors to crawl spaces. A keypad is the device that provides a control interface for the residents to arm and disarm the system using a programmed access code and also to monitor the status of the system.

Door Functional Design Specifications

Doors are protected by installing small magnetic switches inside the frame. Figure 3.1 illustrates the location of the magnetic switch. A magnet is installed in the top of the door that keeps the switch contacts open as long as the door remains closed. An alarm is caused when this switch is disturbed by opening the door. The magnetic switch completes a circuit that is connected to the control panel. Recessed mounted models use magnets that are fitted into drilled holes, and when properly installed, recessed mounted magnetic switches are hard to notice and blend in well with the door.

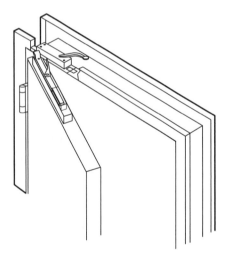

Figure 3.1 Magnetic door switch.

Window Functional Design Specifications

A perimeter security system must include a glass protection system because magnetic switches do not protect against an intruder entering through a broken window. Glass protection systems are available in two categories: vibration and acoustical. The vibration system is mounted on the glass or on a nearby wall and detects movement of the glass. *Acoustical* systems, or *sound discriminators*, sense the sound of breaking glass. The unit can be tuned to react only to the specific frequency of glass breaking, typically 4KHz–6KHz, or it can react to any loud noise. Some manufacturers have combined vibration and sound detectors into one unit that does not activate unless both are detected. These units can be used where the normal conditions would cause a single technology detector to generate false alarms.

Control Panel Functional Design Specifications

The control panel shown in Figure 3.2 is an enclosure that contains all the electronic components, wire termination points, backup battery packs, and telephone termination wiring.

Figure 3.2 Control panel.

Each sensor receives power and is managed from the control panel. It monitors the health and operating status of the total system and sends a signal to the siren when an alarm condition exists. The panel should be mounted in a location that is out of plain view and near a 120v AC outlet where a plug-in transformer can be used to supply low-voltage power to the total system. If a phone line is planned for connection to an external monitoring facility, access to the location where the phone line enters the home must be considered when locating the control panel. Preferred locations are utility rooms, basements, and closets.

Keypad Functional Design Specifications

The best location for a keypad is an area that is both convenient to the family users and secure for the system. The homeowner must ultimately determine how many keypads are desired and where they are to be located in the home. Typically, keypads are placed at the main entry/exit door, in the master bedroom, or in the main hallway of the home. In a multilevel home, keypads are commonly placed on each level.

Physical Devices

The six major physical components of a home security and surveillance system are described in this section. Although some designs include additional components for enhanced performance or custom installations, the physical devices described here are considered the basic building blocks for a security and surveillance system that will provide adequate protection from intrusion. The essential physical assemblies include

- Keypads
- Sensors
- Security panels
- Cameras
- Monitors
- Switchers

Keypads

The keypad is the device used by the home residents to initiate commands for control options and observe the status of the security system. As shown in Figure 3.3, it usually contains an alphanumeric keypad and LED displays that indicate the status of the alarm system.

Figure 3.3 Keypad.

The keypad is used to arm and disarm the system and often includes a panic switch by which the alarm can be triggered instantly in an emergency situation. The alarm can be silenced by the owner by entering the correct coded sequence of numbers on the numerical keypad. The keypad is typically installed inside the home near a door that is most frequently used by the residents. A programmed delay is included as a feature of most systems to enable the users to enter and disarm the system within a fixed delay period (normally 30–45 seconds). The same fixed delay is also used to allow the user to arm the system and exit the home within the fixed delay period. Keypads can also be used to bypass certain areas.

 The bypass function is used to arm the system but disable selected zones or motion detectors inside the home when the family is present. Residents often desire to secure the perimeter area of doors and windows after retiring for the evening but need to bypass interior area motion detection sensors that are activated only when the home is not occupied.

Sensors

Sensors are designed to protect both the perimeter and the open spaces inside the home. As mentioned earlier, perimeter devices primarily protect doors and windows. The most common perimeter sensors are magnetic door switches, window vibration detectors, and window acoustical detectors. Space protection sensors called motion detectors cover interior rooms and hallways and can detect an intruder who has been able to defeat a perimeter device. Exterior motion detectors and motion-activated security lights are also used. The following paragraphs describe the types of sensors required in basic home security and surveillance systems.

Door Switches

Door switches work on a basic principle of a two-part magnetic switch. A switch that is sensitive to a magnetic field is mounted on the fixed structure (frame), and wires from the switch are routed through the wall to the control panel. A magnet is mounted on the door in a position where it is in close proximity to the switch when the door is closed; this also keeps the switch closed. Opening the door moves the magnet away from the switch and causes the switch to "open," which is sensed by the central control panel and activates an alarm. Magnetic switches are available as normally open (NO) or normally closed (NC) to accommodate different wiring designs and controller options.

Window Acoustic and Vibration Detectors

As you learned earlier, window sensors are used to detect the sound of breaking glass. Large glass doors such as patio doors are usually protected by magnetic switches. Windows in the home can also be protected by magnetic switches if they can be opened; however, glass break detectors are recommended by most home security contractors to protect against an intruder entering through a broken window.

Glass break detectors are available in either vibration type or acoustical type. The vibration type is mounted on the glass or on a nearby wall.

Security system magnetic switches are designed for both normally closed (NC) and normally open (NO) options. An NC switch opens when the magnet is moved near the switch, and an NO switch closes when the magnet is moved in close proximity to the switch. This enables magnetic sensor-type switches to be used with various types of security system designs. Sensor door switches wired in parallel use NO contacts so that any closure of the contacts in the circuit activate an alarm condition. Series circuits use NC magnetic switches where any opening of a switch results in an open condition for the circuit, which triggers an alarm condition by the controller.

Motion Detectors

Motion detectors work by detecting the changes in the infrared energy in an area. Because these devices do not emit any energy, they are called *passive infrared (PIR) detectors*. PIR detectors use a lens mechanism in the sensor housing to detect a change in infrared energy across the horizontal sectors covered by the sensor. This type of detector is insensitive to stationary objects but reacts to rapid changes that occur laterally across the field of view. They are the most common and economical type of motion detectors and are available in standard, pet-friendly, and harsh-environment (outdoor) models. An example of a motion detector for interior use is shown in Figure 3.4.

Figure 3.4   Motion detector.

Security Panels

Security panels provide several functions to coordinate the operation and management of a security system. They can include an integrated keypad or LED indicators. Most designs include a power transformer for converting the AC voltage to a DC voltage for the sensor loop and contain a rechargeable battery for backup if the commercial power fails. A terminal strip provides for the connection of the wiring that connects all the sensors to the controller as well as the external telephone line. Most designs include a printed circuit board containing all the electronics and a microprocessor. It also connects to and controls the siren that is activated when an alarm condition exists.

> Security panels are known by numerous names, such as central control box, control panel, alarm panel, and interface panel. They all perform similar functions including controlling and monitoring sensor status, providing power to the system, connecting the telephone line to the monitoring station, and handling the programmable options for the system.

Cameras

Surveillance systems for the home use video cameras that convert the image into a video composite or S-video signal for display on a video monitor. The best type of camera for home systems uses charged coupled device (CCD) technology. These cameras have high resolutions, low operating light, less temperature dependence, and high reliability. A typical CCD camera used in video surveillance systems is illustrated in Figure 3.5.

Figure 3.5 Video surveillance camera.

Camera Resolution and Sensitivity

The two important specifications for cameras are the light sensitivity rating and the number of lines of resolution. The *resolution* of a camera is a measurement of the horizontal lines it is capable of generating. Most standard TVs and VCRs have a resolution of fewer than 300 lines, but video monitors can have a resolution as high as 800 lines. Surveillance cameras come in a range of costs and are available with 300–500 lines of resolution. Higher resolutions make distinguishing fine details and recognizing people at a distance easier.

The amount of light required to obtain a reasonable image is called the *lux rating*. 1 lux is approximately the light from one candle measured from 1 meter. Typical camera ratings range between 0.5 and 1.0 lux.

Monitors

Monitors are cathode-ray tube display systems similar to computer display systems. They are used to display video information processed by the camera. Coaxial cable is used to connect the camera to the monitor that can be located in any area selected by the user. Monitors do not have a TV tuner and usually have better video resolution than standard television receivers. They can also be connected to programmable switchers that receive input from several cameras and show multiple images on a single monitor.

Closed circuit TV (CCTV) monitors are available for black-and-white or color display depending on the resolution and camera selection. Black-and-white monitors have resolutions in the range of 700–1000 lines, whereas color monitors are available with 350–400 lines. CCTV monitors are designed for extended 24-hour-per-day operation.

142 Chapter 3

> Video surveillance systems for both home and commercial business use are referred to as *closed circuit TV (CCTV)* systems. The name is derived from the type of the system used for transmission over a closed circuit or private transmission circuit rather than a standard television broadcast system. CCTV is also used in a wide variety of applications for schools, business video conferencing, retail store surveillance, and gambling casinos.

Switchers

Switchers are devices used with multiple camera systems. Although primarily used in commercial building security and surveillance systems, they can be scaled to fit the needs of a home security system. They enable several cameras to be used with a single monitor. The switcher can be programmed to cycle through all the cameras in a surveillance system or dwell on each camera for a specified length of time, usually in the range of 1–60 seconds. Exterior sensors can detect movement and cause cameras to start recording the image on a VCR.

Quads are special devices that enable the viewer to simultaneously record and monitor four cameras. It does this by splitting your screen into four sections. The normal configuration for connecting a quad switcher with a sensor and a VCR is shown in Figure 3.6, which illustrates the connections between a quad, a monitor, and four surveillance cameras. The monitor can view all four images at the same time.

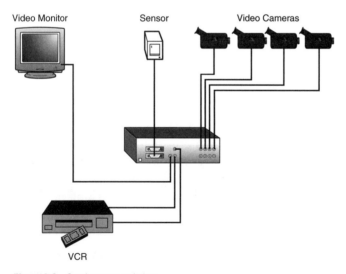

Figure 3.6 Quad camera switcher.

Configuration and Settings

The next section examines the location and configuration settings for each type of sensor and the video surveillance components. You will discover the features and requirements for planning zone layouts, setting up passwords, locating keypads, as well as locating video surveillance cameras.

Zone Layout

Home security and surveillance systems are usually designed with certain areas of the home designated as zones. A *zone* can include interior motion detectors or certain rooms or hallways. The use of zones has several purposes. It enables the user to arm only portions of the system, such as the perimeter doors and windows, while bypassing the interior motion detectors when retiring for the night. When leaving the home, all zones, including the interior, can be armed as required.

A zoned security system layout is also used by the external monitoring service to know which sensor in a designated zone is causing the alarm.

If a sensor were to be reported as just sensor 3 zone 5, this could mean just about anywhere at first. But if the sensor were reported as sensor 3 zone 5 *perimeter*, this would give the operator a better understanding that the violated area is on the outside of the premises. Another reason for using a zoned security system layout is the ease of troubleshooting if a sensor is reported as a bad sensor. For example, if a bad sensor is reported as being in zone 3 perimeter, there is no need to troubleshoot sensors that are located in the interior of the system.

Passwords

Passwords are used as a confirmation tool by most professional alarm monitoring service companies. They are used to avoid a police response when a false alarm situation has been triggered accidentally while still ensuring that the homeowner is safe. The passwords are known only by the residents and the monitoring station personnel. When an alarm condition is received from the home at the monitoring center, the person on duty calls for a police response to the home. However, this occurs only if the alarm is not reset in a predefined length of time. If the alarm is reset at the keypad with the correct code by the owner prior to the time limit, the monitoring station simply calls the subscriber and asks for a password. The password response has two agreed upon formats: One password is used by the homeowner to indicate the alarm condition was accidental and no action need be taken by the monitoring station.

Another secret password is used if the owner is in peril or possibly held hostage by a burglar. This password, when passed after an alarm condition has been reset, results in a police response. This is the option used if the homeowner is asked to respond to the monitoring station phone call by an intruder who (hopefully) does not know the distinction between the two passwords. Passwords are used only if the monitoring station needs to authenticate the cause of an alarm condition when the residents are at home. Law enforcement agencies can penalize homeowners for excessive false alarms.

 Common sense should guide the selection of a security system password used with a monitoring service. Never use family or pet names, birth dates, street names, or any word found in the dictionary. The preferred password should include both letters and numbers, and a password should be memorized and never be written down. Numeric keypad codes should also avoid the use of birth dates or home address numbers.

Keypad Locations

Keypads are located by most contractors on the inside wall near the door most often used by the occupants. They are also frequently located in the bedroom to provide easy access by the users when retiring or arising. Keypads used for home security perimeter protection are not recommended for outside mounting because this would allow tampering, vandalism, or attempts by an intruder to search for a correct code. Some exceptions to this rule are applications such as exterior security gates and gated community entrances.

Sensor Locations

Sensors perform all the functions required for detecting and reporting an intrusion in the area they are designed to protect. Window and door sensors are installed on the door or window frame. Window sensors can be protected by either a magnetic switch that detects an opening of the window frame or other types mounted on the window to detect the sound or vibration of broken glass.

Interior surveillance cameras and motion detectors are usually wall mounted and are used to protect open areas where an intruder might be able to defeat the perimeter security system sensors.

Exterior surveillance cameras and security lighting systems should be located in areas near the front, side, and rear entrances to detect intruders before they reach the perimeter security sensors.

Motion Detector Locations

Motion detectors are usually mounted in the corner of a room. This enables the PIR type of motion detector to cover a 90° field. As illustrated in Figure 3.7, they are sensitive to movement across the sensor field of view.

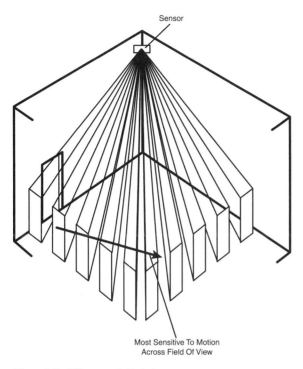

Figure 3.7 PIR sensor field of view.

Window Sensor Location

The location of a window sensor can vary with different types of windows. Vibration detectors are mounted on the glass, whereas acoustical window sensors are typically mounted on an adjacent wall, as illustrated in Figure 3.8.

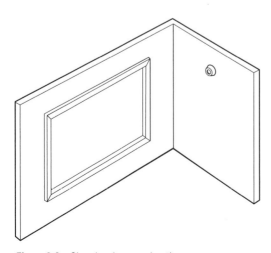

Figure 3.8 Glass break sensor location.

Camera Locations

Surveillance cameras can be mounted outside the home to provide recognition of someone wishing to enter the security perimeter area, such as a front door or driveway gate. They should be located where no blind spots exist or where it is not practical to use other types of sensors. Cameras can be mounted on any surface area of the home or garage where coverage is required as long as the area is illuminated sufficiently at all times after dark. They are often used in the interior of the home to monitor a child's playroom or nursery. Although not usually a problem in a residential application, certain legal implications are involved in using video surveillance cameras. They are not to be used where there is a reasonable expectation of privacy by individuals who are not aware that they are in a location where a camera is installed. This obviously does not apply to a person breaking into a business or home.

Surveillance cameras can be used in most exterior residential locations; however, you should remember that a surveillance camera cannot be located where there is a reasonable expectation of privacy.

Exterior Security Lighting Systems

Exterior lighting is often used to illuminate dark areas or areas to be protected using motion detection sensors that activate security lighting to deter an intruder. Dark areas surrounding the home where there are trees or shrubs are specific areas that need to be illuminated with security lighting systems. Motion detector-activated lights are also popular for exterior lighting. The most common and reasonably priced style is the two-floodlight design with the motion sensor mounted beneath them. The dual-lamp type exterior security lighting system is shown in Figure 3.9.

Figure 3.9 Motion detector light.

Exterior security lighting should be used near landscaped areas around the home, such as shrubs, bushes, foliage, and trees. Security lighting systems used during dark hours avoid the problem of allowing an intruder to enter the area surrounding a home and attempt entry while unobserved in a dark area.

Device Connectivity

Connecting all components of a security system requires compliance with national and local building and fire codes. Each internal and external interface involves selecting and identifying the correct type of wire, coaxial cable, and termination points.

Low-Voltage Wire Connectivity

Security system wiring is rated as low-voltage, limited energy circuit wiring by the local building codes in most states. Almost any gauge wire can be used for connecting sensors to the main control panel, but #22-gauge is the smallest gauge wire recommended for any home security wiring. Quad wire is a type of security wire that is used by many installers; it contains four wires enclosed in a plastic sheath and is used for interconnecting security system components.

As indicated earlier, some types of smoke detectors and heat detectors receive power from the home AC power wiring. They are not classified as low-voltage devices and must be wired in accordance with high-voltage wiring in compliance with local electrical building codes.

Although almost any gauge wire can be used for connecting sensors to the main control panel, #22-gauge is the smallest gauge wire that should be used for wiring home security systems.

Wireless System Connectivity

Wireless systems provide the easiest and most economical method for connecting components in a home security system.

Each sensor in a wireless security system provides its own power using internal batteries. When a sensor detects an intrusion, it transmits the information to a control panel using radio transmissions to signal an alarm condition to a central controller.

Telephone

The telephone line is used by a security system to automatically call a central monitoring facility when an alarm condition exists. The alarm system controller box must be connected to the phone line with an RJ-31x phone jack. This connector takes priority when an alarm is triggered, disconnects the home phones, and dials the number of the monitoring service. This prevents a burglar from taking the phone off the hook to prevent the system from dialing out. The proper wiring connection for an RJ-31x connector is described later in this chapter in the section "Installation Plans."

Coaxial Cabling

Coaxial cable is used to transport video signals between video cameras and monitors in home video surveillance systems. The quality of the video signal is affected if poor-quality cable or the wrong type of cable is used. Structured wiring standards recommend the use of RG-6 coaxial cable and F-type connectors for connecting cameras and monitors.

Termination Points

The main termination points for a home security system are contained in the control panel. All the sensors installed throughout the home are connected with low-voltage wiring and routed through walls to the central controller termination strip.

The telephone line also has a termination point where it enters the control panel enclosure. An RJ-11 telephone jack is retrofitted with a RJ-31x jack. The four telephone wires are then terminated on the controller main connection bus with the other external cabling harness.

In-house Services

Several types of security and surveillance services are available to residential users that have been previously used only in larger commercial business systems. Smaller, scaled-down types of home video surveillance and monitoring systems provide an economical option for homeowners. Two types of systems that provide internal services to residential users are discussed in the following paragraphs.

Video Surveillance and Monitoring Services

Video surveillance and monitoring is a type of service that allows viewing of selected areas on the exterior and interior of the home to be made available to the residents. Identification of visitors can be verified using video cameras and monitors before opening a gate or front door.

Interior video surveillance and monitoring cameras can be used to observe infants in a nursery or small children in a playroom area. Hidden cameras can be linked to a VHS recorder to record the identity of any intruder who might have eluded the security system sensors.

Alarm Types

Alarm sounders not only attract the attention of others outside the home, but also create a sufficiently high level of sound to discourage an intruder. Various types of sirens, horns, buzzers, klaxons, and bells are used to attract attention when an alarm condition is activated. Different types have varying levels of volume for various locations. Older alarm systems used bells typically mounted on an outside wall. Today bells are rarely used except in commercial buildings for fire alarms. They have been replaced in home security systems by solid-state electronic sirens, such as those shown in Figure 3.10. These sirens provide a higher level of sound output as well as a variety of tones and sound pitch.

Figure 3.10 Electronic siren.

> **NOTE** Interior sounders installed in concealed areas in the home operating at maximum sound levels are designed to frighten an intruder into making a fast exit because the sound masks any opportunity to hear an outside approaching police siren. Interior sirens are available from several vendors that operate at sound levels in the 110dB–120dB range, which is near the threshold of pain.

Smoke Alarms

A smoke alarm is designed to alert the residents of a home that a fire is either in progress or in the initial stage.

Smoke alarms have built-in sounders designed to awaken the occupants of a home and alert them of the danger. Many areas now have building codes that require all smoke alarms installed in the home to sound the alarm if any single unit is activated. Most smoke alarms have a test button that checks the sounder as well as verifies that the battery is operational.

External Services

External security and surveillance services are available from private companies which are connected by dial-up phone service to the home security system. Another type of service is available for residents who want to check on the status of the security system when they are away.

External Alarm Monitoring Service

Professional alarm monitoring companies provide 24-hour service for residential security systems for a monthly fee. Monitoring service attendants have a computer system and display with a database containing information about each subscriber. As discussed previously, security systems are designed to send a digital message via phone lines to a central monitoring facility. The attendants contact police or a private armed response guard to go to the location where the alarm was activated. Customized response scenarios are established with the owner and the monitoring service for the use of passwords to verify the identity of the resident when an alarm is triggered. Each monitoring service offers a set of response options discussed earlier to verify false alarms or special distress code word if an intruder forces the resident to cancel an alarm.

Remote Access

External access to the home security system by means of telephone lines is called *remote access*.

Remote access can be designed into a system that permits the user to call from a remote site and set system parameters or obtain voice-synthesized status messages concerning any sensor that has detected a value outside programmed values such as heat, cold, water leakage, loud noises, alarm history, or other custom features.

Industry Standards for Home Security and Surveillance Systems

You will need to know the key standards and codes involved in the installation and configuration of home security and surveillance systems. Installers use these standards when configuring the security system component location and cabling plan. Although numerous standards are available for guidance, local building codes are the final authority when seeking inspection approvals during the construction phase.

The National Electrical Code

Two sections of the National Electrical Code (NEC) describe requirements for security and surveillance systems. NEC Article 725 describes class 1, 2, and 3; remote control; signaling; and power-limited circuits, and NEC article 760 describes fire protective signaling systems.

The local building codes are the final authority for determining construction and installation requirements for wiring, smoke detector location, and connectivity.

The Telecommunications Industry Association and the Electronic Industries Alliance

The ANSI/TIA/EIA standards organizations have only recently addressed the need for residential security standards. In 2002, work was completed by the Telecommunications Industry Association (TIA) 42.2 working group on ANSI/TIA/EIA-570A Addendum 1, which covers residential alarm and security cabling. A summary of the important points covered in this standard are as follows:

➤ Security system wiring should be installed while the building is under construction and prior to dry wall installation.

➤ All low-voltage wire runs that are run parallel to AC power cables should be separated by at least 12".

➤ All wiring must terminate in an alarm or a control panel grounded to a true earth ground.

➤ An RJ-31x jack is required for connection between the off-premise telephone line and the alarm panel to provide priority for the alarm system.

➤ Home run wiring is required from all sensors/detectors to the control panel.

➤ Passive sensors need only two wires, and active sensors require four wires.

➤ 22-gauge or larger wire should be used for connecting sensors to the control panel.

➤ When security wires are crossed with power wiring, they must cross over at a 90° angle.

➤ Video surveillance systems require RG-59 or RG-6 coaxial cable with 95% copper braid.

➤ The ANSI/TIA/EIA 570A addendum for security cabling, points out that the location of sensors and cabling devices must align with the requirements of the National Fire Protection Association, the National Electrical Code, and the National Fire Alarm Code.

 The ANSI/TIA/EIA 570A standard requires at least 12" of separation between parallel runs of security wire and AC power wiring. You should also remember that, when security wire and cabling is crossed with AC power wires, the crossover must be at a 90° angle.

Underwriters Laboratories

The Underwriters Laboratories (UL) is a nonprofit organization that has established standards for all components of security systems and their installation. If the product has the UL label, the device meets or exceeds UL's requirements.

Installation Plans

Installation planning should follow the guidelines mentioned earlier in ANSI/TIA/EIA-570A Amendment 1 and NEC articles 725 and 760. Important requirements are summarized in this section for RJ-31x jacks, motion detectors, and video surveillance systems.

Installing an RJ-31x Telephone Jack

A special type of telephone jack called an RJ-31x is required with all security systems that use a central alarm monitoring service. The home central monitoring panel sends a message to the monitoring station when the alarm is activated. If the telephone line is in use when an alarm message starts or a phone instrument is left off the hook, there needs to be a way to clear the line and override all other call activity. Otherwise, the alarm message could be seriously delayed or blocked completely. Most security systems are wired to the phone line using an RJ-31x phone jack. This 8-pin telephone jack is specially designed to seize the phone line and send the alarm message. As indicated in Figure 3.11, the RJ-31x jack must be installed between telecommunications equipment in the home and the demarcation point where the phone line enters the home.

Customer Premises Equipment and Wiring

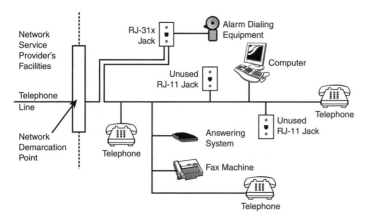

Figure 3.11   RJ-31x jack location.

All alarm systems that use the telephone lines in a home to aromatically dial a central monitoring station must have an RJ-31x jack installed. This enables the alarm system to seize the phone line when an alarm condition exists.

Installing Motion Detectors

Motion detectors should be installed in open areas that cannot be protected by window or door sensors. These detectors require two wires for 12v–16v DC and two wires for the alarm circuit. Care should be taken if pets are to remain in the area when the system is alarmed, although special motion detectors are available that avoid false alarms caused by pets.

Camera and Switching Equipment Installation

Video surveillance cameras can be installed using the supplied mounting hardware facing the area of interest. For outdoor use, a location should be selected where the full view of the area can be observed within the field of view of the camera. RG-6 cable and F-type coaxial connector hardware is used to run the cable from the camera to the monitor or switching equipment. If the camera has a motorized scan feature, be sure to allow for full coverage of the area.

Surveillance System Monitor Installation

Monitors can be mounted in a location convenient to the user. For observing front gates or entrances to the home, such as a front door, a small monitor should be located near the entrance used most frequently by visitors.

Maintenance Plans and Procedures

The final step of an actual installation is to test the system to ensure that it is working properly. When you document the testing procedures, include which devices you tested, how you tested each device, the results from the tests, who tested them, and the date the tests were performed. Always test the system and each individual device to ensure that all devices are working properly.

Troubleshooting can be made a lot easier by keeping good, detailed notes on what the problem is, what the system or device is doing, and what you did to correct the problem. If you or another installer run into the same problem later, fixing it the second time around will be easier if you documented exactly what you did the first time.

Exam Prep Questions

Question 1

> What is the name of the device that allows four surveillance camera images to be viewed on a monitor at the same time?
> - A. A video splitter
> - B. A modulator
> - C. A quad switcher
> - D. A photoelectric sensor

Answer C is correct. A quad switcher is designed to provide simultaneous viewing of four cameras on one monitor using a split-screen display. Answer A is incorrect because a video splitter is used with TV antenna systems and is not an active component required with four-camera surveillance systems. Answer B is incorrect because a modulator is not required for viewing images. Answer D is incorrect because a photoelectric sensor is used to monitor ambient light conditions for security systems and is not required as a component in a video surveillance system.

Question 2

> Which type of jack is required to be installed to allow the security system to seize the line when an alarm condition exists?
> - A. RJ-11x
> - B. RJ-45
> - C. RJ-12
> - D. RJ-31x

Answer D is the correct answer. The RJ-31x jack is used to disconnect all the phones in the home when an alarm condition is initiated. This enables the alarm panel to have priority over all uses of the phone when it needs to call the outside monitoring facility. Answers A and C are incorrect because both types of RJ-11 jacks are used for standard telephone four-wire connections and are not applicable to security system connections. Answer B is incorrect because RJ-45 connectors are used for four-pair UTP cabling in local area network networks and are not required for security system applications.

Question 3

Which TIA/EIA standard describes the requirements for residential security system cabling?

○ A. 568-A
○ B. 568-B
○ C. 570-A Addendum 1
○ D. 570-A Addendum 3

Answer C is correct. Addendum 1 to 570-A is the standard that references residential security system cabling. In 2002, Addendum 1 to the 570-A standard was released covering residential security system wiring. Answers A and B are incorrect because the 568-A and B standards cover commercial building telecommunications wiring standards. Answer D is incorrect because Addendum 3 to the 570-A standard is applicable to whole home audio cabling for residences and is not used to specify security system wiring.

Question 4

What is the term used to describe the amount of light required to obtain a reasonable image with a surveillance video camera?

○ A. Lux rating
○ B. Candlepower rating
○ C. Resolution
○ D. Pixels

Answer A is correct. Video surveillance cameras are usually designed to work in low-light conditions and are rated in lux. 1 lux is approximately the light from one candle measured from 1 meter. Typical camera ratings range between 0.5 and 1.0 lux. Answer B is incorrect because the candlepower rating is the measure of light intensity and is not a specification parameter used with video cameras. Answers C and D are incorrect because resolution and pixels are terms relating to video imagery quality and are not applicable to light measurement for a video surveillance camera.

Question 5

How many wires are normally required for connecting a passive sensor to the control panel?

○ A. 1
○ B. 4
○ C. 2
○ D. 3

The correct answer is C. A passive sensor requires two wires because no power is applied to a passive security system sensor. An active sensor requires four wires because, by definition, an active sensor requires a power source. Answers A, B, and D are incorrect because they each represent the incorrect number of wires required to connect a passive sensor to a control panel.

Question 6

Which type of sensor is used for door installations?

○ A. PIR
○ B. Contact sensor
○ C. Active
○ D. Magnetic switch

Answer D is the correct answer. Magnetic switches are used for doors because of their low cost and capability to be hidden.

Question 7

Which types of cable are used when installing video surveillance components?

○ A. RG-6 and RG-58
○ B. RG-8 and RG-59
○ C. RG-59 and RG-6
○ D. RG-59 and RG-58

Answer C is correct. RG-6 and RG-59 are both 75-ohm types of cable. RG-6 is a higher grade used for low loss applications, whereas RG-59 is satisfactory for short runs. RG-8 and RG-58 are rated at 50 ohms of impedance and are not compatible with most CCTV equipment impedances of 75 ohms, making answers A, B, and D incorrect.

Question 8

What is the minimum separation required between security wiring and AC power wiring when they are installed in parallel wire runs?
- A. 18"
- B. 2 feet
- C. 6"
- D. 12"

Answer D is correct. The ANSI/TIA/EIA-570-A Amendment 1 standard indicates the minimum spacing for security and power wiring is 12" when installed in parallel runs. Answers A and B are incorrect because any spacing between security wiring and power wiring in excess of 12" would be in excess of the minimum requirement. Answer C is incorrect because the answer is less than the minimum spacing of 12" required between power wiring and security wiring.

Question 9

Which type of residential location is typically classified as a perimeter security location when designing a security system? (Select two.)
- A. Windows
- B. Basements
- C. Driveways
- D. Doors

Answers A and D are correct. The sensors typically connected to a single monitoring circuit that surrounds the perimeter of a residence are placed on the doors and windows. This provides protection for any opening in the primary structure. Answer B is incorrect because the basement is not normally included as a perimeter protected area but would use interior motion detectors. Answer C is incorrect because driveways are normally protected by exterior surveillance monitors such as video surveillance or motion detector systems.

Question 10

> Which of the following is true regarding the use and installation of smoke detectors? (Select all that apply.)
>
> ❏ A. For all new residential construction, battery-powered smoke detectors are acceptable for some locations.
> ❏ B. For all new home construction, fire alarm sensors must be powered by the home AC power electrical wiring.
> ❏ C. For homes built prior to 1979, battery-powered smoke alarms are permissible.
> ❏ D. Smoke detectors must be installed near windows and doors to provide alerts for escape routes in case of fire.

Answers B and C are correct. The NFPA rule 72 requires smoke detectors for new residential construction to be powered by AC wiring as the primary power source of power with battery power used only as a backup source in case of a power failure. Answer C is correct because battery-powered smoke detectors are permitted for homes built prior to 1979. Answer A is incorrect because the NFPA rules make no allowance for smoke detectors powered by batteries in specific locations in new residential construction. Answer D is incorrect because smoke detectors are designed to alert the residents of a potential fire condition. Smoke detectors are not designed to aid in locating an escape route from the home in case of fire.

Need to Know More?

 Cumming, Neil. *Security: A Guide to Security System Design and Equipment Selection and Installation, Second Edition.* St. Louis: Butterworth-Heinemann, 1994. (This book includes additional information on selecting the best components and installing home security systems.)

 Traister, John E. and Terry Kennedy. *Low Voltage Wiring: Security/Fire Alarm Systems, Third Edition.* New York: McGraw-Hill Professional, 2001. (This book describes the basic principles involved in planning the low-voltage wiring layout for a home security system.)

 Trimmer, William H. *Understanding and Servicing Alarm Systems, Third Edition.* St. Louis: Butterworth-Heinemann, 1999. (This book is written for the installer. It will help those employed in the home security industry gain additional knowledge relating to various types of security systems.)

4.0—Telecommunications Standards

Terms you'll need to understand:

- Hybrid fiber coax (HFC)
- PBX
- KSU
- Insulation displacement connection (IDC)
- Punch-down tool
- 66/110 connection blocks
- Digital subscriber line
- Splitter
- Registered jack
- USOC
- Multiplexing
- Automatic call distribution
- TDR
- PSAP
- Remote access

Tasks you'll need to master:

- Selecting the correct type of cable for a home structured cabling plan
- Planning the installation of telecommunications wall outlets
- Listing the types of emergency response systems for the home
- Describing the tool used to install wiring on a punch-down block
- Preparing a list of the features offered by a PBX and key telephone system
- Making a drawing that describes a simple home run telecommunications wiring plan
- Making a list of the features of a call blocking service
- Describing the features of DID and extension dialing

Telecommunications Design Considerations

The design for a home telecommunications system must consider the needs and work requirements for all members of the family. The telecommunications design must also include solutions for home network devices and other home automation systems that can be included in the overall telecommunications plan. In this section, you will gain an understanding of the choices available for the home integrator who is planning the design and installation of telecommunications equipment for a modern automated home.

Hybrid Systems Design

The wide range of telecommunications equipment available for home users includes several types of cabling as well as both digital and analog voice systems. Hybrid system design solutions include fiber-optic patch cables discussed in Chapter 2, "2.0—Audio and Video Fundamentals"; coaxial cable; and unshielded twisted pair (UTP) cable used in home structured wiring. Telephone systems discussed later in this chapter include digital and analog systems, PBX and key telephone systems, as well as fax, dial-up analog modems, and digital Internet access equipment. As you can see from these examples, a hybrid telecommunications system design has become standard practice for the professional home integrator.

Hybrid is a loosely defined generic term used in the telecommunications industry to describe any system that uses a mix of media and transmission standards. A *hybrid fiber coaxial (HFC)* network is a commonly used term you might encounter. It consists of a network in which optical fiber cable and coaxial cable are used in different portions to carry broadband mixed content such as video, data, and voice. Both cable TV (CATV) and telephone companies are using HFC in new and upgraded networks to carry both video information and voice conversations in the same system.

Analog Communications Systems

Almost everything in the world can be described or represented in one of two forms: analog and digital. The principal feature of *analog* representations is that they are continuous. In contrast, *digital* representations consist of values measured at discrete intervals. Analog communications systems are those systems designed to transport voice information. The electrical representation of the human voice is called an analog signal. Most of the intelligence for the human voice is contained in a range of audible frequencies between 300 hertz (Hz) and 3300Hz.

Telephone networks are designed to transport all audio sounds including those generated by voice, analog modems, and fax machines that fall into this range of audible frequencies. Analog signals, by definition, can have an infinite range of values at any specific point in time. Voice frequencies are converted to a continuously variable voltage level and sent over copper wires. As shown in the Figure 4.1, telephone handsets perform the conversion between human speech sounds and the varying electrical signals sent over the telephone circuits.

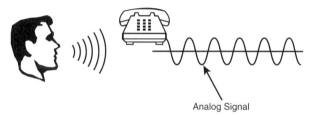

Figure 4.1   Analog signals.

The majority of telephones, modems, and fax machines used in the design of home telecommunications systems are analog devices. Although dial-up modems and fax machines are used to send digital information, they use transmission signals that are in the audible range to be transported over the public switched telephone network (PSTN).

Digital Communications Systems

Digital communications systems use an advanced technique for connecting office telephone systems, computers, and cellular telephone networks. A digital system converts analog voice signals into a digital format. This provides several advantages over analog communications systems described here:

➤ A digital telecommunications system can be easily integrated with computers and digital mass storage equipment because computers handle information in a digital format.

➤ Digital signals are less vulnerable to noise, distortion, and interference than analog signals.

➤ Digitized voice systems are becoming the preferred technology for office telephone systems, answering machines, and interactive voice recognition systems.

➤ Digital communications systems enable several circuits to be multiplexed on a single high-speed circuit. This provides a major savings over having a separate twisted-pair cable for each analog circuit.

- ▶ Digital voice systems are a basic feature of new computer telephony integration (CTI) technology products.

- ▶ Digital voice systems can be used to improve the quality and performance of telephone systems that are not possible with analog systems. Examples include technologies such as forward error correction, voice compression, synthesized voice response, digital repeaters, and voice over IP (VoIP).

Remember that multiplexing is a method for combining two or more information channels onto a common medium. One of the major advantages of using digital transmission is the capability to use multiplexing. This enables several circuits carrying digitized voice or data information to be multiplexed onto a single high-speed circuit. This type of equipment is used to synchronize and combine many digital circuits into a hierarchical system of high-speed digital transmission systems.

PBX Systems

A *private branch exchange (PBX)* is a telephone system within an enterprise that switches calls between enterprise users on local lines while allowing all users to share a certain number of external phone lines. The main purpose of a PBX is to save the cost of requiring a line for each user to the telephone company's central office. The PBX is owned and operated by the enterprise rather than the telephone company (which can be a supplier or service provider). Private branch exchanges used analog technology originally. Today, PBXs use digital technology (digital signals are converted to analog for outside calls on the local loop using plain old telephone service). A PBX includes

- ▶ Telephone trunk (multiple phone) lines that terminate at the PBX
- ▶ A computer with memory that manages the switching of the calls within the PBX and in and out of it
- ▶ The network of lines within the PBX
- ▶ Usually a console or switchboard for a human operator

In some situations, alternatives to a PBX include centrex service (in which a pool of lines are rented at the phone company's central office), key telephone systems, and (for very small enterprises) primary rate Integrated Services Digital Network.

Typically, businesses with more than 50 employees prefer a PBX telephone system because it can control a large number of lines. Mid- to large-size firms wanting to manage their own private communications network usually choose PBX rather than a key system because of its larger telephone set and line capacities.

PBX telephone systems are available in a wide range of costs and features suitable for the small home office as well as large business organizations.

Many use digital switching technology. Typical design features and specifications for a home PBX system include the following options:

➤ The number of internal phone extensions and remote extensions outside the local office can be selected. For a small business PBX, this can range from 8 to 32 extensions.

➤ The number of incoming lines offered typically ranges from 4 to 16.

➤ An auto attendant feature for automatic voice response for inbound calls is available with most PBX products. Voice mail response and music on hold for each extension is included.

➤ Paging, call forwarding, and call conferencing are typical services offered with most home office PBX systems.

➤ Automatic call distribution (ACD) services for the organization are available. When customers call a company that has an auto-attendant PBX system, they typically hear a welcome greeting and instructions to dial the extension of the desired employee or, if the extension is not known, to use the company directory by spelling the employee's name.

Key Systems

For the home office or small business that does not require the services and capacity of a PBX, key systems are a popular choice. Most key systems are supported by a central processing unit called a key service unit (KSU). KSU systems typically support advanced business telephone features such as call forwarding, extension dialing, and voice mail options. The KSU supplies power and intelligence to all the telephone sets.

When planning the design of a KSU key telephone system, a star topology wiring plan is required in which each telephone set has a separate cable between the telephone set and the KSU unit, as shown in the Figure 4.2. Figure 4.2 shows a typical KSU unit using three outside telephone lines.

A less expensive alternative is a KSU-less key telephone system. Each phone is wired in a daisy-chain fashion, as illustrated in the Figure 4.3. A KSU-less design requires a separate cable to each handset for each outside line.

The main difference between a PBX and a key telephone system is you must first dial "9" to access an outside line when using a PBX system. If a business is growing rapidly and needs more than 40 incoming lines, a PBX system is the best option for a privately owned telephone system. Key systems are a less expensive option for the home office.

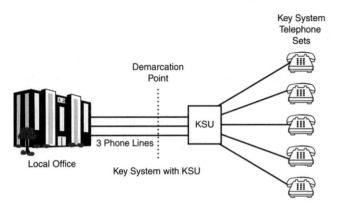

Figure 4.2 Key system with KSU.

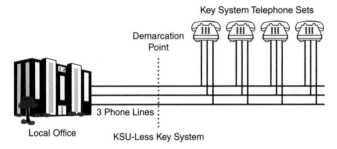

Figure 4.3 KSU-less key system.

A key system supported by a KSU is a popular design choice for an average home office or small business. On the other hand, a non-KSU phone system is the best choice for small businesses where only two phone lines are used and there is no requirement for Web hosting. Remember that the example in which the local telephone company cabling separates from the residential owner's wiring (service wiring from premises wiring) is called the *demarcation point*.

Voice Over IP

Voice over IP (VoIP) is a term used in IP telephony for a set of facilities for managing the delivery of voice information using the Internet Protocol (IP). In general, this means sending voice information in digital form in discrete packets rather than in the traditional circuit-committed protocols of the public switched telephone network (PSTN). A major advantage of VoIP and Internet telephony is that it avoids the tolls charged by ordinary telephone service. VoIP, now used somewhat generally, derives from the VoIP Forum. This forum is made up of major equipment providers, including Cisco, VocalTec, 3Com, and Netspeak, and promotes the use of ITU-T H.323, the standard for sending voice (audio) and video using IP on the public Internet and within an intranet. The forum also promotes the use of directory service standards, so users can locate other users, and the use of touch-tone signals for automatic call distribution and voice mail.

Modern voice coders used for many years in computer sound cards and cellular telephones provide the capability to convert human analog voice information into a digital format using compression to reduce the number of bits transmitted. Small voice coder microchips are inexpensive and have made VoIP an affordable technology. VoIP software formats the compressed voice digital information into IP packets of data and sends it over the network. The IP network handles the voice packets the same as any other packets. Voice communication is therefore possible over any public or private IP network. Because the Internet is a global network, the possibility of worldwide toll-free computer-to-computer Internet telephony has created widespread interest for home network users.

Remote Access Methods, Standards, and Protocols

Remote access is a type of telecommunications service that provides users access to a computer or a network from a location away from the home or work location. Employees on business travel often need access to the corporate private network to read email or access shared files. Remote access is also a valuable tool for home network users who need to access automated home features from a remote location.

A remote access *server* is the computer and associated software that is required on the home or business network. The server might also include a firewall either integrated with the server or as a separate processor and a router. A remote access server can also include or work with a modem pool manager so that a small group of modems can be shared among a large number of intermittently present remote access users. A remote access server can also be used as part of a virtual private network (VPN). The advantage of a VPN remote access connection is the capability to use a public network such as the Internet rather than a dedicated line.

The following three protocols have become popular with the growing use of VPNs for remote access to local area networks (LANs). You will need to become familiar with each of these protocols:

➤ *Layer 2 Tunneling Protocol (L2TP)*—Used to encapsulate a message when it is transmitted over the Internet using a VPN. An L2TP tunnel is created by encapsulating an L2TP frame inside a User Datagram Protocol (UDP) packet, which in turn is encapsulated inside an IP packet.

- *Point to Point Tunneling Protocol (PPTP)*—A network protocol that encapsulates Point-to-Point Protocol (PPP) packets into IP datagrams for transmission over the Internet or other public TCP/IP-based networks. PPTP can also be used in private LAN-to-LAN networking.
- *Point to Point Protocol (PPP)*—A remote access protocol used in PPTP to send multiprotocol data across TCP/IP-based networks. PPP encapsulates IP packets between PPP frames and sends the encapsulated packets by creating a point-to-point link between the sending and receiving computers.

Home Environmental Factors

Several home environmental factors must be considered when designing a telecommunications system. They play an important role in determining how the system will meet the objectives planned by the home users. Important issues relating to the home environment are as follows:

- If a key telephone system is planned, it should be consistent with the room plan of the home, known locations of existing wired phone jacks, and the topology (star or daisy-chain) of exiting telephone cables.
- The telecommunications system should be designed for compatibility with planned or existing home computer networks, Internet access options, home security, surveillance systems, and home audio/video systems.
- If a home security monitoring service is being used or planned in new home construction, the telecommunications system must support an internal telephone connection to the security and surveillance system.
- The home topology and structured wiring plan must be compatible with planned telecommunications systems. As an example, if a key telephone system is planned, it should be consistent with the room telephone cable wiring plan of the home.
- The telecommunications system should support the remote access features of a home network. You must determine whether routers, firewalls, and remote access servers are required and whether the home network will use remote access as a future option or installed during new construction.

Telecommunications Equipment Location Considerations

Locating various components and telephone sets in a home are dependent on the type of phone system selected, the location where the phone lines enter the home (the demarcation point), and the type of phone cabling that might be already present in an existing home. Availability of some of the services, such as DSL and dial-up and cable Internet, are related to certain home environmental conditions discussed in the following paragraphs.

Telecommunications System Characteristics and Restrictions

Telecommunications systems for the home user covered earlier include analog and digital telephone systems, PBX, key systems, and KSU-less telephone systems. Although PBX and KSU key phone systems are seldom selected for an average home design, such systems are practical for a small office or home office location. The KSU should be located in an area close to the demarcation point to minimize the number of cables required. Where KSU-less systems are used in a residential setting, the phones can be located anywhere. A KSU-less system does not require a star home run type wiring topology; therefore, it is less expensive for an existing home. Each phone in this case shares the same number of telephone lines with all other phones in the home in a multiline installation.

Industry Standards and Practices

Numerous telecommunications industry standards groups have developed and published information related to the wiring and location of residential telecommunications systems. The organizations involved in the development of home telecommunication standards that you will need to remember are the Federal Communications Commission (FCC), American National Standards Institute (ANSI), the Telecommunications Industry Association (TIA), the Electronic Industries Alliance (EIA), and the National Fire Protection Association (NFPA). These standards groups and related publications are covered later in this chapter.

Home Environmental Factors Related to Location

Home telecommunications networks are designed to provide both internal and external services; however, home environmental factors have some influence on the choices and location of components and can also influence the services available. As an example, the location (urban, suburban, or rural) of the home office or residence influences the type of Internet access available in a particular geographic area. Consideration must be given to the availability of toll-free dial-up access to an Internet service provider (ISP) in a suburban or rural area.

Location of the residence or home office also determines whether cable or DSL Internet access service is offered in the local geographic area. Unlike standard dial tone services, DSL services are distance limited and typically available to only those homes served by copper wire circuits less than 18,000 feet in length. The percentage of homes within DSL range varies by central office (CO) and can range from less than 50% for some rural central offices to as much as 100% for some urban COs.

Physical Telecommunications Products

Telecommunications products for the home provide a wide range of capabilities and services. They include simple analog phones, key systems, and top-of-the-line PBX systems. Fax machines are now used in the home as well as the business office for transmitting and receiving documents. Videoconferencing is now possible between PCs for the home office and individual users using software, low-cost sound cards, and PC video cameras. You will also learn the names of several caller line identification services as well as the newest methods for terminating telecommunications wiring in the home.

Telephone Fundamentals

The most common type of telephone system for the home is a single line with multiple extensions where all members share the single line/single number–type analog telephone system.

As indicated earlier, most small business and small office users select a key telephone system. Key systems are available with a central controller called a key service unit. Another type of key phone system is available that does not have a KSU and instead distributes the switching functions to individual phones. This type of system is called a *KSU-less* system because the phones are similar to a traditional key phone, with its busy lamps and direct station select keys, but lack the key system's main switch (the KSU). Unlike PBXs and key systems, KSU-less telephone systems avoid structured star or home run wiring, in which separate wire segments link phones to a central switch. In a KSU-less phone system, a single segment of existing in-wall two-pair telephone wiring can generally support two trunk lines for outside calling, while simultaneously carrying intercom communication.

A KSU key telephone system has additional capabilities but is also more expensive. A KSU distributes power to all the phones, sends dial tones to them from the phone company's incoming lines, controls the ring, generates DTMF dialing tones, enables intercom and paging capabilities, and provides features such as music-on-hold. Phone systems using KSUs generally require home-run wiring, with a direct path from the KSU to each phone jack.

 It is important to know the basic differences between PBX, KSU, and KSU-less phone systems. Remember that a KSU-less system is ideal for businesses with two outside lines. Additional outside lines for a small business are usually better served with an upgrade to a KSU phone system. PBX systems are for large businesses with approximately 50 or more employees.

FAX Machine Communications

Fax (abbreviation for facsimile) machines are devices used in home offices or businesses to send and receive documents using standard telephone lines or digital dedicated lines. Faxing is a less expensive alternative to overnight package delivery and can respond to situations in which a document is needed the same day for an urgent business transaction.

Many of the newer fax machines as illustrated in the Figure 4.4 can be connected to a computer to serve as a printer and a computer scanner. This type of device is called a *multifunction* fax machine.

Figure 4.4 Multifunction fax machine.

 In addition to the standard fax functions, a multifunction fax machine is also a copier, scanner, and printer. All these features can be integrated and controlled with a computer.

Fax machines are manufactured to comply with a series of transmission protocols and standards such as transmission speed, compression, and resolution. This feature ensures that fax machines from different vendors are interoperable. The standards for fax machines and protocols are described at the end of this chapter.

Fax Speeds

Most fax machines transmit at speeds of 9600 bits per second (bps); 14,400bps; or 33,600bps. Sending a fax at 9600bps takes 15–60 seconds per page depending on the density of information (text and graphic content). A higher-speed transmission at 14,400bps reduces the time for one page to 10 seconds. At 33,600bps, the time per page drops to less than 5 seconds.

PBX Systems

A PBX is a telephone switching system that provides the physical connection between the outside line used by the caller and the employee's phone on his desk. It is private because it is owned by the business or organization and is located on its facility, typically in a wiring or phone closet. It is called a *branch exchange* because it makes connections inside the business in the same way that a normal telephone exchange connects all the phone numbers served by the phone company.

Advanced PBX phone systems often include auto-attendant, voice-mail, and ACD services for the organization.

An ACD handles incoming telephone calls by routing them according to a pre-defined set of rules. Additionally, the ACD plays announcements to encourage the caller to wait for the next available operator if all lines are busy. When a number does not answer, the caller is routed to the phone's voice mail.

Videoconferencing

Videoconferencing is a telecommunications process that supports real-time video and sound interaction between people in different locations.

Videoconferencing is now possible between two individuals using desktop PCs equipped with software, cameras, microphones, and sound cards.

Videoconferencing offers many advantages for a home office or business. It can often reduce or eliminate the costs for business travel. Instead of sending several employees to meetings in different cities, they can attend the meeting while in their own offices. By connecting each site with faxes, phones, and videoconferencing terminals, both parties are able to hear, see, and share documents in a real-time environment.

Caller Line Identification

The public telephone system has the capability to identify the number of the caller to the receiving (called) party. This type of service is one of many listed under the title of *caller line identification* or *calling line identity (CLI)* services. Features such as caller display and call return use a feature of modern telephone networks that transmits the number of the caller as each call is set up. Called customers, using these new CLI services, can now have access to this information. With caller display, a customer can see the number on her phone before answering the call, and with call return, she can dial 1471 at any time to find out the number of the last caller to her number—whether it was answered or not.

66/110 Connection Blocks

Connection blocks are a type of telecommunications patch panel used to connect incoming phone lines on one side and telephone wires going to each room in the house on the other side. These blocks provide a convenient method of interfacing all telephone wiring at a central location where the outside phone lines enter the home. Wires are connected to each side of the block using a punch-down tool, which inserts the wire into a terminal on the block. As the wire is seated, the terminal cuts through the insulation to make an electrical connection, and the spring-loaded blade of the tool trims the wire flush with the terminal. This type of connection is referred to as an *insulation displacement connection (IDC)*. The two common families of punch-down blocks are 66 blocks, which are usually used for phone work, and 110 blocks, which are used for both data and voice circuits.

Punch-down Tool

The punch-down tool illustrated in Figure 4.5 is also known as an *impact* tool. It is usually supplied with two types of blades: One is for the 66-type block, and the other is for the 110-type block. Both of these blades have two sides to them. One side is for terminating and cutting, and the other side terminates without cutting the wire to allow the wire to be looped to another terminal.

The impact tool itself has two settings, called Hi and Lo. The Hi setting is used to terminate wire on a 66-type block, whereas the Lo setting is used to terminate wire for the 110-type block.

Figure 4.5 Punch-down tool.

The special type of tool required for inserting wires into the connectors on a 110 or 66 block is called a punch-down tool. Most integrators select the less expensive 110 block, which also supports higher-speed data circuit terminations.

Type 66 Connection Blocks

The type 66 block shown in Figure 4.6 contains four columns of 50 pins each, onto which the 50 wires of the 25-pair group are placed. Each wire is placed in a pin and then punched into place, stripping the insulation in the process. The pins in columns 1 and 2 are shorted together, and the pins in columns 3 and 4 are shorted together. This creates an input side and an output side of the block. To connect the two halves, bridging or shorting clips are used between pins 2 and 3 of any particular row. The 66 punch-down block provides easy access to each wire and is used to terminate most twisted-pair cable. Most 66 blocks do not support Category 5 UTP cable.

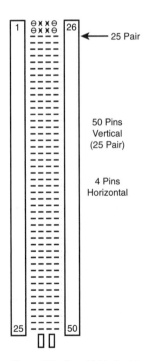

Figure 4.6 Type 66 block connections.

Type 110 Connection Blocks

Type 110 connector blocks are newer than 66 blocks and occupy less space than 66 connector blocks. These blocks are rated for Category 5 UTP cable and often have RJ-45 connectors already attached to them. 110 connector blocks are less expensive and provide the best options for future expansion.

You will need to know the difference between the 66 and 110 termination blocks identified in this section. The 66 blocks are designed to connect 25-pair cables. The 110 connectors, also known as M-type blocks, are rated for higher bandwidth data signals and are used for both voice and data cable terminations.

Standard Configurations and Settings

Standard configurations for home telecommunications settings include the techniques for connecting telephone handsets and broadband digital modems. Expanding on information you have learned in the previous section about the physical components of telecommunications systems, this section includes methods for connecting phone extensions to a connection block using home run wiring.

Voice mail and intercom systems can be configured to adapt to most requirements defined in the design for a home or small office. Many of the optional configurations for each are also included in the following paragraphs.

Phone-line Extensions and Splitters

Phone-line extensions enable two or more phones to use the same line or line group. For home wiring, extensions are easy to install if the home was wired for modular telephone connectors during new construction. For business and home office wiring, extensions should be wired using a star configuration with separate wires run to each outlet, as shown in the Figure 4.7. This is the preferred type of extension wiring because it makes future growth both practical and less time-consuming. Each extension has a dedicated wire from the room phone jack to a punch-down block.

The type of splitter used with a telephone extension referenced in this section enables a phone jack to serve as a multiple port for the devices mentioned, such as a modem or fax machine. This should not be confused with a DSL splitter discussed in the section "DSL Modems an Splitters" in Chapter 1, "1.0—Computer Networking Fundamentals." The phone jack splitter discussed and illustrated here is a simple port expander that enables only one device to use the phone line at one time. If the phone line is shared by a modem or data device, such as a fax with a phone, only one device can use the phone line at one time. The DSL splitter has a very different and more sophisticated role: It acts as a filter to divide the usable spectrum of the phone local loop telephone circuit into a data frequency band and a voice frequency band. The DSL splitter enables the DSL modem and a voice phone call to share the phone line at the same time with no interference from each other.

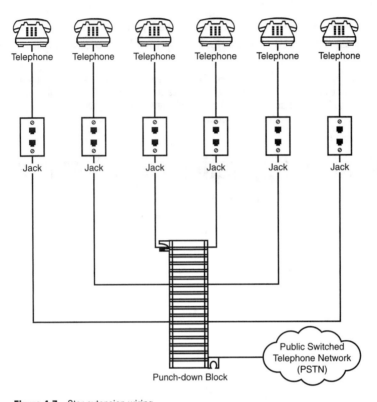

Figure 4.7 Star extension wiring.

Phone jacks can also use a splitter to divide a single RJ-11 phone jack into two or three ports. This works well with locations where a phone, modem, or fax is used with a single wall jack.

Figure 4.8 shows an example of a dual-port splitter that might be installed by the user.

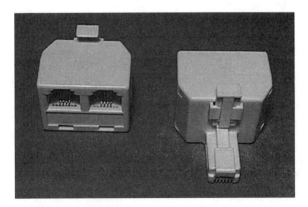

Figure 4.8 Phone-line splitter.

Voice Mail

Voice mail is a telecommunications service that has become a standard feature for home users as well as business organizations. Voice mail can be implemented using an external service provided by a local telephone exchange carrier, using a privately owned PBX with internal voice mail service, or through a privately owned answering machine. Newer voice answering machines use digital storage technology that has replaced older tape cassette answering machines.

Voice Mail Configuration Options

Privately owned digital answering machines offer convenient, time-saving features that enable the user to skip, erase, or repeat a message as he's listening to it. They also usually have a digital LED display to show how many messages have been received and stored, and each message is tagged with the date and time received. These machines enable the user to record a greeting that is played when the phone is not answered. They can also be programmed to answer after a selected number of rings, and the messages can be played back from a remote location by entering the correct code on the keypad.

Some home digital systems provide a feature for individual voice mailboxes. A home-based business can use one mailbox while maintaining the same service for selected family members. A recorded message instructs the caller to press an assigned number for each mailbox before leaving a message so the incoming messages are kept private and organized.

Local telephone companies also offer voice mail service. The phone is answered automatically with your structured voice answering message, and calls are maintained by the local exchange carrier. This type of service is used by businesses and home users who do not want to use a personal home answering machine.

Call Restriction

Call restriction features are available to home and business phone subscribers who want to restrict the phone from being used in an unauthorized way. The home or business phone system owner can restrict calls to specific area codes or to any preprogrammed four digits that restrict calls to any number that matches the four numbers.

Only the owner can override the restriction by means of a password. The owner can then make and take calls by dialing her password.

Call restriction options vary according to individual proprietary features offered by numerous vendors and local telephone service providers. Some examples of how most systems offer call restriction options are listed in Table 4.1.

Table 4.1 Call Restrictions

Type of Call Restriction	Restriction Feature
900 number block	Prohibits the capability to dial any 900-number service
800 number block	Prohibits the capability to dial any 800-number service
011 number block	Blocks the use of an international prefix calling number
Collect call block	Stops the user's phone from accepting collect calls
Long-distance restriction	Blocks the phone from dialing "1" or "0" to initiate a long-distance call

Intercom Systems

Intercom systems are primarily used in the home or home office for communicating with other members of the family or workers in various locations. Intercom systems can be configured using several optional features, outlined in Table 4.2.

Table 4.2 Intercom Features

Intercom Feature	Description
Background music	An AM/FM radio and CD player integrated with the intercom allows music at any or all stations.
Doorbell	Alerts the family when a doorbell is sounded.
Door speakers	Provides the ability to speak with a visitor before opening the door.
Monitoring	Selected stations, such as a nursery, can be monitored from the central controller.

An example of a wall-mounted intercom control station with AM/FM radio and alarm clock is shown in Figure 4.9.

Figure 4.9 Intercom radio.

Methods for Connecting Telecommunications Equipment

Telecommunications systems for the home can be connected using both wireless and wired technologies. The physical structured wiring grades 1 and 2 definitions are described in this section, together with methods for connecting standard telephone sets and telecommunications wiring termination points.

Wireless Connectivity

Wireless connectivity for telecommunications systems can be described as either privately owned systems or public cellular/PCS wireless networks. The three types of private wireless systems used in homes and business organizations are

- *Wireless local area networks*—These use components manufactured in compliance with the IEEE 802.11x standard family.

- *Cordless telephones for the home consumer market*—These use the unlicensed radio spectrum.

- *Wireless PBX (WPBX) telephone equipment*—This includes base stations and handsets that connect users to PBX systems using the unlicensed radio spectrum.

Cellular service is a type of public wireless service that enables mobile subscribers to make phone calls from most urban locations using portable wireless phones that connect to radio cell sites, which in turn connect callers to a mobile switching center. The mobile switching center then makes the connection to the wire-line public switched telephone network.

Telephone Connectivity

Telephones are connected to wall outlets and routed to a termination point where cross-connects or terminal blocks provide the connection to external phone company wiring. For home offices or business locations that have their own PBX or key systems, the phone wiring connects all the telephones to the central PBX controller or KSU.

The standard wall outlet for telephones is called an RJ-11 connector. The detailed wiring plan for an RJ-11 connector is discussed later in this section.

Physical Wiring Types

The standard for new residential telecommunications cabling is called *structured cabling*. The basis for this new concept in home wiring is described in the ANSI/TIA/EIA 570-A, "Residential and Light Commercial Telecommunications Wiring Standard." When properly designed and installed, structured cabling can be used for home networking; audio/video systems; and home security, control, and automation solutions in all areas of the home.

The physical structured cabling system is comprised of three major components: the communications center, the in-wall cabling, and the outlet/patch cord.

The communications center is typically installed close to the point where cable TV, phone, and Internet services enter the home. The center is usually contained within an outer shell and encloses cable management, a cable patching system, a 66 or 110 push-down block, a video amplification unit, and a local area networking hub.

The ANSI/TIA/EIA 570-A standard has established minimum home run wiring requirements in which each room in the home has outlets connected to the communications center for video, data, and voice. Two grades of structured cabling are specified for installation, as follows:

➤ *Grade 1*—For each cabled location, grade 1 provides a generic cabling system that meets the minimum requirements for telecommunication services. This grade provides for telephone, TV, and low-speed data services. A minimum of one Category 3 unshielded twisted pair (UTP) cable and one 75-ohm coaxial cable is required for each location.

➤ *Grade 2*—For each cabled location, grade 2 requires two Category 5 (5E recommended by the FCC ruling) UTP cables and two 75-ohm coaxial cables to each location plus, as an option, optical fiber cabling.

Category 5E cable is recommended for both data and voice circuits in a structured wiring design. Grade 1 provides sufficient capacity for up to two telephone lines plus an Ethernet LAN port for each location.

Remember that *home run wiring* is another name for a star wiring topology in which all cabling is routed from each room to a common termination location. Memorize the grade 1 and grade 2 residential telecommunications wiring requirements. The 75-ohm coaxial cable requirement is often specified as type RG-6 (quad shield recommended).

RJ-11 Connections

RJ-11 is the type of jack used for terminating a single-line telephone cable. The connector has six pins, but only two are used in this example. RJ-11W jacks are wall-mounted, and RJ-11C jacks are flush-mounted. In both RJ-11C and RJ-11W connectors, only two terminals are wired, as indicated in Figure 4.10. The red wire is connected to the number 3 terminal (ring), and the green wire is connected to pin 4 (tip).

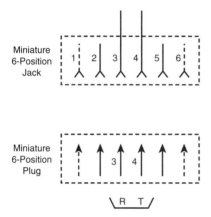

Figure 4.10 RJ-11 plug and jack.

Telephone jacks and plugs are wired to conform to the Uniform Service Ordering Code (USOC) numbers, originally developed by the Bell System and later endorsed by the FCC. One specific type of registered plug and jack hardware can be wired in different ways and have different USOC numbers. USOC has become an acronym that is pronounced "you-sock," and jack wiring schemes are generally referred to as USOC codes. The designation *RJ* means registered jack. The six-position jack shown in Figure 4.11 can be wired for different RJ physical configurations. For example, the six-position jack can be wired as an RJ-11C (1-pair), RJ-14C (2-pair), or RJ-25C (3-pair).

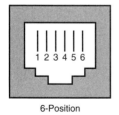

6-Position

Figure 4.11 Six-position jack.

The USOC code designations identify the type of use for the jacks as follows:

- *RJ*—Denotes a registered jack
- *C*—Denotes a flush- or surface-mounted jack
- *W*—Denotes a wall-mounted jack
- *X*—Denotes a complex multiline or series jack

RJ-45 Termination Standards

Both ANSI/TIA/EIA 568A and T568B standards define the two color codes used in wiring RJ-45 eight-position modular plugs. The two standards allow both of these color codes. The only difference is that the orange and green pairs are interchanged, as illustrated in Figure 4.12. The ANSI/TIA/EIA 568-A standard is the preferred standard for home telecommunications cabling.

The rationale for maintaining two standards for registered jacks reflects an earlier LAN standard. Some versions of early LANs were previously developed to use telephone cable as the network media; examples are SynOptics' LattisNet and AT&T's StarLAN. The 568B standard matches the older AT&T 258A color code and is still permitted within the standards. When the preferred 568-A standard pairs are used for Ethernet, pair 2 (orange) and pair 3 (green) are used for the connection to the network and pairs 1 (blue) and pair 4 (brown) are reserved for two telephone lines so there is no conflict.

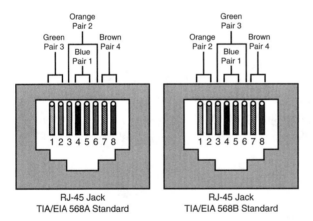

Figure 4.12 RJ-45 color codes.

You will need to memorize the RJ-45 plug and jack wire terminations for both LAN and telephone connections in a structured wiring home design configuration. Pins 4 and 5 (blue, pair 1) and pins 7 and 8 (brown, pair 4) are reserved for voice (telephony). Pins 1 and 2 (green, pair 3) and pins 3 and 6 (orange, pair 2) are used for connection to an Ethernet network.

Termination Points

Telecommunications wiring schemes for homes are usually designed to route all internal wiring to a common termination panel where all low-voltage wiring is terminated in a central distribution panel. The types of termination points discussed earlier for telephone systems are called type 66 or type 110 termination blocks. This is the location where external telephone wiring enters a home office or business and establishes the demarcation point between internal and external telecommunications wiring.

Telecommunications In-house Services

In-house services include all the enhanced telephone features that are available using privately owned voice mail, PBX, and key telephone systems.

Voice Mail

As discussed earlier, mail services are available from local telephone exchange carriers and personal home voice mail systems. In-house voice mail services include personal answering machines, integrated key systems, and PBX voice mail systems.

Personal answering machines are owned by the subscriber, are low in cost, and require no service fee for operation. They are often integrated with a wireless phone and can also be used as an intercom. Key systems have more voice mail features with a choice of port configurations. In key telephone voice mail terminology, a *port* is one connection to the voice mail system. The number of ports on a voice mail system is the number of simultaneous connections to voice mail possible at any one given time. This includes both callers leaving voice mail, as well as a user retrieving her personal voice mail.

Intercom Services

Intercom systems are a convenient and low-cost internal communications system that can be privately owned and installed by the user with a variety of custom features. They provide the capability to communicate with other members of the family from any location in the home or other employees in a small business office. Larger businesses with more than 50 employees typically use a key phone system or PBX with internal extension dialing features for internal voice communications.

Home intercom systems are either hard-wired using twisted-pair wiring in the walls or wireless systems that use the 120-volt AC power wiring in the home to communicate with other stations that are also plugged in to a power wiring outlet. Wireless intercom stations do not actually use the wireless radio medium for connecting to other units. They use the power wiring to connect to other intercom stations and are classified as wireless because no special wiring beyond the power wiring is required for operation.

Hard-wired systems are normally used for new construction, but wireless systems are the most economical option for existing homes.

Call Conferencing

Call conferencing with a privately owned phone system enables multiple users to be connected to other extensions for in-house conferencing. This feature previously was available only on large PBX systems, but many vendors now offer call conferencing on less-expensive key or non-KSU telephone systems.

Most key phone systems support three-way call conferencing, which enables any user to call a second in-house extension. The first called party is placed on hold with a switch hook action. The phone system then supplies three beeps and a second dial tone to the initial caller. Next, the second party extension is dialed. When the second party answers, the initial caller depresses the switch hook again and announces that the conference is ready to begin.

Call conferencing is a feature that can also be used to connect an outside caller to multiple internal extensions. This type of call can often save time when the outside caller might need to order a product but also needs to have specific product specifications or technical information. The person who takes the order can have a technical support member on the line at the same time, thereby making it unnecessary for the outside caller to be placed on hold or switched between different departments.

Extension Dialing

Most PBX and key systems support extension dialing on internal telephones. This feature is also valuable for small, non-KSU phone systems for the home office. For larger business organizations, the direct inward dialing feature of many business phone systems enables an employee's extension to be dialed from an outside phone number. Direct inward dialing (DID) is a service of a local phone company that provides a block of telephone numbers for calling into a company's PBX system. Using DID, a company can offer its outside customers individual phone number extensions for each person or workstation within the company without requiring a physical line into the PBX for each possible connection.

For example, a company might rent 100 phone numbers from the phone company that could be called over 8 physical telephone lines (these are called *trunk lines*). This would allow up to 8 ongoing calls at a time; additional inbound callers would either get a busy signal until one of the calls completed or be able to leave a voice mail message. The PBX automatically switches a call for a given phone number to the appropriate workstation in the company—a PBX switchboard operator is not involved.

Telecommunications External Services

External telecommunications services include the various offerings from local telephone companies made possible by the integration of advanced switching and database management systems in the telephone network infrastructure. Caller ID, voice mailboxes, call blocking, three-way calling, call waiting, and emergency response systems are typical of the external services discussed in this section.

Caller ID

Caller ID is a popular service offered by local telephone companies. It provides a display of the calling party number before the called party picks up the phone. Caller ID requires a connection to the phone line of a small caller ID display similar to the type shown in Figure 4.13. Caller ID also retains a list of the most recent caller phone numbers, including those received while you were away.

Figure 4.13 Caller ID display.

External Voice Mail Services

Business organizations and home users often elect to purchase voice mail services from an external source such as a service provider or local exchange carrier. These services are usually available for a monthly fee based on the type of service desired and the options selected. This type of service is desirable for individuals who work away from the home or business office, such as a sales or field support staff. Voice mail from a service provider enables the subscriber to reach a voice mailbox from a toll-free number anywhere in the United States.

Most voice mail service organizations have added features that include storing a given number of messages for 30 days, paging when someone leaves a voice mail message, multiple voice mailboxes on a single line, and mail forwarding to a single mailbox for multiline subscribers.

Wireless cellular carriers offer voice mail services for mobile phone subscribers. This is another external voice mail service that is a popular choice for travelers who also have cellular phone service. It offers the same advantages as service providers where the subscriber is alerted when a voice mail message has been received. Mobile phone voice mail is also available to the user from any location away from home.

Call Blocking

A number of custom telephone services are collectively called *call blocking* services. The types and features are not always available from all carriers. The instructions for turning any of the features on and off can also vary among different local service providers.

As discussed earlier in this chapter, call restriction is a type of call blocking service designed to stop some specific inbound calls and prevent certain prefixes from being dialed on a phone.

Another type of call blocking is designed to reject calls from numbers included in a subscriber call block list. When a call is received from any blocked number on the list, a recorded message is received by the caller that says, "We're sorry, the number you have reached is not accepting calls at this time."

Outbound call blocking prevents calls to any pay-per-use service and entertainment numbers beginning with the prefix 900 or 976. These prefixes are commonly used by entertainment or informational services that carry considerable per-call or by-the-minute charges.

Another type of call blocking service provides the capability to prevent the calling party phone number from appearing on the caller ID display of the person receiving your call.

This type of service can be initiated on a per-call basis using the following procedure:

1. Lift the telephone handset and listen for a dial tone.

2. Press * 67.

3. Dial the outbound call number.

4. The person you've called will not be able to see your number displayed on his caller ID unit. Instead, a "P" or "Private" will be displayed.

If you have requested call blocking service from your service provider, you do not need to dial a code number sequence to block your number each time. Your number will always appear as "Private." Some providers allow the caller to override caller ID blocking (allowing your number to be displayed) on an individual call by dialing * 82 before placing the call.

Three-Way Calling

Three-way calling is an added service offered by local exchange carriers, long-distance carriers, and mobile wireless cellular carriers. It enables a subscriber to talk to two people in two different locations at the same time.

Three-way calling is initiated by calling the first party, pressing and releasing the switch hook to get a second dial tone, and dialing a second number. When the second party answers, the switch hook is pressed and released again. The three-way connection is then in place and the call can proceed with all three parties connected.

Most local exchange carriers have pricing options for three-way calling that offer a monthly flat fee or pay-per-use fee. Other long-distance carriers offer the service free under some plans. Some plans also allow the subscriber to use the three-way calling feature even if he didn't initiate the call.

Call Waiting

Call waiting is a telecommunications service feature that provides an indication to a subscriber conducting a call that one or more other phones are trying to make a connection to her number.

With call waiting service, the second call is announced by a soft beep. The user can ask the first caller to wait while she answers the second call. She then has the option to return to the first call or disconnect from the first to talk to the second caller.

Emergency Response System

Emergency response using the 911 system has been a free telephone service for several years. Numerous lives have been saved with the numbers 911 being used to provide help to callers during an emergency.

The 911 system connects callers to a facility called a public safety answering point (PSAP). The dispatch operator at the PSAP attempts to obtain as much information as possible from the caller and send the appropriate help to the caller's location. The caller might be in distress and unable to provide location information; the caller's phone number, however, is available to the dispatch operator through a system called automatic number identification (ANI). ANI is a service that provides the receiver of a telephone call with the number of the calling phone. The method of providing this information is determined by the local exchange carrier, but the service is often provided by sending the dual tone multifrequency (DTMF) tones along with the call. This process is also used for caller ID service. The 911 operator has the ANI feature available at the PSAP to assist in identifying the number and the location through an automatic directory service.

Local governments are responsible for the costs of the circuit, emergency response staff, installation charges, and public safety answering point equipment.

Enhanced 911 for Wireless Carriers

Wireless enhanced 911 (E911) is a program mandated by the FCC to improve the effectiveness and reliability of wireless 911 services.

Location information for regular wire-line 911 service calls can be traced using the ANI feature. Wireless calls to the PSAP can be initiated from anywhere in the cellular service area; therefore, the ANI feature is of no benefit to the PSAP for assisting a mobile caller in distress. The caller might be unable to provide the dispatcher with specific location information. The wireless E911 program is divided into two parts, called phase I and phase II. Phase I requires carriers, upon appropriate request by a local PSAP, to report the telephone number of a wireless 911 caller and the location of the specific cell site antenna that received the call. Phase II requires wireless carriers to provide far more precise location information, within 50–100 meters in most cases. Phase II also offers several technologies available to the individual cellular service providers. Some cellular companies use network triangulation between cell site antennas to locate the caller, whereas other carriers use global positioning system (GPS) equipment integrated into the mobile subscriber's phone to provide position location information to the PSAP.

The deployment of E911 service requires the development of new technologies and upgrades to local 911 PSAPs, as well as coordination among public safety agencies, wireless carriers, technology vendors, equipment manufacturers, and local wire-line carriers. The FCC established a four-year rollout schedule for phase II, beginning October 1, 2001, and to be completed by December 31, 2005.

Personal Emergency Response System

A personal emergency response system (PERS) is an electronic device designed to let the user summon help in an emergency. It has three components: a small, battery-powered radio transmitter with a help button; a console connected to the user's telephone; and an emergency response center that monitors calls. When the button is pressed, it signals the console, which automatically dials one or more preprogrammed numbers. Most systems can dial out even if the telephone is off the hook. When its button is pressed, a radio signal prompts a machine connected to the telephone to call the monitoring center for help. The monitoring center usually tries to call back to find out what is wrong. If the center is unable to reach the person or help is needed, the center tries to reach a designated person (such as a friend or family member) to follow up the call. If a medical emergency appears evident, an ambulance or other emergency provider is dispatched.

Telecommunications Industry Standards

Since 1990, the videoconferencing industry has built up specifications, standards, and practices governing audio, video, and call control components, whether over switched circuits, packet data networks, or even dial-up modems.

Fax Transmission Standards and Protocols

Fax transmission speeds and protocols are specified by the International Telecommunications Union – Telecommunications Sector (ITU-T). All the fax and modem standards were previously defined by the CCITT French standards group called the "v.dot" series, familiar to most modem buyers.

Fax machines that meet the ITU-T standard for group 3 fax are for machines that use standard analog telephone lines. The group 3 encoding format is the most commonly used fax in the home and office environment. It supports one-dimensional image compression (compression within the line only) of black-and-white images. The one-dimensional compression scheme uses run-length and Huffman encoding and can achieve compression ratios of 10:1 for office documents and 15:1 for engineering drawings. The resolution of group 3 encoded images is 200 dots per inch (dpi). Group 3 is used on standard speed fax machines connected to the public telephone network.

The ITU-T Fax Group 4 encoding format is a two-dimensional encoding scheme (within the line and from line to line). It can achieve compression ratios of 15:1 for office documents and 20:1 for engineering drawings. Unlike group 3, group 4 assumes an error-free transmission medium, such as ISDN networks. Group 4 fax images can be transmitted using X.400 electronic mail, and the connection options available with Group 4 are much more flexible than with the earlier fax standards. The standards are based on the ITU v.dot standards group. V.29 specifies a speed of 9600bps, V.17 is the standard for transmitting at 14,400bps, and V.34 supports 33,600bps. You can use the high speeds only when transmitting to other fax machines that are supported by the same protocols. But like modems, V.17 and V.34 machines are equipped with a fall-back mechanism that permits them to send and receive at 9600bps to ensure that they are backward compatible with older fax machines.

ANSI/TIA/EIA Standards

The primary installation planning and procedures for residential wiring are described in TIA/EIA 570A, "Residential and Light Commercial Telecommunications Wiring Standard." This is an important standard for home technology integrators.

> Remember the title of the ANSI/TIA/EIA 570-A: "Residential and Light Commercial Telecommunications Wiring Standard." It might appear as the short title "TIA 570" in the exam. This standard is the source document for most of the cabling installation requirements included in the HTI+ Residential Systems exam.

Two grades of structured cabling are specified for installation as follows:

➤ *Grade 1*—For each cabled location, grade 1 provides a generic cabling system that meets the minimum requirements for telecommunication services. This grade provides for telephone, TV, and low-speed data services. A minimum of one Category 3 unshielded twisted pair (UTP) cable and one 75-ohm coaxial cable is required for each location.

➤ *Grade 2*—For each cabled location, grade 2 requires two Category 5 (5E recommended by the FCC ruling) UTP cables and two 75-ohm coaxial cables to each location plus, as an option, optical fiber cabling.

Some new terms introduced in the 570A standard that you will need to remember are defined here:

➤ *Network interface device (NID)*—The NID is the demarcation point from the local exchange carrier. The NID is usually a nonmetallic box found on the side of a single-family house.

➤ *Auxiliary disconnect outlet (ADO)*—The ADO provides a means for the homeowner or tenant to disconnect from the local exchange carrier. Preferably, the ADO is co-located in the distribution device panel.

➤ *Distribution device (DD)*—The DD is a cross-connect facility located within each home or tenant space. Outlet cables from wiring points throughout the house are terminated here.

Prior to the publication of ANSI/TIA/EIA 570-A for residential telecommunications, the ANSI/TIA/EIA standards 568 and 569 are the most widely used standards for telecommunications systems. The 568 standard outlines specifications for a generic telecommunications cabling system. The purpose of the 569 is to standardize design and construction practices within and between buildings that are in support of telecommunications equipment and media.

Telecommunications System Installation Plans and Procedures

Installation procedures for telecommunications outlet boxes and various types of cabling must be in compliance with local building codes, which always have authority over any standards. The location of outlets and selection of cable types are described in the ANSI/TIA/EIA 570-A standard. This document should be used by installers when not in conflict with local building codes.

Telecommunications Wall Outlets

The ANSI/TIA/EIA 570A standard for grade 1 and grade 2 telecommunications residential wiring recommends the installation in each room of RJ-45 eight-position jacks and coaxial cable F-type connectors of the type shown in Figure 4.14.

Figure 4.14 Telecom outlets.

A common procedure is to place both types of jacks and connectors in the same outlet position using a duplex wall jack. This enables a computer and phone to be plugged in at the same location. It is a good practice to place more than one jack location in a room to allow flexibility in furniture arrangements.

Jacks are normally located at the same level on the wall as power outlets, but not closer than 12" to the power outlets. Wall phone brackets with an integral jack are usually mounted near eye level, or a bit lower, on the wall.

Telecommunications Cable Installation Tips

Telephone wire is usually 24 American Wire Gage (AWG) and comes in twisted or untwisted pair. The more twists per inch, the better. Telephone wire is rated by Category 2-5; no Category 1 wire exists. Category 3 wire has fewer twists than Category 5 wire per the same distance of wire. The best wire to use for telephone connections is Category 5E UTP as described in ANSI/TIA/EIA 570-A. The bare minimum number of wires for a residential telephone system is two wires. Category 6E is the latest high-performance UTP cable standard to be approved by the TIA and EIA standards group. It has also been identified in an international standard. The second edition of the International Standards Organization/International Electrotechnical Commission (ISO/IEC) 11801 standard includes Category 6 components as well as cabling. In ISO/IEC 11801, Category 6 cabling is referred to as "Class E Cabling," similar to the notation for Category 5E cable. The Category 6E specifications in the ISO/IEC 11801 document are essentially the same as in ANSI/TIA/EIA-568-B.

The following list provides best-practices tips associated with basic telephone cabling concepts:

➤ Do not pull four-pair cable with more than 25 lbs. of pulling energy.

➤ Never put a splice behind a wall or in an area where it cannot be accessed.

➤ Always test the cable runs after the installation is complete.

When installing any type of four-pair UTP cable, remember the maximum pulling force that should be applied is 25 lbs.

Telecommunications Troubleshooting and Maintenance

Many of the problems that occur in home telecommunications systems can be solved by following a few simple troubleshooting procedures. More complex problems might require calling a professional installer. Most cable problems can be isolated using a time domain reflectometer (TDR), which is designed to locate data cabling problems. The following tips are typical for the level of troubleshooting that should be followed by the home integrator:

➤ *Problem with one phone in a multiphone key system (no dial tone, and all others on separate extensions are working okay)*—The trouble is most likely with your phone itself or with the wire going to it. Check whether all the other phones have dial tones. If the installation is new, check the termination at the wall outlet. Change the line cord going to the wall outlet or check the termination with the proper cable tester. If the phone still does not work, change it with a phone that is working properly. This isolates the problem to the phone or the wall outlet wiring.

➤ *Problems are present with all phones in a residential installation*—Unplug all the telephones, answering or fax machines, caller ID units, and computers from the wall or baseboard jacks inside your home. If any of this equipment is powered by an AC adapter, unplug it from the AC outlet also. Then try each one (one at a time) in each phone jack or outlet. If one phone does not work anywhere, the trouble is probably in the telephone set. If none of the phones works in a particular jack, the trouble is with that jack.

Know that a time domain reflectometer (TDR) is a type of handheld tester that can pinpoint the location of a break in a cable. When all phone lines except one pair have dial tones in a new telephone cabling installation, you should check the termination wiring and the connecting cable between the phone and termination point.

Exam Prep Questions

Question 1

> Which type of phone system design requires a separate cable to each handset for each outside line?
> - A. KSU system
> - B. KSU-less system
> - C. PBX system
> - D. Cordless system

Answer B is correct. Each telephone handset that is a component of a KSU-less system (unlike a KSU) requires a connection to each phone line in the home. Answer A is incorrect because a KSU system uses a star topology wiring plan in which each telephone set has a separate cable between the telephone set and the KSU unit. All phones are controlled by the KSU unit. Answer C is incorrect because a PBX phone system manages the interface between the outside lines and the individual phones, eliminating the need for any direct connections between the phones and the outside line. Answer D is incorrect because cordless phones can be daisy-chained with no requirement for direct connection to an outside line for every phone.

Question 2

> Which feature of the public telephone network provides the automatic identification of the calling phone number to a 911 emergency dispatch operator?
> - A. PBX
> - B. ANI
> - C. DTMF
> - D. Call blocking

Answer B is correct. The automatic number identification (ANI) feature supports 911 answering centers and assists in the identification of the calling party. Answer A is incorrect because a PBX is a private telephone switching system and is not a public telephone system feature. Answer C is incorrect because the DTMF is the signaling system for electronic telephone switches that selects the correct local office switching path when a call is initiated. Answer D is incorrect because call blocking is an option that enables subscribers to prohibit certain types of call prefixes to be dialed on a phone.

Question 3

Which type of telecommunications technology is used to conduct voice communications over the Internet or a private intranet?
- A. PBX
- B. VoIP
- C. ISDN
- D. KSU

Answer B is correct. VoIP stands for Voice over IP. It is also referred to as Internet telephony. Answer A is incorrect because PBX is a private exchange telephone system and is not used primarily to conduct voice communications over the Internet. Answer C is incorrect because ISDN is a digital broadband Internet access method used to support both voice and data, but it is not a VoIP technology. Answer D is incorrect because KSU is a device for managing small customer-owned telephone systems and does not directly provide VoIP services.

Question 4

Which type of cable is recommended for data and voice communications using a structured cable design?
- A. RG-6
- B. Category 2 UTP
- C. Quad shield
- D. Category 5E UTP

Answer D is correct. The ANSI/TIA/EIA 570-A standard recommends Category 5E type cable for both voice and data throughout the home. Answer A is incorrect because RG-6 is a type of cable used for video applications in structured wiring but is not used in this type of application for voice and data communications in the home. Answer B is incorrect because Category 2 cable is no longer supported or recommended in any type of structured cable standard. Answer C is incorrect because quad shield is a type of outer insulation used in the construction of coaxial cable.

Question 5

When connecting home telephone lines using RJ-45 plugs with Category 5E cable, which connections are reserved for telephony?

- ○ A. Pins 4 and 5, and 7 and 8
- ○ B. Pins 1 and 2, and 3 and 6
- ○ C. Pins 2 and 3, and 4 and 5
- ○ D. Pins 5 and 6, and 7 and 8

Answer A is correct. Pins 4 and 5 (pair 1) and pins 7 and 8 (pair 4) are the pairs and pin numbers used for telephone connections when using an RJ-45 connector with UTP cable. Answer B is incorrect because pins 1 and 2 and 3 and 6 are used for data, not voice telephony. Answer C is incorrect. Although pins 4 and 5 are used for telephony, pins 2 and 3 are conductors in different pairs and would not be a correct connection for either telephony or data. Answer D is incorrect because pins 5 and 6 are not wire pair numbers but are connections used in two different pairs.

Question 6

Which standard was developed specifically to support Residential and Light Commercial Telecommunications Wiring?

- ○ A. ANSI/TIA/EIA 606
- ○ B. ANSI/TIA/EIA 568-B
- ○ C. ANSI/TIA/EIA 570-A
- ○ D. ANSI/TIA/EIA 569-A

Answer C is correct. The standard for new residential telecommunications cabling that describes structured cabling. The basis for this new concept in home wiring is described in ANSI/TIA/EIA 570-A, "Residential and Light Commercial Telecommunications Wiring Standard." Answer A is incorrect because ANSI/TIA/EIA 606 is the Administration Standard for the Telecommunications Infrastructure of Commercial Buildings. Answer B is incorrect because ANSI/TIA/EIA 568-B is the standard for Commercial Building Telecommunications Cabling. Answer D is incorrect because ANSI/TIA/EIA 569-A is the standard for Commercial Building Standards for Telecommunications Pathways and Spaces.

Question 7

Which device is required to avoid interference on the analog voice line from the DSL data transmissions on the same line?
- A. A gateway
- B. A multiplexer
- C. An RJ-11 jack
- D. A splitter

Answer D is correct. The splitter is essential for proper operation of a DSL modem over regular analog phone lines. Answer A is incorrect because a gateway is a type of device used to connect external communications systems to a home network. Although the gateway can contain an integrated DSL modem, its main purpose is not to act as a DSL splitter. Answer B is incorrect because a multiplexer is a device used to synchronize and combine many digital circuits into a hierarchical system of high-speed digital transmission systems and is not directly associated with a DSL service. Answer C is incorrect because an RJ-11 jack is used to provide the connection for a standard voice telephone handset.

Question 8

Which telecommunications network protocol is used to encapsulate Point-to-Point Protocol (PPP) packets into IP datagrams for transmission over the Internet?
- A. PPTP
- B. TCP
- C. UDP
- D. SMTP

The correct answer is A. The Point-to-Point Tunneling Protocol (PPTP) is an encapsulation protocol used to hide the IP packet header when transmitting packets over the Internet. Answer B is incorrect because TCP is a Transport layer protocol used with IP packet headers and does not perform encapsulation of packets. Answer C is incorrect because UDP is also used with the IP header for Transport layer services. Answer D is incorrect because SMTP is a TCP/IP suite protocol used to transfer email.

Question 9

Which types and quantities of cabling are required for a grade 2 structured wiring design?

- ○ A. One Category 3 cable plus one 50-ohm coaxial cable
- ○ B. Two Category 5 (5E recommended) UTP cables and two 75-ohm coaxial cables to each location plus, as an option, optical fiber cabling
- ○ C. One Category 5 (5E recommended) UTP cable and one 75-ohm coaxial cable
- ○ D. One Category 3 cable (Category 5 recommended) and one 75 ohm coaxial cable

Answer B is correct. The ANSI/TIA/EIA 570-A standard states that for each cabled location, grade 2 requires two Category 5 (5E recommended by the FCC ruling) UTP cables and two 75-ohm coaxial cables to each location plus, as an option, optical fiber cabling. Answer A is incorrect because 50-ohm coaxial cable is not recommended for home structured wiring designs. Answer C is incorrect because grade 2 requires two each UTP and two each coaxial cables. Answer D is incorrect because Category 5 (5E recommended) cable is required for grade 2 structured wiring, not Category 3.

Question 10

Which standard applies to fax machines that use standard analog dial-up telephone lines?

- ○ A. Group 4 fax
- ○ B. ANSI/TIA/EIA 568-A
- ○ C. Group 3 fax
- ○ D. DSL

Answer C is correct. The ITU-T standard for group 3 fax is for machines that use standard analog telephone lines. Answer A is incorrect because group 4 fax is the standard for digital error-free transmission circuits such as ISDN. Answer B is incorrect because the ANSI/TIA/EIA/ 568-A standard is for commercial building wiring and does not specify fax transmission standards. Answer D is incorrect because DSL is a high-speed Internet access service offered by local exchange carriers and is not associated with analog fax transmission standards.

Need to Know More?

 http://www.hometoys.com/articles.htm. This Web site contains a compilation of home automation and networking articles published in *EMagazine*.

 Held, Gil. *Voice and Data Internetworking.* Berkeley, CA: McGraw-Hill, 2000. (This book provides a detailed description of the latest voice technologies and data transmission protocols.)

 O'Driscoll, Gerard. *The Essential Guide to Home Networking Technologies.* Upper Saddle River, NJ: Prentice Hall PTR, 2001. (This book provides detailed information on the new home networking technologies related to power-line and phone-line networks.)

5.0—Home Lighting Control and Management

Terms you'll need to understand:

- ✓ Kilowatt-hour
- ✓ Three-way switch
- ✓ EMT
- ✓ CEBus
- ✓ GFCI
- ✓ Romex
- ✓ Ballast
- ✓ Armored cable
- ✓ MC cable
- ✓ Ohms
- ✓ BX cable
- ✓ X10
- ✓ TRIAC
- ✓ THHN

Tasks you'll need to master:

- ✓ Describing the connection for a three-way switch
- ✓ Identifying the uses for NM cable
- ✓ Identifying three types of luminaries
- ✓ Listing the requirements for completing a load analysis
- ✓ Explaining the theory of operation for a GFCI
- ✓ Describing the concept of zone lighting
- ✓ Describing how to conduct a test to determine whether an AC outlet is wired incorrectly
- ✓ Identifying the type of tester used to measure AC current in a cable

Home Lighting Design Considerations

The initial step in the design of a home lighting system for both new construction and a remodeling project is to determine the overall load (*current*) requirements for all the lighting systems. This is only one component, though, of the total electrical load projection because the other appliances and home heating systems must be included.

Wiring runs and wireless components must be identified as well as any possible control systems or zoning features. This section identifies the essential items in a lighting system design plan.

Load Requirements

Calculating the total electrical power required for a home lighting system is an essential part of the total power management plan that the electrical contractor must consider. The total peak power used at any instant in time is called the *load requirement* that is imposed on the home power wiring and the power source supplied by the electric utility company. The unit of power used is the *watt*, which is equal to the voltage times the current used by an electrical device. New home design considerations for electrical wiring must consider not only the power required for lighting, but also the power required for other electrical systems. Examples of other equipment included in the total power load requirements are

- Kitchen appliances
- Heating, ventilating, and air conditioning (HVAC) system
- Home entertainment systems
- Garage door opener
- Home computer and office equipment
- Washer, dryer, and dishwasher
- Refrigerator and freezer
- Workshop and outdoor AC power tools
- Home security and low-voltage lighting
- Ceiling fans

The initial step in calculating load requirements is to determine the watts/hour of all lighting and electrical devices. Electrical contractors use a table in which each appliance, light fixture, HVAC, entertainment systems, and so on is listed with the power in watts/hour. Some items might be used for less than 1 hour at a time, such as a garage light. In this example, the garage light would be calculated for fractions of a watt-hour. A 120-watt light multiplied by .2 hours (120 × .2 hours = 24 watt-hours). A watt-hour is a relatively small amount of power; therefore, expressing power consumed in kilowatt-hours (1000 watt-hours) is easier. Kilowatt-hours are abbreviated as kWh. Electrical power companies charge consumers for the amount of kilowatt-hours consumed during a month.

The final step is to calculate the combined load of all planned home lighting and appliances over an extended period. Calculations are made in this case for fractions of a watt-hour. The example of a lighting system load calculation is included in Table 5.1 for each device. The graph in Figure 5.1 also illustrates an example of a 7-day load plotted in kilowatts (watts × 1000) per hour. Note how the load varies with time of day as well as the day of the week.

Table 5.1 Home Lighting Load Requirements Calculation Table

Electric Component	Watts (Volts × Amperes)	Hours/Day	Average Watt Hrs/Day
Garage light	120	.4	48
Security lights	340	4.2	1428
Kitchen (fluorescent)	160	3.2	512
Den lights	140	4.4	616
Dining area	120	2.2	264
Bedroom areas	255	1.8	459

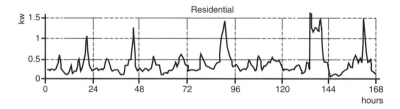

Figure 5.1 Kilowatts used per hour (168-hour summary).

The watt is the unit of power a device uses when connected to a power source. The unit of measure for the amount of electricity used over a unit of time is called the kilowatt-hour (kWh). It is the amount of electrical energy that a 100-watt light bulb uses in 10 hours. kWh is the basis for the charges shown on a monthly utility bill, which are determined by a reading of the electric power meter. The dollar amount of the bill is based on the number of kilowatt-hours used times the cost per kilowatt-hour charged by the utility company.

Grounding

Electrical ground in an AC power system is a wire connected to the earth, hence the name *ground*. Part of the electrical wiring of every home is a wire that is connected to a large metal rod driven deep into the ground thereby ensuring a good connection to earth ground. The alternating current (AC) power entrance service for residential users is shown in Figure 5.2.

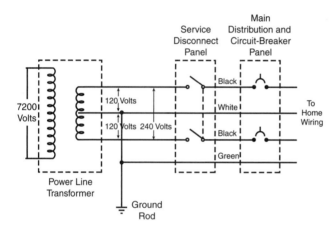

AC Power Distribution

Figure 5.2 AC power distribution.

Figure 5.2 shows the power-line transformer and ground rod location. The 240v service, neutral wire, and ground wire with appropriate color codes are shown. The neutral wire is the return path for home lighting and wall outlets. The ground (green) wire is the safety wire that provides the common ground reference point for all circuits in modern residential wiring.

> Hot and *neutral* are terms frequently used to describe the black-and-white wires, respectively, in residential wiring. The green wire referred to as the *equipment grounding conductor* is returned to the ground rod and normally does not carry any current unless a fault occurs in the wiring. The National Electrical Code (NEC) indicates that the equipment grounding conductor used in residential power wiring must be green or bare.

In electrical equipment and lighting fixtures that have a safety ground connection (as evidenced by a three-prong plug), the safety ground is always connected to any exposed metal parts of the equipment in case of a wiring fault inside the appliance or lamp fixture. If a fault were to occur, such as an accidental connection between the black (hot) wire and the metal case, the safety ground connection would cause the hot connection to be directly connected to earth, which would result in the activation of protective devices such as a GFCI (covered later), fuse, or circuit breaker shutting down the power circuit.

In the past, earth grounds for home service entrances used water pipes as the earth ground reference point. Because increased use of plastic water pipes and nonconducting fittings has made the effectiveness of grounding to plumbing systems questionable, the method does not meet current safety standards. Although water lines still must be used in most circumstances, the NEC now states that the home also requires one or more supplemental grounding electrodes buried in the house foundation or in the earth outside the home. One or more copper-clad grounding rods approximately 8 feet in length is often recommended. Local code enforcement might also require that grounding rods be added to existing homes when new electrical work is done or when the home is sold.

Wireless/Wire Runs

A wireless light control system is designed to use radio frequency signals or infrared transmitters to carry the commands for light control direct to the module that controls the lights. Installation of a wireless wall controller involves replacing a conventional wall switch with the wall-mounted transmitter (connecting to hot and neutral) and mounting power controllers directly to the first fixture in a group of lights or remotely on a junction box. Because the controls communicate via radio frequency, the wall transmitters do not require any wiring directly to the power controllers. This makes the installation of additional wall controls in an existing room very easy.

Wire Runs

Local and national codes require wire that is insulated and is the most efficient size for the appropriate application. The most commonly used interior wiring is a 12- or 14-gauge nonmetallic (NM) sheathed cable, sometimes called *Romex*. Within the cable are plastic-coated copper wires, colored for each function.

Home-run

Home-run wiring is a method used in modern automated home systems where all the wiring is routed from a central point in the house, usually in a wiring closet. At the central point, the wires terminate at various types of patch panels or cross-connect terminals. This point becomes the interface between home wiring and external telecommunications wiring. Home-run wiring is the basic concept used in structured wiring. As you learned in Chapter 1, "1.0—Computer Networking Fundamentals," and Chapter 2, "2.0—Audio and Video Fundamentals," the structured wiring approach is to run a full bundle of wire and cable to every significant room in a home. The full-structured bundle consists of two 4-pair Cat 5 cables and two coax cables (usually quad shield RG-6) and, optionally, two multimode optical fibers. This provides a method for having a home prewired for future automated home features.

Daisy-chain

Daisy-chain wiring is used in home lighting systems where load requirements for individual lighting systems do not require a dedicated home-run connection to the main power distribution panel. The hot wire from the main distribution panel might be rated at 20 amperes and is used to provide power to several lighting circuits, as illustrated in the Figure 5.3.

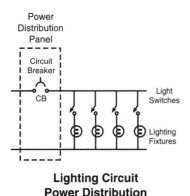

Figure 5.3 Lighting circuit power distribution.

Power-line Controls

Home lights can be controlled with plug-in modules that use the existing power wiring to send control information to any light in the home. The two well-known power-line protocols for managing home lighting are the X10 and CEBus power line protocols.

Both X10 and CEBus modules communicate between transmitter and receiver modules using radio frequency (RF) signals transmitted over the existing AC wiring that carries digital information. The RF signals must be generated by a module or controller plugged in to the AC power. Wireless handheld remote controllers are also available to control the modules inserted in power outlets. This means the wireless controllers (transmitters) available are not directly sending an X10 or a CEBus signal to the receiver modules; instead, a transceiver module is plugged in to the wall that has an antenna that receives the over-the-air RF signal from the wireless controller and then translates the signal into an X10 or a CEBus signal and transmits the signal over the house wiring.

Both systems use plug-in modules and are available as single-room systems and whole-house systems. The transmitter modules are used in standard wall switch boxes. The light fixtures plug in to receiver modules that use standard wall AC power outlets. They can be used to control lights in a variety of ways, including

➤ To extend an existing light with a single light switch to be controllable from two, three, or even more locations with no additional wiring.

➤ For a single light control or whole-home control options.

➤ To switch lights on and off at set or random times.

➤ To control groups or zones of lights or appliances from a single switch point.

➤ To enable lights and appliances to be controlled from outside the house. This is possible from a remote location using dial-up or Internet access via a computer X10 or CEBus interface.

X10 and CEBus are the two power-line control protocols you will need to remember. They are used to send signals through handheld wireless keypad devices and the home power lines to remotely control the lighting scenes and illumination levels in any area of the home.

Conduits

A *conduit* is a type of enclosure for wiring installed in locations that are classified as hazardous due to potentially explosive air environments, underground wiring, outdoor conditions, and moist locations. Conduits are available in a variety of types for various environments. They range from rigid metal conduit, intermediate metal conduit, thin-walled conduit called electrical metallic tubing (EMT), and nonmetallic types called polyvinyl chloride (PVC) tubing. Local building codes specify how and where conduit must be used. As a general rule, conduit is used to prevent wiring to be exposed when installed outside of walls or where wiring would be subject to damage or contact with harmful chemicals.

Thin-wall Conduit

Electrical metallic tubing, commonly called thin-wall conduit, is metallic tubing that can be used for exposed or concealed electrical installations. Its use should be confined to dry interior locations because it has very thin plating that does not protect it from rusting when exposed to the elements or humid conditions. It is the most popular choice for use among all conduit types, is less expensive than rigid conduit, and is much easier to install.

Zoning

Residential lighting can be programmed for different zones. A *zone* is one group of lights controlled by one switch or dimmer module. Automated zone management of lighting allows different light levels to exist in different areas. Zone lighting can be set to lower or raise the light level during particular hours of the day, and zone systems can be programmed to remember specific zone settings.

Remote Access

Remote access to home lighting systems is achieved through numerous products available for home automation. Access is possible using the telephone in the home or by dialing in to the home automation system from a remote location. Security system status and temperature sensors can be used to adjust lights, appliances, and thermostats and monitor activity and remote events by entering key commands on the telephone from a remote location. Lighting system management is one of the options available on most total home remote access systems. When the telephone in the home is accessed from a remote location, an automated answering system prompts the caller for a correct password entry code to access home environmental status. Clear voice menus lead the user through steps to turn lights on and off or select any programmed lighting scene for day or night conditions.

Home Lighting Equipment Location Considerations

The locations of home lighting fixtures can be either wall mounted or portable depending on room location and access to natural light. Wall outlets should be located on every wall to provide maximum flexibility for portable lamp fixtures and control units for remote access.

Planning the Locations for Home Lighting

Lighting systems are usually planned and designed during the development of architectural drawings for the home. The planning for the location of various light fixtures such as in-ceiling lights and switched-wall outlets for portable lamps are all essential to proper light distribution for a home.

The number of lighting circuits depends on the size of the home in square feet and the number of residents in the home. Standard practice and lighting circuit codes indicate a requirement for at least two 120v circuits, staggering the outlets for lighting so that, if a breaker trips, adjoining rooms will still have lights. Also, if the home has a wired smoke-detector (it uses the AC power line instead of batteries), it should be wired into the lighting circuits. This aids troubleshooting because the homeowner is more likely to be aware of a tripped breaker that operates lighting; therefore, it helps in determining whether there is power to the smoke detector.

For locating the outlets for lighting fixtures, duplex receptacles are normally placed 6–8 feet apart in newer homes, depending on local building codes. The idea is to have enough circuits so that extension cords are not necessary. Ideally, each room should be on a separate circuit. Some outlets could be switched so that a table or floor lamp could be turned off at a wall switch. As indicated earlier in the wall outlet configuration for split operation, the duplex outlets are separated by the breakout fins so that one of the dual outlet sockets is switched and one is not switched.

Physical Lighting Products

Lighting products for home systems include wall outlets, portable lamps, light fixtures, modules for controlling brightness, and automated window treatments for controlling natural light. In this section, the basic characteristics and physical features of lighting products are reviewed.

Power Outlets

Power outlets are designed to provide AC power to lights and other appliances located throughout a home. The National Electrical Manufacturers Association (NEMA) has developed standards for the physical appearance of receptacles. Figure 5.4 shows the physical features and terminology for standard wall outlets.

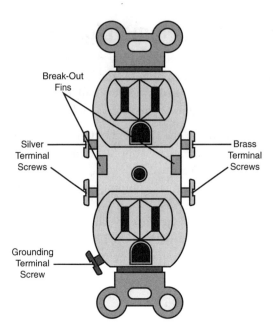

Figure 5.4 Power receptacle configuration.

The large slot on the left in Figure 5.4 identifies the neutral (white) wire and the narrow slot on the right is connected to the hot side. The round hole is connected to the ground wire. Outlets have two terminal screws on each side—one pair is black or brass in color and the other screws are silver.

The black or red (hot) wire always connects to the brass terminal located on the same side as the narrow slot, whereas the white wire always connects to the silver terminal located on the same side as the wide slots. The ground wire connects to the green screw. Note that two screws are used on each side. When the break-off link between the brass and silver screws is removed, the power is connected to only one of the outlets. The two screws can then be used independently for power and control of the upper and lower receptacles. This is often used where it is desired to use one of the receptacles for a switched outlet controlled from a light switch for turning on or off an appliance or light from another location while leaving the other receptacle wired for permanent AC voltage service.

A back-wired option is available on most outlets. This option has openings in the rear that provide openings for inserting the wires. The same color code should be observed for inserting the black and white wires to the proper side of the receptacle. This type of electrical connection might not be acceptable for some local building codes. When contemplating the installation of this type of outlet, you should first check the local electrical building code requirements.

The correct method for connecting house AC power wiring requires the hot, neutral, and ground wires to always be connected to the correct terminal on a wall receptacle. The black (or red) wire is always connected to the terminal, and the white wire is always connected to the silver terminal. The green wire (ground) is always connected to the green screw terminal. When a receptacle is wired correctly, the narrow slot is always the hot connection and the wide slot is always the neutral connection.

Ground Fault Circuit Interrupters

The ground fault circuit interrupter (GFCI) is another type of wall outlet used to protect people from electrical shock. The GFCI detects an unbalanced condition in the current flowing into and out of the load connected to its hot and neutral sockets. An unbalanced condition occurs when a faulty tool or appliance is shorting to the metal case and possibly leaking electrical current to ground through a person coming in contact with the device. The GFCI detects the condition and quickly removes power from the outlet.

Dimming Modules

Dimming modules are used to vary the voltage to an incandescent light fixture. This enables the user to control the amount of light from a lighting system or light fixture. In many residential lighting systems, various lighting areas and zones might not need to be operated at full brightness levels. They normally are required to fade in and out and be used at different light levels at different times. Dimmer modules are not usually used with florescent light fixtures.

Older types of dimmers use a variable resistor to limit the current and resulting voltage available to the lamp. More modern electronic light dimmers utilize a variable phase control principle where the voltage is switched on and off with special components inside the dimmer module. The type of components most commonly used are called silicon controlled rectifiers (SCRs) and thyristor for AC applications (TRIACs). Variable phase control, mentioned earlier, refers to allowing only portions of the AC cycle to go through the load. It is usually performed with a TRIAC or two back-to-back SCRs. AC current sources have a zero-crossing two times every cycle; this happens 120 times per second for 60 hertz (Hz) AC service.

The TRIAC turns off 120 times per second. By varying the turn-on period, you can vary the amount of power applied to a lighting circuit. This produces the waveform shown in Figure 5.5.

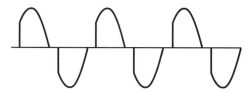

Figure 5.5 80% power setting.

To deliver only half the power, the TRIAC is fired between each zero-crossing, producing the AC waveform shown in Figure 5.6.

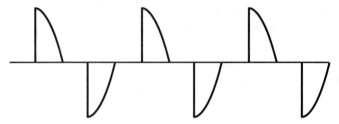

Figure 5.6 50% power setting.

By firing the TRIAC just before each zero-crossing, you can further reduce the total power to the load to 20% with the waveform shown in Figure 5.7.

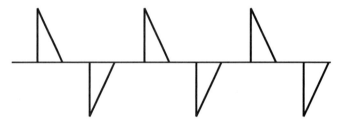

Figure 5.7 20% power setting.

This type of module is normally used for incandescent light dimming because lights are resistive loads and will not flicker even if they get only voltage spikes, as illustrated in Figure 5.7.

Automated Dimmer Modules

Automated dimmer modules are available for direct replacement of standard manual dimmer switches. Automated dimmers have the capability to turn the light on and off, plus dim it to any level, at the switch location or from any X10 transmitter. The light can be dimmed by holding the switch until the light dims to the desired level. In addition, some automated dimmer switches have a feature that gently fades lights on over a 2-second period and a resume feature that returns the light to the dim level it was set to before the light was turned off. X10-compatible two-wire dimmer switches can be used only with incandescent lights.

Light Switches

Light switches are used to control lighting circuits from locations that are convenient to home residents. The types of switches available for residential installations vary according to where they are installed, the type of circuit they are controlling, and the current rating of the load connected to them.

Standard Switch Configurations

Light switches are categorized according to the number of contacts available. The four primary design configurations for switches are shown in Figure 5.8 and are described in the following:

► *Single pole single throw (SPST)*—This type of switch is used for operating a single light fixture. The incoming hot wire is hooked to one terminal screw, and the outgoing hot wire is connected to the other screw.

► *Single pole double throw (SPDT)*—This type of switch is referred to as a *three-way* and is used for controlling a light from two locations. It has a double throw capability and can change the current path between two wires. A description of a three-way installation for controlling a light fixture from two locations is discussed later in the following paragraphs.

► *Double pole single throw (DPST)*—A DPST switch is another type of on-off control and is used to interrupt a two-wire circuit.

► *Double pole double throw (DPDT)*—This type of switch is another changeover type that is normally used to control 220v two-phase circuits.

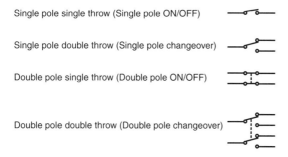

Single pole single throw (Single pole ON/OFF)

Single pole double throw (Single pole changeover)

Double pole single throw (Double pole ON/OFF)

Double pole double throw (Double pole changeover)

Figure 5.8 Light switch configuration.

Three-way Light Switches

Three-way switch circuits are used frequently in residential installations. They are used to control a light fixture or lighting array from two locations with one three-way switch located at each position. An example of this type of installation is where a light needs to be controlled at the top and bottom of a stairway. A four-way switch is used with a circuit that controls a light from three locations; this type of switch is used less frequently than the three-way switch is.

At a glance, three-way switches look the same as the common single pole switch does, but instead of having only two screws on which to make your connections, they have two connections on one side and two on the other side (see Figure 5.9).

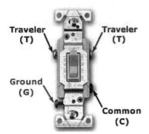

Figure 5.9 A three-way switch.

As indicated in Figure 5.9, one terminal is referred to as the *common* terminal, one terminal is the ground terminal, and the other two are known as *travelers*. This is because the electrical connection either goes from the common screw, to one or the other travelers, depending on the switch position. This is also illustrated in Figure 5.10.

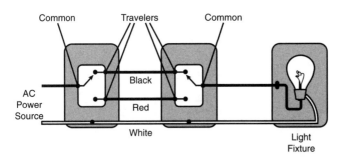

Figure 5.10 A three-way switched circuit.

Study Figure 5.10 and you will discover that the light fixture can be turned on or off by moving either switch to the opposite traveler position. Also, the red wire is always used as the traveler.

As indicated in this section, a three-way switch is used when two switches are required to control one light from two locations. The term can be misleading because the switch controls a single light from two locations. Memorize the wiring diagram for a three-way switch and remember how it is used.

Photoelectric Switches

Photoelectric switches are a special type of switch used for controlling lights depending on the light intensity (*lux value*). This type of switch contains a special sensor that operates a switch when the light intensity or darkness reaches a certain value. Photoelectric switches are typically used in outdoor locations for landscape and security lighting. Most photoelectric switches are integrated in a single fixture that includes the sensor and lamp fixture, and most have adjustments to turn off lights after a number of hours when having lighting is unnecessary from dusk to dawn. They also have a time delay of approximately 15 seconds to prevent the fixture from turning on or off in cases of short-duration intensity changes.

Clock Switches

Clock switches are used to turn lights on and off at given intervals. They are often used to control lights when the members of the family are away and want the home to appear occupied. They can be inserted into a standard wall outlet; the clock motor obtains power from the AC receptacle. Lamps or appliances can be plugged in to the clock timer switch, and programmable tabs can be arranged on the dial to turn the light on or off at any time. Clock timers are rated for the amount of power, current, and voltage that can be delivered to the load connected to the outlet on the timer.

Remember the following types and characteristics of various special purpose light switches. Three-way switches are needed to control a light fixture from two locations. Switches that operate on the amount of light present are called photoelectric switches, whereas clock switches are used to turn lights on and off at preset time intervals during a 24-hour cycle.

Light Fixtures

Light fixtures can be selected for the home in a wide range of styles, including wall mounting, ceiling, track lighting, table lamps, recessed ceiling lights, and a large variety of custom designs to fit any home decor. Generally, two types of lighting are used: *ambient* and task. Ambient lighting typically comes from a ceiling fixture that evenly distributes light throughout an entire room. *Task* lighting, on the other hand, is located on or near your work surface and is switched on when required to illuminate specific areas. Recessed lighting installation requires special attention to avoid heat problems to surrounding objects.

Incandescent lighting, along with halogen and fluorescent lighting, is the most popular type of illumination because of its low cost and ease of installation. You should become familiar with the details of these types of light sources for the home.

The term *luminaries* might appear in exam objectives for residential lighting systems. The term refers to a complete lighting unit using any type of illumination, such as halogen, fluorescent, or incandescent.

Incandescent Lighting

The standard incandescent light bulb has changed only marginally over the past several years. Incandescent lamps are the least energy-efficient electric light source and have a relatively short life (750–2500 hours). Light is produced by passing a current through a tungsten filament, causing it to reach incandescence and glow to produce light. With extended use, the tungsten material in the filament slowly evaporates, eventually causing the filament to fail.

Three-way bulbs provide three settings for brightness and contain two filaments. They must be installed in lamp fixtures that have a switch that can select one or both of the three-way bulb filaments. When the switch is operated, it selects the lowest wattage filament (30 watts as an example). The next position selects the second filament (70 watts), and the third position of the switch connects both filaments to the line voltage to produce a 100-watt light output.

Halogen lamps are another type of incandescent lamp. They have a longer life than conventional light bulbs, but they are only slightly more efficient. Halogen lamps are best suited for lighting areas where a direct focus of light is required.

Fluorescent Lighting Fixtures

Fluorescent lamps are a type of low-pressure or low-intensity discharge lamps. These lamp consist of a closed tube that contains two cathodes, an inert gas such as argon, and a small amount of mercury. When the correct voltage is supplied to a fluorescent lamp, an electrical arc is formed between the two cathodes. This arc emits energy that the phosphor coating on the lamp tube converts into visible light.

A fluorescent lamp tube contains argon combined with a minuscule amount of mercury. At the low pressure within the lamp, the mercury vaporizes, even at temperatures only slightly above room ambient. An electrical discharge ionizes the mercury vapor, which emits ultraviolet (UV) radiation. The UV radiation stimulates phosphors that coat the interior of the lamp's glass envelope, and the phosphors convert essentially all the UV radiation to visible light. The conversion of electrical energy to light is much more efficient than in an incandescent lamp, and a considerably smaller fraction of the input energy is converted to heat. Generally, fluorescent fixings give out approximately three times as much light per watt as a halogen light. The color of the light that a fluorescent lamp produces depends on the composition of the lamp's phosphors.

Fluorescent Ballasts

A florescent fixture includes an assembly called a *ballast*. It serves as a current-limiting device and is used during the startup phase. When voltage is applied to the fixture, the starter allows current to flow through filaments of the tube. The current causes the starter's contacts to heat up and open, thus interrupting the flow of current. This causes a great increase in voltage in the ballast due to its inductive properties, which ionizes the gas mixture and causes the lamp to emit visible light. Because the gases in the tube have what is referred to as negative resistance, the ballast now plays the role of a current limiter and stabilizes the operation.

Rapid Start Florescent Ballasts

Most modern fluorescent fixtures use a rapid start ballast and fluorescent bulbs designed for this type of ballast. With this type of ballast, a small current always flows through the filaments. This type of bulb has a gas content that ionizes quickly, and it is available in various designs, such as magnetic, electronic, and hybrid types that have a high operating efficiency.

Automated Window Treatments

Windows can be enhanced in an automated home design with the addition of motorized control of blinds. With automatic window control designs, you can open, close, or adjust window coverings from a wall switch or by handheld remote control units. They are frequently used with home theater systems where sunlight entering an area can be shaded for best viewing of movies or television programs.

The ability to coordinate the operation of window coverings with a security system so that shades are opened or closed to create a lived-in look when the family is away is also useful.

With motorized shades, the user can easily operate window treatments that are very large or are located in places that are difficult to reach. The motors that move the drapes or shades are normally concealed within the valance at the top of the window treatment.

Standard Configurations and Settings

Home lighting systems can be customized to meet the needs of any design with innovative systems such as lighting scenes, exterior and interior security lighting, and lighting zones. The various security configurations and customized settings can be used to make the home appear occupied when the family is away. The following paragraphs illustrate some examples of custom home lighting options.

Lighting Scenes

Scene lighting is one of the features offered by several home automation vendors. Selecting a specific lighting scene is similar to a scenario used by a stage lighting director in which a unique lighting condition is planned or programmed in a memory chip or computer and implemented at the touch or command of a single switch action.

Whole-home light automation systems use a central processor to communicate with all keypads in the house. Pressing a "scene" button on the keypad in the foyer could set not only the lights in that area, but also in the kitchen, family room, and patio. The same command might also lower the temperature and disarm the security system.

Some systems offer a touchscreen display that can command the lights to adjust to various scene settings. Lighting scenes can be changed for entertaining, evening time, viewing a movie, or decorative lighting in the daytime hours. The lights can also be programmed to automatically shut off after the residents go to bed, saving energy and time. They similarly turn on in light a path when anyone gets out of bed in the early morning hours.

Security Lighting

Security lighting can be integrated with automated whole-home lighting systems, or they can be standalone lighting fixtures such as motion detector lighting modules located outside the home. Outdoor lighting is used on a nightly basis by many homeowners for safety, as well as security and aesthetic reasons. Motion sensors instantly turn lights on when motion is detected and off when there is no activity for a desired period of time. Photoelectric sensors discussed earlier in this chapter measure the amount of daylight present and turn the lights on or off accordingly for added security.

Motion Detector Security Lighting

Motion sensors (discussed in Chapter 3, "3.0—Home Security and Surveillance Systems," detect infrared energy that radiates from objects warmer than the surrounding environment. When the detector senses a warm object moving across its field of view, it automatically turns on the security lights. The light stays on anywhere from 1 to 20 minutes, depending on a preset timer. The detector automatically turns off the light unless it continues to sense movement, and a photocell deactivates the light during daylight hours.

Motion detector security lighting systems also support operation of the light by a manual switch. Double-flipping the switch turns the light on manually; repeating this switch action returns the light to the normal automatic sensing mode.

Security lighting systems operate with two types of sensor systems for controlling lights. Photoelectric sensors operate by sensing the absence of natural sunlight. Infrared sensors operate security lights by detecting heat in the form of infrared background energy caused by humans moving through an infrared sensor pattern.

Lighting Zones

Lighting zones can be configured in a home to automatically control groups of lights called zones. Zone lighting is usually included as one of the modes available in whole-home lighting systems configured to control lights from multiple locations. Zone lighting is a concept used in lighting management to group lights together for single-switch control. As an example, kitchen lighting might have work area lighting, track lights, and recessed lighting that can optionally be programmed for single-zone single-switch control.

Device Connectivity

Lighting systems are connected with several types of wire and cable. The choices are dependent on the type of environment in which the wiring is to be installed. Building codes and electrical codes also govern the types of wire and cable that can be used in home lighting systems. In this section, you will become acquainted with all the names, ratings, and locations for the various wire types.

Nonmetallic Cable

Nonmetallic (NM) cable is the most widely used cable for indoor wiring. Romex is a brand name for a type of plastic insulated wire that is often called nonmetallic sheath; however, the formal electrical codename is NM cable. This type of wire is suitable for stud walls, on the sides of joists, and places that are not subject to mechanical damage or excessive heat. Newer homes are wired almost exclusively with NM wire.

NM cable is an example of cable in which several wires are wrapped together. When a cable such as NM has two number 14 wires, it is known as 14-2 (fourteen-two) cable, or in the case of three number 12 wires, it would be identified as 12-3 (twelve-three).

MC Cable

Metal clad (MC) cable includes its own flexible metal covering, as shown in Figure 5.11.

Figure 5.11 MC cable.

MC cable is composed of THHN soft-drawn, copper wire and is composed of conductors and an insulated grounding conductor. It is suitable for branch, feeder, and service power distribution in commercial and industrial applications as well as in multifamily buildings, theaters, and other populated structures. The letters *THHN* refer to heat resistant thermoplastic (90° C) for dry locations. This is a reference to the type of insulation on wiring, coded per the NEC. All the NEC designations for wire insulation letter codes are referenced later in this chapter.

MC cable can be installed in cable trays and approved raceways and as aerial cable. It is suitable for use in wet locations (when it has a PVC jacket on it) or in dry locations, at temperatures not exceeding 90° C.

Many installers prefer the use of MC cable instead of conduit because it installs quickly and does not require the level of experience necessary for installing rigid conduit.

MC cable has an overall Mylar wrap enclosing the conductors. Only MC cables can be used in places of assembly of more than 100 persons, per NEC Article 518.

Remember the distinction in the standards between NM and MC cable for home lighting. National Electrical Code Article 334 defines NM cable as "a factory assembly of 2 or more insulated conductors having an outer sheath of non-metallic material." Article 330 of the NEC describes MC cable as "a factory assembly of one or more insulated circuit conductors with or without optical fiber members enclosed in an armor of interlocking metal tape or a smooth or corrugated metallic sheath."

Armored Cable

Armored cable is identified by the code name AC cable. AC cable is often referred to as BX, although this is the trademark of a specific manufacturer that has become a generic name for armored cable. Armor clad and metal clad might sound like the same cable, but they have some important differences. AC cable uses the interior bond wire in combination with the exterior interlocked metal armor as the equipment grounding means of the cable. MC cable, on the other hand, is manufactured with a green insulated grounding conductor, and this conductor—in combination with the metallic armor—comprises the equipment ground.

AC cable can have up to only four insulated conductors; a fifth insulated conductor is allowed by UL if it is a grounding conductor. Each conductor in AC cable is paper wrapped.

Low-voltage Wiring

Low-voltage wiring is a term used in this book and most of the residential wiring industry to include all wiring for audio/video components, network data cabling, security sensor wiring, phone lines, telecommunications systems, and low-voltage landscape lighting. This encompasses all residential wiring with the exception of 120v power wiring. Whole-home structured wiring features call for the use of RG-6 cable for audio and video distribution and Category 5 cable for telecommunications systems and components.

For lighting systems in the residential home area, low-voltage wiring is used for landscape lighting and sensor connectivity for home security systems.

Most exterior low-voltage lighting systems use 12v derived from transformers that convert standard 120v AC power sources to 12v for exterior lighting.

Low-voltage landscape lighting systems have several advantages. Most important, low-voltage landscape lighting equipment is easier to install. Except for the transformer that's connected to a 120v circuit, the entire system runs on harmless 12v distribution cabling. Low-voltage wiring is lightweight and can be laid on the ground or buried just below the surface.

In contrast, a 120v system runs on the same power that supplies your home and requires the same precautions and expertise that normal house wiring does. Outdoor 120v lighting needs to be installed according to code and can require buried conduit. Once installed, 120v systems are relatively permanent.

Termination Points

Termination points for home lighting systems are established in the circuit breaker panel. Each circuit is identified by the circuit breaker that protects the outlets and lighting circuits connected to it. The hot wires (black) are connected to the circuit breakers on the 120v bus, and the neutral (white wires) are connected to the neutral bus.

Industry Standards

This section describes the standards organizations and related standards documents that you will need to know concerning the home lighting and management industry.

Wiring Types and the National Electrical Code

The National Electrical Code sets standards for lighting circuits and home wiring types. A set of standard terms used to identify different wire types are listed here. Electrical contractors use these designations when selecting the correct wire type for each environment. They are

- *T*—Thermoplastic (60° C) for dry locations
- *TW*—Moisture-resistant thermoplastic for wet locations
- *THHN*—Heat-resistant thermoplastic (90° C) for dry locations
- *THWN*—Heat-resistant (75° C) for dry and wet locations
- *W*—Moisture-resistant
- *H*—Heat-resistant
- *HH*—More heat-resistant
- *NH*—Non-halogen for wiring where human life might be at risk due to fumes from toxic gases in a fire

The *N* in types THWN and THHN means the wire has a nylon outer jacket (which helps reduce abrasion damage when pulling through conduit.)

Wire Gauge and the Number of Conductors

Multiconductor cable is labeled according to the number of active conductors and the wire gauge. As an example, 12-3 indicates a cable with 3 each number-12 gauge conductors. This is typically the black, white, and green wires. A 14-3 cable is a 3-conductor cable containing number-114 gauge wire. 14-2 indicates a 14-gauge cable with 2 conductors; it can also have a bare copper ground wire.

Standards and Organizations

The following standards and related organizations publish various documents relating to home lighting control and management:

- *The Electronic Industries Alliance (EIA) Standard 600*—The "Introduction to the CEBus Standard" is the authoritative documentation on the power-line transmission protocol used for home automation and lighting management.

➤ *The National Electrical Manufacturers Association (NEMA)*—An organization of manufacturers of electrical equipment, including, but not limited to, wiring devices, wire and cable, conduit, load centers, pressure wire connectors, circuit breakers, and fuses. NEMA is the voice of the electrical industry, and through it standards for electrical products are formulated. Generally, these standards promote interchangeability between products of one manufacturer with like products made by another manufacturer. In some cases, standards relating to product performance are also formulated by NEMA, but these are the exception rather than the rule. NEMA standards are certainly not compulsory, but generally they are accepted by those manufacturers that help write them as a way of making their products more saleable and acceptable.

National Electrical Code

The National Electrical Code is a document that basically describes recommended safe practice for the installation of all types of electrical equipment. The NEC is not a legal document unless it is so designated by a municipality as its own statute for safe electrical installations. It is revised and published every three years. The NEC is "national" only in the fact that it is the only document of which all or part is accepted by all states as an electrical guide. It is also the only document of its kind written with national input supplied by 20 panels of advisors containing several hundred experts in the electrical field from all parts of the country. The sponsoring agency of the NEC is the National Fire Protection Association (NFPA). In this chapter, the following documents were used in the development of material on cable types:

➤ NEC Article 333 covers armored cables. Armored cables are manufactured in accordance with UL 4.

➤ NEC Article 334 covers metal clad cables. MC cables are manufactured in accordance with UL 1569.

Underwriters Laboratories, Inc.

Underwriters Laboratories, Inc. (UL) is an independent, nonprofit product safety testing and certification organization. Each year, more than 17 billion UL marks are applied to products worldwide. In this chapter, the following UL standards were referenced in the topic on wire types:

➤ Armored cable (AC) is manufactured in accordance with UL 4.

➤ Metal clad (MC) cable is manufactured in accordance with UL 1569.

National Fire Protection Association

The NFPA develops, publishes, and disseminates timely consensus codes and standards intended to minimize the possibility and effects of fire and other risks. Virtually every building, process, service, design, and installation in society is affected by NFPA documents.

The NFPA focus on true consensus has helped the association's code-development process earn accreditation from the American National Standards Institute (ANSI). NFPA is the sponsoring agency for NFPA 70, NEC.

Installation Plans and Procedures

As discussed earlier, electrical metallic tubing, commonly called thin-wall conduit, is metallic tubing that can be used for exposed or concealed electrical installations. Its use should be confined to dry interior locations because it has very thin plating that does not protect it from rusting when exposed to the elements or humid conditions. It is less expensive than rigid conduit and much easier to install because the process of bending requires less effort and the ends do not have to be threaded. In comparison to the other wiring systems, it ranks behind rigid conduit but ahead of the other types of wiring when considering the quality and durability of the installation. For this reason and because of the decreased cost in materials and labor, it is typically specified for home-building construction. It can be installed in the same manner as rigid conduit, except it uses pressure-type couplings and connectors instead of threaded units.

 Remember that EMT is the acronym you will often hear for a popular type of conduit. EMT is recommended for dry installations primarily to protect wiring from physical damage.

Installation Procedures for Single-Pole Switches

A single-pole switch has two brass screw terminals. Both are hot leads for one incoming and one outgoing line. Those are all the wires that connect to the switch.

The neutral wires tie together separately, and the ground wires tie together separately in the box.

Many new switches include a ground screw; others might not have one. If you have a choice, select a switch with a ground screw terminal, which is where the bare copper or green wire connects.

Troubleshooting and Maintenance Plans

Troubleshooting home lighting systems requires some basic knowledge about the home wiring layout and some experience installing and maintaining electrical wiring systems. You should not troubleshoot any electrical problem without taking the following safety steps:

➤ Always turn off the circuit breakers or remove the fuse that controls power to the circuit on which you are working.

➤ Play it safe and get an inexpensive voltage tester that lets you know whether it's okay to work on a wall outlet or light fixture. Place the tester on the wiring or in the slots in the AC outlet. Simple testers provide an LED or other indication that the voltage is present. This lets you know whether you've turned off the wrong circuit breaker or fuse or if there's a problem with the breaker.

➤ If you want to perform more extensive tests or troubleshooting steps, obtain a digital volt-ohmmeter. This type of tester can measure voltage, check for proper connections, and give a direct reading of the amount of voltage on a wire or an AC outlet receptacle.

➤ Use a tester that connect to an AC outlet and trace the circuit to the breaker panel. The tester can identify the circuit breaker that controls the power to the circuit on which you want to work. Plug-in testers can also troubleshoot outlets for reversed neutral/hot wires, improper connections, or missing grounds. It is not uncommon to find errors in electrical outlet wiring that were overlooked during the installation.

➤ Select a type of tester(s) that can read voltage, current, and resistance in ohms.

You will need to remember that an ohmmeter is a type of tester used to measure the resistance of an electrical circuit or component. The unit of measure for resistance is ohms. Testers for measuring resistance normally include the capability to also measure AC and DC volts. Measuring current in a circuit is usually performed by experienced electricians using a clamp-on type tester that measures current flow without disconnecting any wires.

Exam Prep Questions

Question 1

What is the correct terminal connection for the neutral wire when terminating power wiring in a standard AC wall outlet?

- ○ A. The brass terminal
- ○ B. The silver screw
- ○ C. The green terminal
- ○ D. The black terminal

Answer B is the correct answer. The silver terminal located on the left side near the wide slot of a receptacle is the proper correction for the neutral (white) wire. Answer A is incorrect because the brass terminal is used for the black hot wire connection. Answer C is incorrect because the green terminal is used for the ground wire and not the neutral wire. Answer D is incorrect because the black terminal is used for the hot wire on some types of wall outlets that use a black screw instead of a brass screw.

Question 2

What is the purpose of a GFCI?

- ○ A. It is used to limit the amount of current delivered to any load to a maximum of 15 amperes.
- ○ B. It is used to protect electrical devices from damage caused by electrical surges and lightning.
- ○ C. GFCIs are used only in locations where dangerous fumes might present a hazard for explosions.
- ○ D. GFCIs are used to protect people from electric shock.

Answer D is the correct answer. Ground fault circuit interrupters are a type of wall electrical outlet designed to protect persons from electrical shock when using plug-in cord-connected portable electrical equipment, especially in wet or damp locations. GFCIs prevent life-threatening contact with power circuits by quickly disconnecting the electrical current if the flow of electricity is detected to be returning via a ground path instead of the neutral return path. Answer A is incorrect because a GFCI is not designed to limit the current to 15 amperes, and answer B is incorrect because a GFCI is not a surge protector. Answer C is incorrect because GFCI outlets are designed to guard against electric shock from grand faults and are not necessarily associated with hazardous area installations.

Question 3

Which type of switch is required to operate a light from two locations?
- ○ A. Three-way switch
- ○ B. DPDT switch
- ○ C. SPST switch
- ○ D. Four-way switch

Answer A is the correct answer. A three-way switch is used to control lighting systems from two separate locations. Answer B is incorrect because a DPDT switch is normally used as a main service entrance disconnect switch and is not used as a three-way type switch. An SPST switch is suitable for controlling a light from a single location, so answer C is incorrect. Answer D is incorrect because a four-way switch is used to control a light from three locations.

Question 4

Which type of cable is sometimes referred to as BX cable?
- ○ A. NM
- ○ B. AC
- ○ C. NH
- ○ D. HH

Answer B is correct. Armored cable is often identified by the code name AC cable, which is also referred to as BX. Although this is the trademark of a specific manufacturer, BX has become a generic name for armored cable. Answer A is incorrect because NM is the designation for another type of cable identified as nonmetallic cable. Answer C is incorrect because NH cable is the designation for non-halogen cable. Answer D is incorrect because HH is used as a subdesignation for thermoplastic nylon-coated wire.

Question 5

> Which type of electrical power cable is required for installation in places that are used for assembly of 100 persons or more?
>
> ○ A. MC cables
> ○ B. NM
> ○ C. Category 3 UTP
> ○ D. Romex

Answer A is correct. Metallic clad (MC) cable is required for installation in locations where there might be an assembly of more than 100 people at one time (per NEC Article 518). Answer B is incorrect because nonmetallic cable is used in residential wiring but is not usually permitted by most building codes in commercial areas with 100 or more people. Answer C is incorrect because Category 3 cable is not classified as a power cable. Romex is a brand name for NM-type cable and is not an MC cable, so answer D is incorrect.

Question 6

> Which type of unit acts as a current-limiting device in a fluorescent light fixture?
>
> ○ A. A starter
> ○ B. A switch
> ○ C. A socket
> ○ D. A ballast

Answer D is correct. The ballast is an inductive component attached to a fluorescent fixture, and it provides an inductive load to the AC current when the lamp is in operation. Answer A is incorrect because starters are used for igniting the firing sequence in older fluorescent fixtures and are not used as current limiters. Answer B is incorrect because a switch is a device used to interrupt or apply voltage to a circuit and is not considered to be a current-limiting device. Answer C is incorrect because a socket is the generic name for an electrical wall outlet and is not a current-limiting device.

Question 7

How many individual conductors are contained in an NM cable sheath labeled 12-3?
- A. 12
- B. 3
- C. 2
- D. 5

Answer B is correct. The designation for NM sheath cable indicates the wire gauge (12 in this example) followed by the number of conductors (3). Answer A is incorrect because 12 refers to the wire gauge and not the number of conductors. Answer C is incorrect because the designation 12-3 refers to a three-conductor cable and not a two-conductor cable. Answer D is correct because the designation 12-3 refers to three-conductor cable rather than five-conductor cable.

Question 8

Which type of lamp is the most practical for locations where direct focus is required?
- A. Incandescent
- B. Fluorescent
- C. Halogen
- D. Three-way

The correct answer is C. Halogen lamps are recommended for close work areas identified in the lighting industry as direct focus lighting. Answers A, B, and D each include a type of lighting that could be adapted for direct-focus lighting applications. However, halogen lighting is a more appropriate type of illumination recommended by lighting experts for direct focus areas. Therefore, answers A, B, and D are incorrect.

Question 9

Which wire insulation color is used for identifying the ground wire in a lighting distribution system?

- A. Green
- B. Black
- C. Red
- D. White

The correct answer is A. The green wire is used for the ground conductor in home AC power distribution. Bare copper wire is also acceptable for use as the ground. Answer B is incorrect because the black wire is used for the hot connection. Answer C is incorrect because the red wire is used as the hot wire or the traveler wire in a three-way switch circuit. Answer D is incorrect because the white wire is used as the neutral wire.

Question 10

Which type of switch is used for controlling lights in a home or at outside locations depending on the light intensity?

- A. Three-way
- B. DPDT
- C. Photoelectric
- D. Clock timer

The correct answer is C. A photoelectric switch contains a special sensor that operates a switch when the light intensity or darkness reaches a certain value. Photoelectric switches are often used for decorative landscape lighting. Answer A is incorrect because a three-way switch is used to manually control lighting from two locations. Answer B is incorrect because a DPDT switch is normally used in a fuse box or circuit breaker panel to disconnect the primary power to a residence. Answer D is incorrect because a clock timer switch is used to control lights at a specific clock interval and does not have the capability to measure light intensity.

Need to Know More?

 Gerhart, James. *Home Automation and Wiring*. Berkeley, CA: McGraw-Hill, 1999. This book covers a number of home technology topics, including lighting and control standards.

 O'Driscoll, Gerard. *The Essential Guide to Home Networking Technologies*. Upper Saddle River, NJ: Prentice Hall PTR, 2001. This book provides detailed information on the new home networking technologies related to power-line technologies used to control and manage home lighting systems.

 http://www.absak.com/design/powercon.html#top. This Web site provides an excellent tutorial and worksheet for conducting a residential load analysis.

6.0—HVAC Management

Terms you'll need to understand:

✓ Air handler
✓ Bimetal elements
✓ Damper
✓ Static pressure probe
✓ Time-of-day programming
✓ Zone settings
✓ Automatic changeover
✓ Condensing unit
✓ Reversing valve
✓ Staging thermostat
✓ Water column
✓ Centralized system

Tasks you'll need to master:

✓ Memorizing the troubleshooting tips for air handlers
✓ Knowing the common checks for an air conditioning system
✓ Remembering the common troubleshooting tips for thermostats
✓ Planning a duct installation plan for a single-story residence
✓ Describing the problems associated with thermostats containing mercury

HVAC is an acronym for heating, ventilating, and air conditioning. HVAC systems cover the full scope of products used to provide seasonal climate control for the home. HVAC systems are installed by trained professionals who are familiar with building codes as well as fire and safety regulations. As a home integrator, you will need to become familiar with the terms; the equipment categories; and the functional relationships between such items as programmable thermostats, zone controls, ducts, whole home fans, damper controls, condensers, furnaces, air handlers, and heat pumps.

As you can see, HVAC systems include more than the heating and cooling units. An automated home can include zones for separate control of heating and air conditioning settings for different areas of the home. They also can include new automated control systems such as motorized dampers. HVAC systems represent one of the more expensive investments to the new home and remodeled home owner; therefore, as an integrator, you will be expected to know how to identify the components and visualize how they are connected to the control interfaces in a modern residential setting.

This chapter provides the information you will need to know as a home integrator for designing, integrating, and identifying the key components for installing a home heating, ventilating, and air conditioning system.

HVAC Design Considerations

In this section, you will explore the design considerations you will need to know concerning heating, ventilating, and air conditioning (HVAC) systems. The topics covered include a discussion of zoned versus non-zoned designs, single or distributed systems, air handlers, water-based systems, and remote access for HVAC control.

Zoned/Non-Zoned Designs

When heating and cooling systems are operated in a non-zoned configuration, temperature differences in various parts of a residence can become much more than simply a minor annoyance.

Zoned system designs should be seriously considered when the residence is multilevel or is designed with wings or sprawling ranch styling. If the home has large, open areas with vaulted ceilings or rooms with many windows, zoning can also be useful. A zoning system is also useful in a home with a finished basement or attic, an indoor hot tub or swimming pool, or an exposed concrete floor.

Zoned Design Considerations

When a zoned system is installed, the communicating thermostats located in various areas of a home feed information to the automated zoning panel. The zoning panel sends commands to open and close dampers located in the heating and cooling ducts to control the temperature in each room and thereby eliminate the problem of hot and cold spots in various areas of a residence.

Used with programmable thermostats, a zoning system can save up to 30% on heating and cooling costs. Each zone can be customized for its own climate, controlling the temperature differences between floors, keeping the bedrooms cooler at night, or adjusting the living room temperature to account for large windows.

 The best solution to eliminate hot and cold spots in a residence is to use a zoning HVAC design.

After the zones have been established, each one must be provided with its own sensor or thermostat. The zone is also given its own control mechanism. In a forced air system, this mechanism is typically a motorized damper used to control the volume of warm or cold air that flows into the zone.

When a thermostat in a particular zone detects a need for heating or cooling, it generates a proper signal that is transmitted to a centralized control panel. The control panel manages the activities of all the zones by driving the motors that control the dampers in the ductwork, as shown in Figure 6.1. It causes the damper motors to open or close the zone damper depending on the demands of the zone thermostat.

Figure 6.1 Controlling the dampers.

For example, if a particular zone requires more heat, the controller drives the motor so that it opens the damper to that zone, allowing more heated air to flow into the zone. It also closes the dampers for all the other zones. In this manner, the full heating/cooling energy of the system is directed into one zone until the thermostat indicates that the desired temperature has been reached.

If a given system is using one unit for heating and cooling, the full output of the HVAC system can be directed into each zone to deliver the maximum capacity when required. In other words, in a residence that has a single 4-ton HVAC system, all 4 tons of capacity can be delivered into each zone. Conversely, if the same residence were heated and cooled by a pair of 2-ton units, only 2 tons of heating or cooling capacity could be applied to each zone.

Single or Multiple Pieces of Equipment

In most homes, the operation of the HVAC equipment represents the largest part of the home's energy costs. This is one of the prime reasons it is in the best interest of the homeowner to control the HVAC system as efficiently as possible. One of the design trade-offs involves the comparison between centralized and distributed HVAC designs. The characteristics of each type are listed as follows:

➤ *Distributed systems*—In distributed systems, multiple heating and cooling units are located throughout the residence. Control of the different heating and cooling units is typically performed individually.

➤ *Centralized systems*—In centralized systems, heating and cooling units occupy a centralized location within the residence and provide heating and cooling to the entire residence, usually through metal ductwork. Control of these systems also tends to be centralized.

In a distributed HVAC system, independent heating and cooling units are located at key locations throughout the residence. The heating units can be electrical heaters, wood burning stoves, or gas stoves, such as the one shown in Figure 6.2.

Figure 6.2 A typical distributed heating unit.

For the most part, the cooling units are window or wall-mounted air conditioning units such as the one shown in Figure 6.3. Other cooling systems, such as fans, can also be used in standalone form.

Figure 6.3 A typical distributed cooling unit.

Typically, each distributed heating/cooling unit is controlled through a local control device such as an on/off or variable selector switch. These devices are used to control the availability of the unit (on/off) as well as its level of operation (warmer/cooler).

Centralized Residential Heating Systems

In a centralized heating system, a master heating source is placed at a centralized location and used to generate heat or cooling that is then distributed throughout the residence via a network of pipes or ducts. The centralized heat source is typically some type of furnace. The furnace shown in Figure 6.4 can use several fuel types to generate heat, including wood, coal, and fuel oil. Other furnace types can utilize electricity, natural gas, or propane.

Figure 6.4 A furnace designed to use wood, coal, or fuel oil.

Furnaces come in all sizes and shapes and are designed to burn various fuels with varying efficiencies. In most modern residential settings, the furnace used is a forced air, natural gas unit. The heat generated by the furnace is used to heat air that is then pushed into different rooms via an installed air duct system. These systems are typically referred to as *forced-air* systems. A motor-driven fan unit, called the air handler, is responsible for moving the heated air out of the furnace and propelling it through a system of ductwork to the individual rooms of the home.

However, in rural areas or areas without access to natural gas pipelines, furnaces often produce their heat using propane or electricity. Yet, all residential furnaces work on the same principle, regardless of the fuel used. The fuel is first burned in close proximity to a heat exchanger, and the resulting heat from the exchanger is used to heat the air that is intentionally passed over it. The heated air is then delivered throughout the residence through the duct system.

Centralized Residential Cooling Systems

The cooling portion of a centralized heating/cooling system operates in basically the same way as the heating side. A cooling device such as an air conditioner, a heat pump, or a fan is used to provide cool air that can be forced through the duct system of the residence, as shown in Figure 6.5. In most cases, the residence employs the same ductwork for cooling and heating functions.

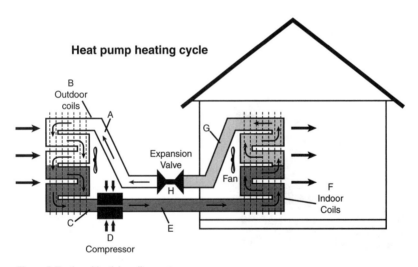

Figure 6.5 A residential cooling system.

Air Handler

The air-handling unit is composed of a cabinet that includes the central furnace, the air conditioner or heat pump, and the plenum/blower assembly. The air handler, commonly referred to as the *blower*, is used to move the heated or cooled air through the residential ductwork. It normally consists of a fan unit that sits inside fan housing, and outside air is pulled into the fan unit and pushed into the ductwork. The fan housing is typically installed directly inline with the distribution ductwork, as shown in Figure 6.6. The fan is driven by an electric motor, whose shaft might be connected directly to the fan, or it might drive the fan through a pulley and belt system, as shown in Figure 6.6.

Figure 6.6 Air handler.

Direct drive units typically place the fan inside the fan unit. Although this makes the motor more difficult to access, these motors do not require any user maintenance activities.

In practice, the term *air handler* has been applied to several variations of the unit just presented. If an evaporator coil is added to the basic air handler, it becomes the indoor unit for a central air conditioner or a heat pump. Likewise, if electric heating elements, circuit breakers, and a control system are added to the air handler, it becomes an electric furnace.

Water-based (Radiant) Heating

Warm air is not the only warming media that can be used. Some home heating systems are water-based radiant heaters. In these systems, the fuel is used to heat water in a central reservoir, which is then piped into heat-radiating devices (*radiators*) positioned in each room. In some systems, hot water is circulated through the system, whereas in others the water is actually converted to steam prior to being forced through the radiator distribution system.

Traditional Radiant Systems

The traditional radiator is a heating unit that transfers heat through radiation into the air around it through conduction. This heat, in turn, warms the air in the room through natural convection processes, as shown in Figure 6.7. The convection current circulates the heated air throughout the room; as the hot air rises, it is replaced by heavier cool air.

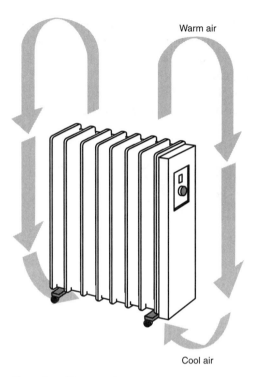

Figure 6.7 The convection process.

Solar Alternatives

The open direct radiant heating system can be converted to solar heating. Its required electricity can be harnessed directly from the sun by using cost-effective solar thermal collectors, as shown in Figure 6.8.

Figure 6.8 Solar thermal collectors.

Because electricity is a premium fuel source that carries with it high financial and environmental costs, simple heating tasks such as space heating or the production of domestic hot water might better be accomplished with solar thermal collectors. The user should consider a design whereby solar thermal collectors are used to create the heat energy required for space heating and domestic hot water.

Geothermal Heat Pumps

A geothermal heat pump is a heating and cooling system that uses the relatively constant temperature of soil or surface water as a heat source and sink for a heat pump. A heat pump moves the heat from the earth (or a groundwater source) into the home in the winter and pulls the heat from the house and discharges it into the ground in the summer. The underground piping loops serve as a heat source in the winter and a heat sink in the summer. Geothermal heat pumps save money in operating and maintenance costs compared to traditional gas heating and air conditioning systems. Although the initial purchase price of a residential geothermal heat pump system is often higher than that of a comparable gas-fired furnace and central air-conditioning system, heat pumps are more efficient and return the investment in a few years.

Remote Access

Various manufacturers have developed access systems that permit the homeowner to control the residential HVAC system remotely, through telephone operations and X10 controllers. Truly convenient HVAC control can be achieved by including a remote access system that enables the owner to make changes to the HVAC system even when away from the residence.

X10 Remote Access Design Features

A typical remote access design employs a standard telephone connection and an X10 telephone transponder to communicate with the residential control system. The transponder receives touch-tone input signals sent to it across the telephone lines and encodes them on the residential power lines using the X10 protocol. The encoded commands can be used to change the temperature set point for the programmable thermostat.

A typical scenario might include the homeowner setting the temperature of the residence to 50° while at work and wanting to raise the temperature to 72° when he arrives home. The user simply calls the residence and the telephone transponder should answer the call. Initially, the transponder requires the caller to enter an access code for security purposes. Afterward, he can program the system through the touch pad on the phone, using X10 codes.

The transponder decodes the X10 commands and places them on the 110Vac power lines. When the X10-compatible thermostat, which is also attached to the power system, receives that signal, it decodes it and changes the temperature setting to the new value, causing the heat to come on.

Programming is accomplished through the touch pad on the telephone. A programming cross-reference chart is typically supplied with the X10 thermostat, so the user can see what each key sequence does to the thermostat.

Equipment Location Considerations

Proper location of system components is essential for reliable and efficient operation of HVAC system. This section covers the exam topics you will need to review, including the location of thermostats and sensors.

Thermostats

In any residence, the locations of the thermostats are very important for the successful and efficient operation of the HVAC system. Another name for a thermostat is a *temperature controller*. Thermostats can be grouped into two major types: legacy (traditional) thermostats and intelligent thermostats. In this section, you will learn more about the location requirements and functional capabilities of each type.

Legacy Thermostats

Legacy thermostats conduct their operations by several means. If they are positioned improperly or located where they will receive invalid input, their usefulness is greatly diminished.

Thermostat Mounting Location

By observing several well-proven precautions, the HTI+ technician can ensure the proper placement of legacy thermostats.

For example, if the thermostat incorporates a mercury switch, it must be mounted completely horizontal for the liquid mercury to be able to use the effect of gravity when making and breaking the contact between its embedded wires. Therefore, always use a leveling tool when mounting a mechanical thermostat. Many types of thermostats can be horizontally mounted by simply holding a small, torpedo-type level against the bottom of the thermostat and tightening the mounting screws at the level point. Most installers, however, are recommending the replacement of mercury thermostats with electronic types.

If the thermostat incorporates magnetic switching contacts or operates electronically, it can probably be mounted upside down or sideways because these types do not depend on gravity to achieve proper operation.

Because outside walls conduct some degree of heat and cold originating from outside of the house, these temperature variations adversely affect the thermostat's accuracy. To avoid this possibility, thermostats should be mounted on interior or partitioning walls.

Placing thermostats near heat sources such as warm air ducts, gas or electric ranges, clothes dryers, fireplaces, and ovens guarantees the improper operation of any HVAC system. Rooms some distance from these devices will feel uncomfortably cool. In addition, locating a thermostat near less-evident sources of heat, such as lamps and electronic equipment, can also become a problem. See Figure 6.9 for some visual guidelines.

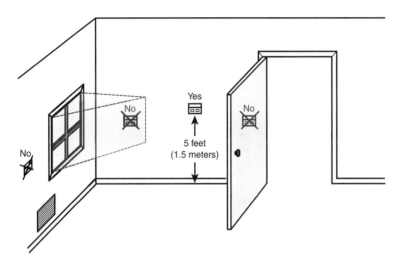

Typical Location of Thermostat

Figure 6.9   A visual guideline for thermostat placement.

As you can see from Figure 6.9, it is not good practice to locate mechanical thermostats in close proximity to doors or vertically sliding windows. The vibrations and jarring produced by the opening and closing of these doors and windows can affect thermostatic operation. In addition, the area behind doors or alcoves does not support enough air circulation to provide accurate temperature readings.

Thermostats should be located well above and to the side of any light switches. This helps to avoid any accidental adjustments being made to a thermostat when the lighting is turned on or off.

An excellent idea is to locate the thermostat near the return air grill because it gains the opportunity to sense the temperature of the air being drawn back into the system. The living or dining room is also a preferred location as long as no cooking or refrigeration units are on the opposite side of the wall.

The kitchen is also not a suitable location for mounting a thermostat because the kitchen is normally warmer or cooler than the other rooms in the house.

 Remember the locations that can adversely affect the operation of a thermostat. Areas that are normally warmer than other areas, such as kitchens, washer and dryer areas, fireplaces, and ovens are likely areas for improper operation of a thermostat. The best area for a thermostat location is near the warm air return grill.

Intelligent Thermostats

Intelligent thermostats are also known as *communicating* thermostats and have found wide application in modern HVAC systems. Operating similarly to a conventional thermostat, they are additionally capable of being controlled either locally or remotely from a home automation system. The intelligent thermostat shown in Figure 6.10 can be integrated into a home automation system or be operated independently with the use of a proprietary system controller.

Figure 6.10 An intelligent thermostat.

Wireless Thermostats

An added dimension in thermostat technology is available with the wireless thermostat. Users appreciate the capability of adjusting the temperature from the lounge chair in the living room, the table in the kitchen, a favorite couch, or the bedroom.

Wireless Thermostat System Operation

The wireless thermostat, shown in Figure 6.11, consists of two parts: a thermostat and a receiver.

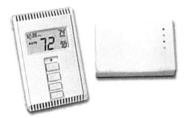

Figure 6.11 A wireless thermostat with receiver.

The receiver contains the wiring connections to the HVAC unit. It can be located using the same criteria used with wired thermostats where the availability of adequate wiring exists, such as at the old thermostat location, inside a rooftop package unit, in a furnace closet, or in an attic or a basement. This makes it no more difficult to install than the standard 24-volt (v) AC thermostat.

Sensors

Sensors are often used with controlling devices such as thermostats. It is through the actions of sensors that controlling thermostats make their decisions about when and how to activate the HVAC system.

Temperature Sensors

Zoned systems commonly employ a single thermostat that operates in conjunction with multiple temperature sensors. To achieve accurate temperature averaging, some thermostat networks are capable of interconnecting with up to six indoor temperature sensors per thermostat. The thermostats connect to each sensor using two-strand thermostat wire.

Leaving Air Temperature Sensors

Leaving air temperature sensors are located in the supply trunk between the bypass damper and the evaporator coil or heat exchanger. They play an important role in the HVAC system's performance. In systems that do not include a bypass damper, a temperature sensor should be located between the zone dampers and the evaporator coil (or heat exchanger).

A discharge air temperature sensor (DATS) should never be located directly above the heat exchanger or strip heater because radiant heat can cause false temperature readings. A good rule of thumb is to keep a DATS out of the heat exchanger's radiation pattern, as shown in Figure 6.12.

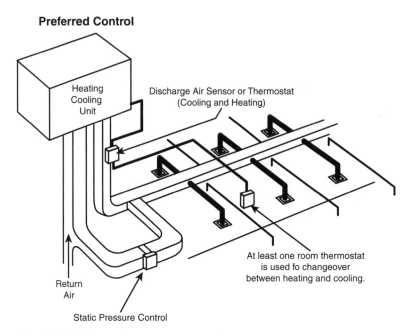

Figure 6.12 Discharge air temperature sensor placement.

Physical Products and Components

This section examines several HVAC control product types that the working HTI+ technician might encounter. The most important HVAC products you will need to know about include thermostats, air handlers, damper controls, furnaces, condensers, and distribution panels.

Thermostats

The thermostat is typically the centerpiece of a residential temperature control system. It acts as a process controller that measures the ambient temperature, compares it to a user-defined set point, and creates an output action to force the two values together. This means the output action calls for the application of heat when the measured temperature is below the set point value established by the user.

Control of centralized heating and cooling systems is generally provided at two levels as follows:

➤ *Overall control*—Overall control is provided by centralized thermostats that measure the temperature at key points within the residence and act to turn the heat/cooling on or off at some predetermined set point.

➤ *Localized control*—Localized control is provided by adjustable airflow vents installed in the registers (exit vents of the ductwork). These devices are used to control the volume of warm or cool air that can enter the room over time. This setting is used to fine-tune the temperature control function by matching air volume against the natural heat up and cool down characteristics of the room.

Many types of thermostats are available to handle a variety of heating and cooling systems. Some thermostats are designed for a single function, such as for heating or cooling, whereas others are designed to handle both heating and cooling functions. Still other designs are used to perform multiple-stage heating and cooling functions.

Thermostat Operation

The most common type of thermostat is based on the characteristics of a bimetallic strip of metal. In these thermostats, two metal strips are bonded together and used to open and close a set of electrical contacts. Because the two metals have different temperature coefficients of expansion, they naturally expand and contract at different rates based on the temperature around them. As the temperature changes, one metal expands or contracts at a faster rate than the other, causing the strip to bend in an arc.

The amount of bend in the strip is dependent on the temperature applied to it. As the strip bends, it moves a set of electrical contacts closer to each other or farther away from each other.

One end of the strip is anchored to the thermostat base, whereas the other is allowed to move freely. A fixed contact is also mounted on the thermostat base, and a movable contact is mounted on the free moving end of the strip.

In a heating operation, when the bimetallic strip is cooled it moves in a direction that brings the two contacts together. When they meet, a low-voltage electrical circuit is closed, which in turn engages the coil of a heat relay or the solenoid of a gas valve. The energized coil closes a set of high-voltage (120v/240v AC) contacts that control the flow of current to a set of resistive heating elements.

 Understand the operation of a bimetallic strip for some types of thermostats. This is the most common type of sensor found in legacy thermostats.

Mercury Thermostats

Mercury-containing thermostats use mercury tilt switches. Due to the toxic nature of mercury, these types of switches are no longer used in HVAC systems. Programmable electronic thermostats are mercury free and are more energy-efficient than the mercury model. The installer should look for programmable electronic thermostats that have the ENERGY STAR label.

Mercury Thermostat Environmental Issues

Mercury switches pose no threat to human health or the environment while they are intact. The mercury is typically encapsulated in a hard plastic casing, stainless steel, or a glass ampoule. Most of these devices are very well-protected in the inside of the thermostat or other appliance, and normal handling of these devices by the HTI+ technician will not disturb the mercury switch. Even during rough handling at a solid waste facility, the majority of mercury switches appear to stay intact. The critical time when these switches can become capsules of poison is when the appliances are shredded by a scrap metal recycler or crushed in a landfill. Both of these processes are likely to compromise the integrity of the switch casing, so you need to keep in mind the local policy for the proper disposal of components that contain mercury.

HVAC systems using mercury switches should be replaced with non-mercury switches. HTI+ technicians should be aware of the local building codes and mercury disposal and recycling programs that apply to each area. Most HVAC wholesalers consolidate thermostats from contractors and send them to recyclers. About 3 grams of mercury are in each mercury tilt switch. Most thermostats have one switch and some have two, but up to six switches are possible.

Staging Thermostats

Staging thermostats are designed to provide greater efficiency and lower operating expense by providing multiple levels of heating and cooling capacity. For example, a staging thermostat causes a multistage HVAC system to operate at a reduced btu level during mild weather and shifts into a higher-capacity mode when the weather turns colder. Likewise, the cooling equipment operates at a higher efficiency level when the weather is mild and shifts into a higher-capacity mode when the weather becomes hotter.

These thermostats typically provide an automatic changeover feature that enables the unit to change from a heating thermostat to a cooling thermostat without human intervention. The homeowner can simply set the desired heating and cooling temperatures for the residence, or zone, and the system will maintain the residence at those points regardless of the season. To avoid creating a situation where the heating and cooling units are working against each other, the HTI+ technician should remember that the cooling setting should not be set below the heating set point and the heating setting should not be set above the cooling set point.

Most thermostats include a fan switch that permits the user to circulate air throughout the system without including the heating or cooling units. When this switch is set in the auto position, the fan runs in conjunction with the heating and cooling systems. However, when the switch is set to the on position, the fan runs continuously, without respect to whether the heating or cooling functions are running.

Electronic Thermostats

Modern HVAC systems are generally equipped with smart electronic control systems. Smart thermostats connect into the HVAC system via the existing residential wiring and communicate with the various control modules through three-conductor control cables. They often include innovative features such as large and easy-to-read displays, ENERGY STAR-compliant circuitry, manual override for all functions, and hold buttons to prevent automatic operation. They also permit the alternative display of the outdoor temperature and the indoor temperature.

ENERGY STAR is a voluntary labeling program sponsored by the U.S. Environmental Protection Agency (EPA) and the U.S. Department of Energy that identifies energy-efficient products. Qualified products exceed minimum federal standards for energy consumption by a certain amount, or where no federal standards exist, have certain energy-saving features. Such products can display the ENERGY STAR label.

Air Handlers

Air handling equipment has become a major segment of the HVAC industry because complete systems can be custom built to any size, shape, or cubic-foot-per-minute (CFM) range specified by the user. The equipment itself can be constructed in single or multiple pieces that can be assembled at the job site.

The air handler typically houses items such as the supply fan(s), return fan(s), heating coil, cooling coil, and filters. Its purpose is to provide the pressure force and conditioning necessary to get the airflow through the entire system, and at the design temperature and humidity condition. Often, the air handler is the place in the system where supply and return air mix, outside air is mixed with system flow, and exhaust air is extracted.

Because of its multifunctional nature, the air handler is the single most critical system component. Essential maintenance of the unit includes the lubricating of fan bearings, periodic changing of filters, and thorough cleaning of the coils.

Whole House Fans

Although air handlers can provide sophisticated solutions for keeping the air inside a residence at specifically selected temperatures, they are relatively expensive to operate when compared to whole house fans.

Daytime attic temperatures during the summer months can easily reach 150° F! This is heat that normally remains trapped in the attic. If the attic remains exceptionally warm during summer evenings due to the trapped warm air, an air conditioner can run around the clock trying to cool things down. An alternative solution to this unwanted expense could be a whole house fan, engineered to efficiently and quietly reduce the indoor temperature of the user's home.

Because indoor air temperatures can often remain higher than outdoor temperatures, particularly in the evening, a whole house fan can help pull this unwanted heat out of the residence. By opening several windows, or a door, cooler outside air can be drawn into the home while the heated air is pushed out through the existing attic and roof vents. Running a whole house fan during the night results in the air conditioning unit not being needed until later in the day.

Attic Fans

Attic fans are another type of HVAC device that can reduce the temperature of the living area on warm days. These are fans placed next to vents in the attic that flush out the hot attic air. This also alleviates the problem of high-humidity levels that accumulate during the day. Moisture migrates through the ceiling toward the roof, where it comes in contact with the cold structure and causes the formation of frost or mold. Although attic fans are used in many older homes, most HVAC contractors favor the whole-house fan for overall efficiency.

You need to remember the advantage of a whole house fan: It helps pull unwanted heat out of the attic during evening hours, resulting in energy savings for the HVAC system. Also keep in mind the purpose and origin of the ENERGY STAR labeling program mentioned earlier and its benefits for residential HVAC systems.

Damper Controls

Located in the distribution ductwork that regulates the airflow, damper controls are movable plates. Typically used in a zoning application, they direct air to the areas that need it most. To do this, they often limit the flow of heated/cooled air to the required areas by permitting the shutting off of airflow to unoccupied or unused rooms. Due to the increased airflow in the occupied areas, the temperature changes more quickly, providing for shorter run times by the heating/cooling system.

Certain rooms are habitually warm or cool, regardless of the settings of the thermostat. If some rooms are too hot or cold, try adjusting the dampers in the registers. If the system has them, adjust the dampers in the warm-air ducts.

When a duct run is so long that the airflow through it decreases to the point where it is no longer effective, it might be necessary to install an inline duct fan to provide the necessary boost to the flow (see Figure 6.13). These inline duct fans are commonly available in diameters of 5", 6", 8", or 10".

Figure 6.13 An inline duct fan.

When the air flowing through a long duct run decreases, an inline duct fan can be employed to increase the airflow.

Damper Heating Adjustments

To make sensible heating adjustments to a residence, you should leave the thermostat at one setting while letting the system run for several hours to stabilize the temperatures. While stabilization is taking place, you should open the dampers wide in the coldest rooms. The next step involves manually adjusting the dampers room-by-room until the temperatures are balanced. As mentioned earlier in the section "Zoned Design Considerations," dampers can also be operated automatically with motors. Thermostats in each zone send commands to the zone panel requesting a change in airflow. The zoning panel sends commands to the motors to open and close dampers located in the heating and cooling ducts to control the temperature in each room. Motorized dampers can be installed in any home that has been designed for zoned temperature control.

Duct Planning

One of the important considerations for planning a residential duct system or, for that matter, determining the performance of an existing system is measuring the system's airflow. After a determination of a given system's airflow has been made, there are also the questions of what the right amount of airflow is and how to obtain it.

Accurate airflow measurements reveal the truth about system performance. For example, low airflow reduces an air conditioning system's total capacity and its efficiency. This is why the airflow should be measured and why it should be remedied if it's too low. Because a variety of ways exist to measure (or infer) a system's airflow, there are bound to be differing ideas about which method is best. One method is to measure static pressure.

Measuring Static Pressure

An accurately performed static pressure test lets you monitor the airflow in a duct system. The test should be performed using a static pressure probe (SPP), several of which are shown in Figure 6.14.

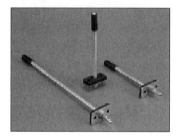

Figure 6.14 A static pressure probe.

The SPP is designed to detect the static pressure in an air handler, a duct, a plenum, or any other HVAC equipment. It contains two orifices situated vertically opposite each other, enabling the canceling out of any errors that might be induced by the airflow.

The probe has a number of holes around its circumference and should be oriented parallel to the air stream while measurements are being taken. This orientation cancels the velocity pressure, so the given reading is pure static pressure.

Furnaces

The furnace is the component of the centralized residential heating system responsible for generating heat to be distributed throughout the residence. As mentioned earlier, different types of furnaces perform this task in different ways.

The major components of a furnace include a heat exchanger, a heat source (fuel burner or heating elements), an air blower, housing, and some type of control system. As mentioned earlier in this chapter, the furnace can be based on coal, fuel oil, gas, wood, or electric heaters for heat sources.

In normal operation, air is drawn into the furnace through the return air inlet by the fan unit. The air is pushed past the outside of the heat exchanger where it is warmed and expelled through the warm air outlet and travels through the residential ductwork.

In nonelectric furnaces, the hot gases created by warming the heat exchanger must be exhausted to the outside through a vent or flue. These gases are created when the fuel source is burned inside the heat exchanger.

The control system is used to monitor the temperature in some predetermined location(s) and adjust the rate of heat generation in the furnace. The most simple temperature control system is an on/off thermostat that is used to turn the heat source, such as an electric heating element, on and off or to open and close a gas valve as the temperature at the thermostat moves above and below the desired set point.

Electric Furnaces

Electric furnaces work by applying electrical current to one or more resistive heating elements that heat the air inside the heat exchanger.

Either 120v AC or 240v AC from the home's electrical power system is applied to the resistive heating elements, causing them to heat up. However, the power applied to the elements is controlled by a low-voltage temperature signal from the thermostat.

Normally, residential temperature control is performed in an on/off manner. When the temperature in the vicinity of the thermostat rises above the setting on the thermostat, an electrical connection in the thermostat opens and the controller removes power from the heating elements, causing them to cool down. Likewise, when the room temperature drops below the thermostat's set point, the connection is closed and power is again applied to the heating elements.

Gas Furnaces

Gas furnaces employ a gas burner arrangement that consists of a main burner responsible for heating the air in the chamber and an automatic igniter. Older furnaces use a pilot light that burns constantly to ignite the gas during a start sequence and a thermocouple assembly that senses the presence of the pilot light. The heat from the pilot light flame generates a signal telling the controller to keep the safety emergency shut-off valve open. If the pilot light goes out or the thermocouple loses the heat from it, the signal disappears and the safety valve shuts to prevent the possibility of the heating system dumping gas into the furnace from the main burner. Newer furnace systems use an automatic igniter that generates a high-voltage spark from the igniter power supply. This eliminates the pilot light assembly and saves the energy used by the pilot light when the furnace is not operating.

Condensers

In an air conditioner, a heat exchanger assembly (called the *condenser*) is used to lower the temperature of the ambient air. After the air has been chilled, it can be circulated throughout the residence by a duct system.

The condensing unit is also known as an *outdoor coil*. As an integral component of a heat pump or a central air conditioning system, it is located in an outdoor casing that is vented. It serves as the transfer point where heat is transferred from inside the home to the outside air.

Alternatively, when heat is desired, a forced air furnace uses a blower fan to draw air from the indoor air return and force it across the heat exchanger. This air is then forced into the supply path to branches of ductwork and out the supply vents.

Evaporators

An evaporator coil is that portion of a heat pump or central air conditioning system that is located inside the home. It is also known as an *indoor coil* and functions as the heat transfer point for warming or cooling indoor air. It consists of a series of pipes connected to a furnace or air handler that blows the indoor air across the evaporator coil, causing the coil to absorb heat from the air. The cooled air is then delivered to the house through the ducting. The refrigerant in the coil then flows back to the compressor, and the cycle starts over again.

Heat Pumps

Heat pumps normally appear in split systems and packaged cooling units designed for mounting on rooftops or windows. They are refrigerated and perform both heating and cooling operations using the properties of refrigeration. As with any system or unit that utilizes refrigeration, heat transfer is continually taking place.

In the example of a household refrigerator, heat is transferred from the inside of the unit to the outside. If the refrigeration unit is being used with an air conditioning system operating in a cooling mode, the evaporator (inside coil) performs the task of removing the heat from the inside air. The heat is transferred to the condenser (outside coil) where it is given up to the atmosphere. When the terms *inside coil* and *outside coil* are used, they almost certainly are referring to the use of a heat pump. In fact, whenever refrigerated air conditioning is being discussed, the evaporator is always known as the coil. When the equipment is operating in a cooling mode, the coil (evaporator or inside coil) is always kept at a very low temperature.

Warm air holds larger amounts of humidity, whereas cool air is much drier and holds very little moisture. This can be demonstrated in a cooling system where the air habitually gives up its moisture when forced across a low-temperature coil (the evaporator). This is a process called *condensation*. In situations where high amounts of surrounding humidity are present, the resulting condensation reduces the amount of available cooling per ton (12,000 btu). For example, in an area of extremely high humidity, a home equipped with 5 tons of cooling (60,000 btu) might receive the benefit of only 4 cooling tons because approximately 1 ton of cooling would be expended in condensation efforts. Geothermal heat pumps, discussed earlier in the section "HVAC Design Considerations," offer many advantages over conventional heating and cooling systems and are the preferred deign for new residential construction.

Reversing Valves

A heat pump uses a reversing valve, such as the one shown in Figure 6.15, to redirect the refrigerant flow. It enables the inside coil and outside coil to act as either an evaporator or a condenser.

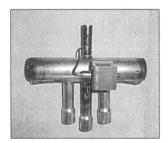

Figure 6.15 A heat pump reversing valve.

For example, in a heat mode the outside coil acts as the evaporator (extracts warmth) and the inside coil acts as the condenser (gives up warmth). Depending on the manufacturer's design, with the reversing valve energized the unit can be in either a heat or cool mode. Normally, reversing valves are energized with 24v AC via a step-down transformer.

 The purpose of a reversing valve in a heat pump is to redirect the refrigerant flow.

Distribution Panels

A zoning control panel can manage the demands of one zone calling for cooling while another calls for heating. It automatically sorts out these conflicting demands and runs the heating or cooling equipment to provide optimum comfort in all zones.

It is also capable of handling auxiliary heat and add-on systems without requiring a fossil fuel kit. This is due to the fact that when the auxiliary heat source is energized, the panel automatically locks out the compressor. It also works with any single-stage mechanical or electronic thermostat, even when used with heat pumps, because auxiliary heat functions are controlled through a fixed and timed upstage program.

A single thermostat located in each zone sets the temperature independently, whereas rectangular or round motorized zone dampers control the air distribution to the zones. The actuators are either two-wire, spring-return types or three-wire, power-open/power-closed designs and are wired directly to the HVAC equipment.

Standard Configurations and Settings

HVAC systems require the settings of various components for proper operation. This section covers the various settings and configurations you will need to remember for the HVAC portion of the HTI+ exam. This section covers the functional purpose of zone programming, system programming, time-of-day programming, and seasonal presets.

Zone Programming

The difficulty of the family reaching unanimous agreement about a comfortable residential temperature can be greatly alleviated with the help of zone programming. For example, the six-zone controller shown in Figure 6.16 is capable of helping the family obtain different heating, ventilation, and air conditioning temperatures in different parts of the home.

By zone programming the HVAC system, the ground floor of a two-story residence can be properly heated without overheating any family members upstairs.

Figure 6.16 An HVAC six-zone programmable controller.

Residential energy management using this integrated zone controller permits individualized programming of six thermostats, dampers, or water valves. For example, one thermostat can provide cooling for a certain part of the house, while another thermostat can be used to keep a specific room heated. An outdoor damper can be programmed to allow outdoor air to enter the system, or an outdoor temperature sensor can permit the user to see remote temperature information on an indoor thermostat.

Mounted near the HVAC equipment or in a utility closet, the controller appears to be a regular thermostat to the existing HVAC equipment. Running properly with as few as two zones, the system requires a wall thermostat and one open duct damper for each. The user retains the option of adding more zones as the residential requirements grow.

The zone controller is capable of requesting the current temperature of each zone, reading and setting the setpoints for each zone, reading and setting the mode for each zone, and issuing many other commands. Also, by changing out a chip on the PC board, the system can be controlled remotely via RS-232.

System Programming

If you are responsible for programming this type of zoned HVAC system, you must be familiar with using an ASCII-based, freeform message format. Successful communications will need to be established between either a remote host or a master system and its family of thermostats and sensors. The required connectivity is achieved using either an RS-232 or RS-485 network interface.

A suitable controlling thermostat for controlling a single mechanical system having a node rather than a zone is shown in Figure 6.17.

Figure 6.17 A controlling thermostat for single-node operations.

This controlling thermostat is powerful enough to not only provide typical thermostat functions, but also to send and receive information via RS-232 or RS-485 communications. This allows the thermostat's setpoint, mode, and fan operations to be remotely manipulated. The remote system can perform status requests regarding the thermostat's temperature, setpoints, and modes.

Time-of-Day Programming

In some of the previous examples of sophisticated HVAC zone programming, the manufacturers included some time-of-day features in several of their products.

One of the main reasons for time-of-day programming is to save money on energy costs. Integrated temperature control is capable of providing optimum comfort while demonstrating residential energy savings of close to 20%. Such systems are designed to automatically reduce the heating and cooling when the residents are asleep or away from home. When the residents wake up or return home, the normal temperatures are automatically restored.

Time-of-day scheduling is defined as automatically turning off equipment when it is not needed, on a predetermined time schedule. Routines must allow for holidays, weekends, and daylight savings time. Override features enable the HVAC equipment to operate whenever the residence is being used for activities outside normal programmed hours and should not interfere with normal HVAC operation. However, users should always be aware that savings are achieved from not running equipment unnecessarily.

Equipment that is usually controlled by time-of-day operations includes ventilation fans, exhaust fans, interior lighting, exterior lighting, and security lighting.

Devices designed to implement time-of-day scheduling can be electromechanical timers, electronic timers, standalone controllers, computer-based systems, lighting controls, or programmable logic controllers (PLCs).

Residential HVAC parameters typically requiring time-of-day control are the temperature and operational settings. This includes the capability of the program to vary the schedule for different days of the week and certain holidays. Time-of-day settings are normally adjusted for cooling and heating so that comfort settings are maintained for human occupancy and comfort during occupied periods.

It is during the unoccupied periods when time-of-day programming works most effectively to maintain specified temperature and humidity levels that not only save energy, but also preserve the integrity of the home's structure and contents. This is accomplished by preventing indoor temperatures from becoming too cold/hot and keeping the humidity from becoming too high.

Smart programmable thermostats are capable of learning the heating characteristics for the space being heated. They can even preheat a given area at a new time for each occupied day. The capability to learn preheat times saves more energy than is possible with other programmable thermostats. Preheat operations can be set not to occur on holidays and weekends, while the setback temperature, occupy/vacate times, time-out delays, and holidays are programmable parameters set during installation.

Seasonal Presets

Timed programming for HVAC systems is often maintained through such situations as extended power outages. Programmable thermostats are capable of keeping track of the day, month, and year, enabling accurate accounting for seasonal occurrences such as leap year and daylight savings time. Custom programming for statutory holidays can also be accommodated and, once programmed, these types of systems automatically update their seasonal presets.

HTI+ technicians should keep in mind that during the summer, every degree that the air conditioning thermostat is raised results in a 5% decrease in cooling energy consumption. Likewise, during the winter, every degree that the thermostat is lowered results in a 3% decrease in heating energy consumption. Thermostats that regulate unoccupied rooms in the home can save substantial amounts of energy by operating at a setting of 55° F in the winter and above 80° F in the summer.

Device Connectivity

HVAC systems require you to be trained in the procedures for connecting the heating and cooling components and understanding the features of automated thermostats and HVAC cabling. This section covers physical connections between controlling devices and the equipment being controlled using communications cable and termination points designed for HVAC systems.

Communicating Thermostats

A communicating thermostat integrates the control functions of a modern automated HVAC system with the external computer interface, such as X10 or CEBus controller software. The connectivity wiring for a typical communicating thermostat is shown in Figure 6.18.

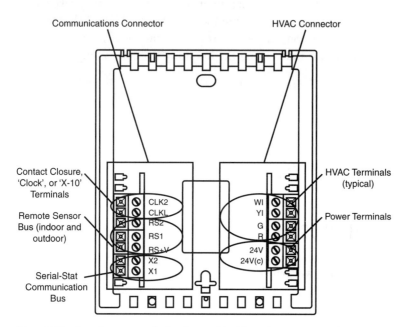

Figure 6.18 A modern communicating thermostat wiring scheme.

The communicating thermostat includes the standard HVAC terminals on its right side and the required communications terminals on its left.

Communications Cable

Communications continually occur between various HVAC systems' components, but the daily routine can be severely disrupted when something occurs to interrupt the connectivity. With newly installed systems, the home user might still be waiting for the equipment to work properly. Without the required connectivity, the wait could be a long one.

Cable Types

Regardless of whether the cabling is being installed in a new home or in a residence that is being remodeled, the correct types of cable must be selected. As you will discover in Chapter 10, "10.0—Structured Wiring—

Low-voltage," structured wiring and cabling standards dictate the use of coaxial cable and unshielded twisted-pair cables to each room in the home. For HVAC management and control, UTP communications cable is the preferred cable for connecting various components of the system.

Computer Networking

When computers are used to run residential HVAC networks, links between the various controllers and the server/computer are accomplished using Category 5 UTP communications cable that complies with the residential structured wiring standards covered in Chapter 10.

The wiring used between the thermostat and the working HVAC equipment is typically a special HVAC type of twisted-pair cabling similar to the wire used for home security systems. Specially designed and color-coded for thermostat/HVAC use, these cables are available in two-, five-, and seven-conductor versions. For cabling runs of up to 250 feet, 20-gauge cable supports loads of up to 4 amps. For cable runs of approximately 200 feet, 18-gauge cable handles loads up to 8 amps.

For systems that will be used only to supply heat, two-conductor cable is all that is needed, whereas systems that provide both heating and air conditioning usually require four- or five-conductor cable.

Termination Points

The typical residential HVAC system uses various termination points for its wiring patterns. These termination points are required for both sensors (temperature or humidity) and for the HVAC control cables. The sophistication of the layout increases substantially for a multifloor residence.

Termination points merely indicate the locations where the wiring ends; they do not necessarily dictate which pieces of equipment will eventually be connected at these points. Although this chapter deals with the HVAC aspects of home automation, wiring terminations involve a host of other systems as well, such as A/V, security, lighting, and residential communications.

Similar to other types of construction drawings, floor plans are usually prepared to indicate where the control cables will be terminated for the heating and air conditioning system.

In a structured wiring design, all the cabling, including the wiring used with other systems, should be terminated (home run) at a wiring closet located in the basement or wiring closet. This is also the logical equipment location for the indoor half of a split HVAC system.

Standards and Organizations

The authority to approve or disapprove the suitability of one product over another has been vested with various technical bodies recognized by manufacturers and consumers around the world. This arrangement has been deemed necessary to achieve some measure of control over the explosion of technical development that has occurred during the last several years and is continuing to occur.

The major organizations listed here have published standards that directly impact the acceptability of both products and labor, with regard to home HVAC control and management system planning and operations. They have been charged with approving or disapproving the suitability of one product over another, and manufacturers and consumers have come to depend on the appropriate standards published by these organizations.

National Electrical Code

Sponsored by the National Fire Protection Agency, the National Electrical Code (NEC) document contains information on which all aspiring electricians must be tested before obtaining their licenses. Having spent so much time studying it, most electricians can easily locate any required information in the code. Other users would be advised to use the index.

The National Electrical Code is merely a guideline, but most states require a permit and an inspection for performing this type of work. Although following the NEC cannot guarantee safe electrical installations, it is the best guide available. Given the many applications available for specific wiring jobs, an electrician should use her own judgment, while keeping above the minimum safety standards set forth by the NEC.

Because each state can differ slightly in its requirements for inspection and code compliance, the local city or town wire inspector should be contacted before any wiring work is undertaken. The local wire inspector is The Authority Having Jurisdiction in most locations and, as such, is responsible for rule interpretation and code enforcement.

TIA/EIA Standards

The Telecommunications Industry (TIA) and Electronic Industry Alliance (EIA) standards are certified by the American National Standards Institute (ANSI). Telecommunications Systems Bulletins (TSBs) are addenda to or explanatory comments about either an industry standard or an interim standard, and they are often integrated into the next standard revision.

Several ANSI/TIA/EIA standards pertinent to residential HVAC control and management systems are as follows:

- *TIA 232-F*—"Interface Between Data Terminal Equipment and Data Circuit-Terminating Equipment Employing Serial Binary Data Interchange" (ANSI/TIA/EIA-232-F-1997) (R2002). This document is applicable to the interconnection of data terminal equipment (DTE) and data circuit-terminating equipment (DCE) employing binary data interchange.

- *TIA/EIA 570-A*—"Residential Telecommunications Cabling Standard" (ANSI/TIA/EIA-570-A-99). This document standardizes requirements for residential telecommunications cabling. These requirements are based on the facilities necessary for existing and emerging telecommunications services.

- *TIA/EIA 570-A-1*—"Residential Telecommunications Cabling Standard - Addendum 1 - Security Cabling for Residences" (ANSI/TIA/EIA-570-A-1-2002). This addendum provides recommendations and specifications for security cabling systems in residences. It contains references to national and international standards.

- *EIA 600 CEBus SET*—"EIA Home Automation System" (CEBus). Provides the necessary specifications for the Consumer Electronic Bus (CEBus), a local communications and control network designed specifically for the home. The CEBus network provides a standardized communication facility for the exchange of control information as data among devices and services in the home.

IEEE Standards

A nonprofit, technical professional association of more than 377,000 individual members from more than 150 countries, the IEEE's full name is the Institute of Electrical and Electronics Engineers, Inc.

Because of the size and scope of its membership, the IEEE has become a leading authority in technical areas including computer engineering, biomedical technology, telecommunications, electric power, aerospace, and consumer electronics.

Underwriters Laboratories, Inc.

As an independent, nonprofit product safety testing and certification organization, Underwriters Laboratories, Inc. (UL) has tested products for public safety for more than 100 years. With each additional year, more than 17 billion UL marks are applied to products from all over the world.

Founded in 1894, it holds an undisputed reputation as the leader in U.S. product safety and certification. However, the UL is becoming one of the most recognized, reputable conformity assessment providers in the world, not just in the United States.

At the time of this publishing, UL services also include helping companies achieve global acceptance for their products, whether it is an electrical device, a programmable system, or a company's quality process.

UL lists various manufacturers for their specific types of installations and equipment, tests the equipment for specific uses, and judges their installations against the governing standards.

National Fire Protection Association

In an effort to reduce the worldwide burden of fire and other hazards on the quality of life, the National Fire Protection Association (NFPA) has developed and advocated a scientifically based consensus of codes, standards, research, training, and education since its founding in 1896.

Today, with its headquarters located in Quincy, Massachusetts, the NFPA is an international, nonprofit membership organization that supports more than 75,000 members representing nearly 100 nations and 320 employees around the world.

In addition to setting installation guidelines for fire equipment, the NFPA also serves as the world's leading advocate of fire prevention. It holds classes to instruct various installers on how to follow the guidelines and is an authoritative source on public safety.

Installation Plans and Procedures

This section lists the installation procedures you will need to know for the HVAC portion of the HTI+ exam.

Thermostat Installations

Specific differences exist when performing various thermostat installations. These differences contrast most sharply between legacy and intelligent thermostat types.

Legacy Thermostats

Regardless of whether the installation involves a single zone or a complete zoning system, the important considerations listed here should be followed for locating and mounting legacy thermostats.

Do not install legacy thermostats in the following locations:

➤ In drafts, dead spots behind doors, or any corners where air simply does not move

➤ Where they will be directly impacted by hot or cold air flowing from the ducts

➤ Where they will be affected by radiant heat from the sun or from major appliances

➤ Near concealed pipes or in close proximity to chimneys

➤ In any area that does not receive any heated or cooled air behind it, such as an outside wall

Intelligent Thermostats

The guidelines for installing intelligent thermostats are similar to those associated with legacy models. In general, do not install intelligent thermostats where they will be affected by the following conditions:

➤ Drafts or dead spots

➤ Dead air behind doors or in corners

➤ Hot or cold air from ducts

➤ Radiant heat from sun or appliances

➤ Concealed pipes or chimneys

➤ Outside walls that create unheated (or uncooled) areas behind the thermostat

As with legacy models, most building codes call for intelligent thermostats to be placed 5 feet (60") above the floor (although some call for 4 feet). Also, thermostats should not be located in direct sunlight or where they cannot sense the temperature of return air (air entering the unit or system) from the conditioned space. Also, they should be located near the return air grill for the conditioned space.

Duct Installations

When installing a residential duct system, it is important to keep these points in mind:

- To minimize the number of dampers the system will require, dedicate branch runs for each zone.

- Ductwork should be designed to handle between 0.08" and 0.10" of water column (WC) of static pressure for each 100 feet of duct.

- To reduce the amount of air turbulence in the duct, do not mount dampers within 3 feet of the plenum.

- Because the dampers provide a direct interface with the system, always mount them in easily accessible locations for situations when servicing will be required at a later date.

- To be able to manually balance the system, be sure that manually balancing dampers are installed in each duct.

Water column (WC) is a common measure of air pressure used in HVAC systems. It is measured by test instrumentation when analyzing duct systems for proper airflow.

The method or system of distributing heating and cooling throughout the home is often called *residential thermal distribution*. Among the popular distribution systems are those that blow air through ducts, circulate heated water beneath floors or through hot water radiators, and use electricity through baseboard heaters. Among these popular systems, research has indicated that forced air systems are potentially the most wasteful. Due to the problems of air infiltration and external installation techniques, forced air systems have the potential for significant energy losses.

The Department of Energy (DOE) and the California Institute for Energy Efficiency (CIEE) have funded supporting research for this conclusion through the Lawrence Berkeley National Laboratory (LBNL). The findings have indicated that forced air distribution systems typically waste 25% of the energy used for heating and cooling through their ducts.

Specific Problems

One of the most widespread problems has to do with where the ducting is located. Often ducts are installed in locations largely outside of heated or cooled residential areas, such as basements, crawlspaces, garages, and attics. This also holds true for the location of the air conditioning coils and the furnace. These locations result in ducting that leaks air to and from the outside environment, as shown in Figure 6.19.

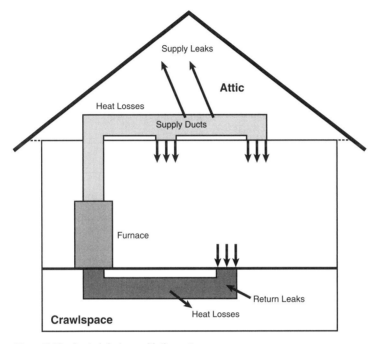

Figure 6.19 Duct air leakage with the system on.

In addition, some of the heat distributed through the system is lost through the duct walls by simple heat conduction. This is all conditioned air being lost to the outside instead of heating or cooling the residence. Losses due to air leakage and heat conduction cost the owner energy and money.

Of course, if a residence is using the basement as living space and locating all the ducts, the furnace, and the air conditioner in the basement, the air leaks or heat conduction losses go to the basement and are not necessarily lost to the outside environment. The energy and financial penalties caused by leaks and conduction losses would be negligible under such circumstances.

One additional consideration for a leaky ducting system is that the furnace or air conditioner is incapable of supplying sufficient heating or cooling to distant rooms. If the residents are uncomfortable, these losses hurt, even if they occur completely within the confines of occupied space.

Air leakage can also occur at points where the ducts connect at various points in the system, such as plenums, or behind registers. This is conditioned air that is not reaching the residents, even though it's being paid for. If additional air is leaking into a residential heating or cooling system, this only increases the amount of outside, unconditioned air that must be heated or cooled. Because the outside air is either cooler or warmer than the air inside the residence, the heating or cooling capacity of the system is wasted treating this air rather than the inside air. Under these conditions, turning off the system, as shown in Figure 6.20, does not provide a viable solution because the duct leakage increases the rate of ventilation. This, in turn, increases the heating or cooling demand.

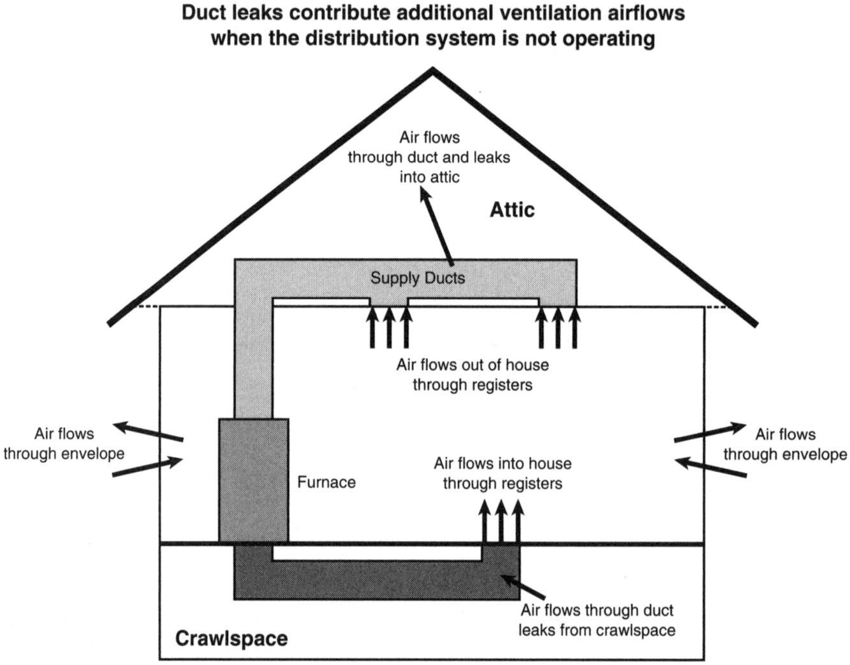

Duct leaks contribute additional ventilation airflows when the distribution system is not operating

Figure 6.20 Duct air leakage with the system off.

This type of leakage is much less likely from water- or electric-based systems, but the damage that might occur would be very expensive to repair. Building codes are extremely sensitive to leakage from water- and electric-based systems, and strict construction and installation controls are enforced.

Duct Locations

In most new and existing U.S. housing, the attic seems to be the most popular location for the HVAC duct system. Yet, the attic is absolutely the poorest choice for duct placement, especially when it comes to keeping a residence cool during the summer months. Beneath a sunny roof, it is not uncommon for attic temperatures to reach 150° F during summer afternoons.

The situation results in heated ducts and heated air inside them. This is particularly true for any air leaking into the ducts from the return side of the system. The air conditioning system might not be capable of cooling the house, and the air coming through the registers might actually be warmer than the house air, turning the air conditioner into a heater!

Building designers often ignore duct locations, leaving the decisions as to where to place them up to the installing contractor. A predictable outcome of such a situation is that the ducting system is difficult to service and maintain. If the homeowner attempts to reduce system losses through retrofit or repair, this problem is only heightened.

A typical view of such an attic is shown in Figure 6.21.

Figure 6.21 A maze of attic ducting.

This attic is filled with ducting that appears to be completely disorganized. The furnace is in there somewhere, but the homeowner might be hard pressed to find it.

Air Leakage

More important than conduction losses are air leakages from the ducting, which is usually constructed of sheet metal in older homes. Newly constructed homes often use flexible, plastic duct. Even though little or no air leaks out through the plastic or metal walls of flexible ducts, it often leaks out at the connection points. For example, substantial leakage can occur at the point where the duct connects to the furnace/air conditioner. Other trouble spots can occur at various branches in the duct system.

Troubleshooting and Maintenance Issues

During warm weather, most homeowners know enough about their HVAC equipment to switch the thermostat to the cool setting, set the fan to its automatic mode, and adjust the temperature setting low enough to operate the AC section. The majority of users also understand the importance of keeping the system filters clean. Beyond these basic measures, however, common residential HVAC problems can go unrecognized. This section lists those areas you will need to know concerning standard HVAC troubleshooting and maintenance issues.

Air Conditioning

Although residential air conditioning systems are complex and include many parts and variables, several suggestions can save the owner from having to place a service call. Of course, if a system is not cooling, qualified service is probably required. However, this does not prevent the owner from checking the obvious.

The air filter is the number-one reason for an HVAC system failing to provide proper heating and cooling. Filters should be checked every 1–3 months depending on the size of the home and climate conditions. As an HVAC technician, you should become familiar with the American Society of Heating and Air Conditioning Engineers organization. It has published standard 52.2-1999, which establishes a test procedure for evaluating the performance of air-cleaning devices as a function of particle size. It is the standard adopted by the industry for grading the effectiveness of various types of HVAC air filters.

If everything is running and the filter is clean, the unit might be low on refrigerant. A quick way to check for this possibility is to follow these steps:

1. Feel the larger of the two pipes going into the condenser.
2. Determine whether the pipe is cold. If it is not, the unit is low on refrigerant and requires professional service.
3. Make sure that all the vents are open. This seems obvious, but closed vents keep a warm room warm, even when the air conditioner is operating properly.

Another important inspection is to visually examine the evaporator coils. Dirty evaporator coils can act like a dirty filter, but they are more difficult to detect because the symptoms occur gradually over a period of time. Telltale indications are an air conditioning system that is not cooling as well as it did the previous year. If the electric bill is unexplainably higher or the unit continually freezes up under normal operating conditions, suspect dirty evaporator coils.

Air Handler Problems

If the air handler is not running, this usually means that the condenser is not running either. The user or HTI+ technician can perform the following checks:

1. Switch the system off at the thermostat.
2. Reset the breaker, or replace fuses if necessary.
3. Switch the system back on.
4. At this point, if the breaker trips or the fuse blows again, the owner should call a qualified service provider.
5. If both the breakers and fuses check out, proceed with additional troubleshooting steps.
6. Switch the system off.
7. Check the condensation drain section.

Heating

Home users do not enjoy calling for professional help every time a simple HVAC problem crops up. You should take some time to review common heating system problems with the owner and provide a list of things to check. Taking this simple approach can help to reduce the frequency of maintenance calls for a seemingly nonfunctioning heating system.

Thermostat

Thermostat problems can often be remedied by following these tips:

➤ Make sure the system selector switch is in the heat position.

➤ Set the temperature at least 3° above the current room temperature.

➤ Check the battery condition of an electronic thermostat.

➤ If the batteries are rechargeable, recharge them. Otherwise, replace them. The battery condition can usually be determined by carefully examining the display. Some displays even provide a battery low indicator.

➤ If the system uses a forced air furnace, check to see whether the thermostat has a furnace fan switch.

- Switch the fan to the on position. Normally, the fan switch is located at the bottom left of the thermostat and is marked on and auto.
- Listen to hear whether the furnace fan comes on.
- Check for any air flowing from one of the supply registers, if you cannot hear the fan.
- Consider the possibility that there might be no power to the furnace if the fan is not running.
- Check whether the thermostat is set properly,
- Check whether there is power to the furnace.

Exam Prep Questions

Question 1

> Which component of an HVAC system is used to move the heated or cooled air it receives from the furnace or air conditioner through the residential ductwork?
> - A. The condenser
> - B. The air handler
> - C. The thermostat
> - D. The evaporator

The correct answer is B, the air handler. The air-handling unit is composed of a cabinet that includes the central furnace, the air conditioner or heat pump, and the plenum/blower assembly. The air handler, commonly referred to as the blower, is used to move the heated or cooled air it receives from the furnace or air conditioner through the residential ductwork. Answer A is incorrect because the condensing unit, which is also known as an outdoor coil, is not used to move air. As an integral component of a heat pump or a central air conditioning system, it is located in an outdoor casing that is vented. It serves as the point where heat is transferred from inside the home to the outside air. Answer C is incorrect because a thermostat is used to measure the temperature of the interior of a home. It sends a signal to the furnace and air conditioning system when the temperature reaches a preset value. Answer D is also incorrect because the evaporator is that portion of a heat pump or central air conditioning system that is located inside the home. It is also known as an indoor coil and functions as the heat transfer point for warming or cooling indoor air. It is not used to move air in an HVAC system.

Question 2

> Which solution can the HVAC installer recommend for certain rooms that are habitually warm or cool, regardless of the settings at the thermostat?
> - A. Replace all the ductwork.
> - B. Clean the condenser coils.
> - C. Install a damper.
> - D. Replace the furnace.

The correct answer is C. A damper is used for directing air where needed and to block its path from areas where it is not needed. In HVAC systems, dampers control air temperatures within a specifically defined zone area. Answer A is incorrect because replacing all the ductwork is not a proper solution. Answer B is incorrect because, although cleaning the condenser coils can be a good maintenance procedure, it will not solve the problem associated with certain areas that are habitually warm or cold. Answer D is incorrect because replacing the furnace is a costly error that would not address the temperature differential problem.

Question 3

Which one of the following locations is a preferred location for mounting a thermostat?

- A. Near the return air grill
- B. Near the kitchen door
- C. 8" from the floor in a hallway
- D. Near the warm air duct

The correct answer is A, the return air grill. It is good practice to locate the thermostat near the return air grill because it has the opportunity to sense the temperature of the air being drawn back into the system. This avoids the problem of locating the thermostat near other sources that are incapable of monitoring the air temperature being returned into the heating system. Answer B is incorrect because the kitchen area can result in extra heat reaching the thermostat area, causing an imbalance in the temperature in other areas. Answer C is incorrect because thermostats should never be mounted close to the floor in any area. Answer D is incorrect because mounting the thermostat near the warm air ducts will establish a temperature range for the area that is less than the desired temperature range because the warm air will adversely affect the true temperature of the area being monitored.

Question 4

Which type of device is recommended for removing warm air from the attic during the high-temperature periods of the summer season?

- A. An air conditioner
- B. A heat pump
- C. A whole house fan
- D. A condenser coil

The correct answer is C, a whole house fan. Although air handlers can provide sophisticated solutions for keeping the air inside a residence at specifically selected temperatures, they are relatively expensive to operate when compared to whole house fans. A whole house fan is therefore the recommended solution. It helps pull unwanted heat out of the attic during evening hours, resulting in energy savings for the HVAC system. Answer A is incorrect because an air conditioner is used to remove warm air from the living area of a residence and does not remove warm air from the attic. Answer B is incorrect because a heat pump is a type of HVAC system designed to heat and cool the living space of a residence. Answer D is incorrect because a condenser coil is a component of a heat pump that is used to transfer heat from the interior living area to the outside atmosphere. It is not used to remove attic air.

Question 5

As a general rule, what is the relative savings in air conditioner energy consumption during the summer months when the thermostat is raised 1°?

- A. 15%
- B. 10%
- C. 20%
- D. 5%

The correct answer is D, 5%. It is generally assumed in the HVAC industry that during the summer months, every degree that the air conditioning thermostat is raised results in a 5% decrease in cooling energy consumption. Answers A, B, and C are incorrect because they are each in excess of the 5% figure accepted in the HVAC industry as the effective cost or savings for each degree of thermostat temperature setting.

Question 6

Which function is performed by an evaporator coil in an air conditioning system?

- A. It is located outside the home and removes cold air from the condenser coils.
- B. It is the portion of the system located outside the home and collects cooled air from the evaporator.
- C. It is that portion of the air-conditioning system located inside the home and functions as the heat transfer point for cooling indoor air.
- D. It is located inside the home and provides for the transfer of humid air from the outside to the internal rooms in the home.

The correct answer is C. An evaporator coil is that portion of a heat pump or central air conditioning system that is located inside the home. It is also known as an indoor coil and functions as the heat transfer point for warming or cooling indoor air. It consists of a series of pipes connected to a furnace or air handler, which blows the indoor air across the evaporator coil, causing the coil to absorb heat from the air. The cooled air is then delivered to the house through the ducting. The refrigerant in the coil then flows back to the compressor and the cycle starts again. Answer A is incorrect because the evaporator coil is located inside the home, not on the outside area. Answer B is incorrect for the same reason. The evaporator coil is not located on the outside of the residence. Answer D is incorrect because, although the evaporator coil is located on the inside, it does not transfer humid air from the outside area to the inside area but performs the opposite transfer of humid air from the interior to the outside air.

Question 7

Which type of thermostat typically provides an automatic changeover feature from a heating thermostat to a cooling thermostat without human intervention?
- A. A staging thermostat
- B. An electronic sensor thermostat
- C. A bimetal thermostat
- D. A cooling thermostat

The correct answer is A, a staging thermostat. Staging thermostats typically provide an automatic changeover feature that enables the unit to change from a heating thermostat to a cooling thermostat without human intervention. The homeowner can simply set the desired heating and cooling temperatures for the residence, or zone, and the system should maintain the residence at those points regardless of the season. Answer B is incorrect because an electronic sensor thermostat does not necessarily have both cooling and heating season programmable settings, as a staging thermostat does. Answer C is incorrect because a bimetal thermostat is a mechanical temperature sensing unit that has no seasonal programming capability. Answer D is incorrect because a cooling thermostat is designed only for operation with an air conditioner and has no capability to monitor or control heating units.

Question 8

Which type of HVAC component is used by a heat pump to redirect the refrigerant flow?

- ○ A. A condenser switch
- ○ B. A collector valve
- ○ C. A thermostat cooling switch
- ○ D. A reversing valve

The correct answer is D, a reversing valve. Heat pumps use a reversing valve to redirect the refrigerant flow. It enables the inside coil and outside coil to act as either an evaporator or a condenser. Answer A is incorrect because a condenser does not have a component called a switch. Answer B is incorrect because a collector is a device associated with solar power heating systems and has no function in a heat pump HVAC system. Answer C is incorrect because a thermostat cooling switch is a component of a programmable thermostat and is not used to redirect refrigerant flow.

Question 9

What is the recommended interval for checking the filters on an HVAC system?

- ○ A. Annually
- ○ B. Semiannually
- ○ C. 1–3 months
- ○ D. Every 2 years

The correct answer is C. Filters should be checked every 1–3 month to determine whether they require cleaning or replacement, depending on the climate and size of the home. Answer A is incorrect because filters need to be inspected more often than once per year. Answer B is incorrect because filters should be examined more often than semiannually. Also D is incorrect because filters need to be checked more often than every 2 years.

Question 10

Which switch position on a staging thermostat enables the fan in an HVAC system to run continuously?

- ○ A. Auto
- ○ B. On or auto
- ○ C. Off
- ○ D. On

The correct answer is D, the on position. A staging thermostat is a programmable all-season thermostat that provides an automatic changeover feature that enables the unit to change from a heating thermostat to a cooling thermostat. A fan switch permits the user to circulate air throughout the system without including the heating or cooling units. When this switch is set in the auto position, the fan runs in conjunction with the heating and cooling systems. However, when the switch is set to the on position, the fan runs continuously, without respect to whether the heating or cooling functions are running. Answer A is incorrect because the auto position on the switch operates the fan only when the thermostat calls for cooling or heating. Answer B is incorrect because, although the correct position for the fan to operate continuously is the on position, the second part of the answer is incorrect. In the auto position, the fan operates only in conjunction with the heating and cooling cycles. Answer C is incorrect because the off position disables all heating and cooling functions as well as the fan.

Need to Know More?

 Breath, Newton and Peter Scott Curtis. *HVAC Instant Answers.* Berkeley, CA: McGraw-Hill Professional, 2002.

 http://www.aexusa.com/pspecah298.htm

 http://www.marinespecialists.com/marineairsystems/products/Blow_Thru_Air_Handlers.html

 http://www.rewci.com/wholehousefans.html

 http://www.espenergy.com/2_speed_professional.htm

 http://www.bsdsolutions.com/hvac/hvacfun.htm

 http://hem.dis.anl.gov/eehem/01/010903.html

 http://www.heil-hvac.com/henergy.html

 http://www.bakerco.com/faq/2.7.htm

 http://www.2.cs.cmu.edu/Groups/AI/html/faqs/ai/fuzzy/part1/faq.html

7.0—Home Water Systems Controls and Management

Terms you'll need to understand:

- ✓ ORP/pH
- ✓ Sump pumps
- ✓ Rotor sprinkler heads
- ✓ Impact rotors
- ✓ Gear-driven rotors
- ✓ Pressurized and nonpressurized pipes
- ✓ Seasonal presets
- ✓ Backflow prevention valve
- ✓ Zoned systems
- ✓ Relays
- ✓ VA rating
- ✓ Still well
- ✓ Low-voltage limited energy circuits

Techniques you'll need to master:

- ✓ Determining the basic requirements for a residential irrigation system
- ✓ Listing the basic features of a programmable water system controller
- ✓ Making a list of the common problems encountered with residential watering systems
- ✓ Listing the features of three common types of sprinkler heads
- ✓ Listing four main advantages for the use of a water system remote controller

Home water system controls and management systems have become necessary additions to a modern automated home. Irrigation systems, pools, spas, and sump pumps have given rise to the need for automated home management systems that can function essentially without human intervention after the initial programming has been performed. Zone programming for electronic irrigation systems can tailor the watering cycle to the different needs of the many varieties of plants and lawns. Irrigation system programming and management systems can also be designed to adapt to local mandates for watering lawns when periods of drought occur.

This chapter has been developed to address the automated water system management tools and components that you will need to know more about to pass the HTI+ exam. Each section in the chapter addresses design, installation, location, component connectivity, and installation tips for water controls, sump pumps, irrigation system, and pool water management devices. You will need to concentrate on the terms and standards related to water management systems discussed in this chapter with special emphasis on exam alerts and practice questions.

Design Considerations

The design of a home and garden water system can include devices to automatically control the times and intervals for watering lawns, shrubs, and flowering plants. Systems can also distribute water in different zones that have different watering requirements. In addition, design requirements must be considered for measuring the quality of the water to ensure safe and proper operation of pools and spas.

You must consider the advantages for including a remote access capability for a water system. Most irrigation system contractors require users to include wireless device control of the system for ease of maintenance and troubleshooting, as explained later in this section.

Timed Systems

Irrigation timers are simple controllers that include a clock unit capable of activating one or more valves of the irrigation system at specified times. Several designs are commercially available with many features and a wide range of costs. Timed systems are used primarily for irrigation management and pool and spa operation.

Irrigation Timers

Users residing in areas of the country where municipal water use is strictly regulated can rely on timed systems to ensure strict adherence to local regulations, especially during the hours between midnight and 6 a.m. Some communities even offer water bill credits if timed systems are equipped with rain sensors, such as the one pictured in Figure 7.1, which keep irrigation systems from switching on during periods of precipitation.

Figure 7.1 A rain sensor.

When setting up automated home water control systems designed for irrigation purposes, questions as to how much to water and when to water must be answered. In areas of the country where the shortage of water is a major concern, overwatering can result in fines. On the other hand, the opposite extreme of underwatering can lead to stunted plants, damaged lawns, and loss of sensitive plants.

Pool, Spa, and Fountain Timers

Timers help to automate several important equipment functions required to keep pools, spas, and fountains in proper condition. They turn the heaters on and off as required to keep the water temperatures comfortable and suitable for normal operations. They can also help to automate the activities of the filter pumps and the filter backwashing system.

Chemical dispensers have become somewhat automated using timers to add chlorine or operate ORP/pH controllers.

NOTE Oxidation reduction potential (ORP) is a measure of the pool water's overall ability to eliminate wastes. High oxidation is present when pollution is low and the water is of high quality. ORP is rapidly becoming the standard means of testing and regulating pool water sanitation. pH is a measurement of the acidity or alkalinity of a liquid. The pH scale ranges from 0 to 14 with a water value of 7 being neutral. Hydrochloric acid has a value close to 0, and potassium hydroxide has a value of 14.

Zoned Systems

Larger irrigation systems typically use a central irrigation controller located in a garage or utility room that operates a set of distributed solenoid valves positioned throughout the area to be irrigated. The controller avoids the need for manual operation for large systems. The controller connects to each solenoid valve with underground low-voltage wiring, and the electrically operated solenoid valves control the flow of water to the sprinkler heads installed in each zone.

Zoned water management systems provide a degree of automation needed to ensure the correct amount of irrigation of selected areas. Irrigation systems should be zoned so that plants with different water needs are irrigated separately. Lawns, for example, might have watering needs different from shrubs and flowers.

Remote Access

Remote access control of a water system is often necessary for a variety of reasons, as follows:

➤ You need to operate all your valves while at the sprinkler head during a test.

➤ You are spending time at the controller manually turning stations on and off.

➤ You want to avoid the added cost of a programmable central controller.

➤ You are an irrigation system contractor and install commercial or residential irrigation systems.

NOTE Water system remote control units are usually manufactured to operate with a type of controller made by the same company. Universal wireless remote controllers are also available for controlling X10 home automation devices.

Equipment Locations

Water control and management systems require control valves to be located in the area where the sprinkler heads can be connected with moderately short runs of irrigation pipes. Backflow and double-check valves are also required for isolating the irrigation system from the main municipal water supply. The location and operation of other water system components such as relays, heaters, pumps, and solenoids are also discussed in the following section.

Irrigation System Control Valve Locations

The control valve shown in Figure 7.2 is designed to control an individual circuit of a home irrigation system.

Figure 7.2   An irrigation control valve.

For moderately large landscaping requirements, the normal approach is to locate separate valve housings for front yard and backyard irrigation, preferably with their covers situated at ground level to avoid becoming a trip hazard, as shown in Figure 7.3.

Figure 7.3   Ground-level valve housing.

Backflow Prevention Valve Locations

Most city and state plumbing codes require the installation of some type of backflow prevention device wherever a permanent irrigation system is connected to a municipal water supply. This device is designed to prevent contaminated water from the irrigation system from getting back into the main water supply.

Most codes require, as a minimum, a double-check valve device be installed at the point of connection. Where a high degree of hazard exists, such as when fertilizers or chemicals are injected into the irrigation water, some codes require a more sophisticated device called a reduced pressure backflow preventer. These devices are expensive but, fortunately, are usually not required for home irrigation systems because fertilizers are not injected.

Keypads

Keypads are used as an integral part of a water management controller to enter commands, set the clock, set parameters for each zone, and manage the overall programmable features of an automated water system. Handheld remote control keypads use wireless transmitters to send commands to the water management controller. Wireless remote keypads are very useful when adjusting sprinkler heads because they allow the same level of control from a remote location that is available at the controller box keypad.

Relays

The magnetic pull of a solenoid coil can be used to open or close an electric switch. Conversely, an electric switch can be used to activate a solenoid coil. Either configuration is called a *relay*. A relay, then, is an electrically operated switch and can be located to isolate one electrical circuit from another. For irrigation purposes, a relay serves to isolate one water distribution circuit from another.

Heaters

Keeping the water in pools and spas properly heated is critical to the user's enjoyment. Proper water temperatures for a swimming pool usually depend on the user's activity in the pool and personal taste.

Water temperatures for spas are typically kept much warmer than those for swimming pools. Before entering a spa, the user should measure the water temperature with an accurate thermometer because the temperature-regulating devices supplied by spa manufacturers vary considerably.

Water temperatures between 100° F (38° C) and 104° F (40° C) are considered safe for a healthy adult, but under no circumstances should the water temperature be permitted to exceed 104° F (40° C). In fact, lower water temperatures are recommended for young children or when spa use lasts longer than 20 minutes.

Heaters can be fueled by various means such as gas, propane, electric, and solar panels.

 The safe maximum temperature range for spas is between 100° F (38° C) and 104° F (40° C). This is the range for adults and should be lower for children.

Pumps

Home water management systems require the installation and use of various types of water pumps. Despite the fact that circuits are being controlled, irrigation systems usually need to boost the house pressure so sprinkler systems can operate properly. Pools and spas use specialized pumping systems to obtain the required flow, filtering, and circulation patterns; fountains and waterfalls use their own types of pumping techniques as well. The location of this equipment varies according to the duties being performed.

Irrigation Pumps

Most home sprinkler systems use the available pressure from the main utility water supply or, in locations near large rivers or lakes, the local irrigation service supply line. With the use of pressure regulators, and the limiting of one active circuit at a time, this is usually sufficient.

If the residence is fairly large, with lawn and garden development spread over a wide area, the sprinkler irrigation system can require the assistance of a booster water pump. In certain higher ground areas supplied by a local water company, the pressure supplied might not be enough because of the higher expenses incurred to build additional pumping stations.

The user, then, is left no choice but to install an irrigation booster pump, such as the one shown in Figure 7.4, to raise the water pressure supplied by the local water utility company.

Figure 7.4 A sprinkler pump.

The pump should be located at the head of the irrigation supply line, before the filters, timers, and manifolds. Pumps running from the municipal water system should be located as close as possible to the water meter or the mainline before it goes into the home.

Pool and Spa Pumps

Pumps designed for pools and spas are usually custom pumps. They fall into two general categories called in-ground and above-ground pumps. *In-ground* pool pumps, such as the one shown in Figure 7.5, are self-priming and are usually mounted below the water level.

Figure 7.5 An in-ground pool pump.

Solenoids

Solenoids are electromechanical devices that are an integral part of an electrically operated water valve or control relay switch. Electrical current is supplied to the solenoid coil and the resulting magnetic field acts on the plunger, whose resulting motion actuates the valve or relay switch. Solenoids are located in automatic irrigation system valves, pool and spa control circuits, and sump pump controllers. They are particularly useful in isolating the controller low-voltage circuits from the device to be controlled, such as a pump motor that uses a 120v or 240v AC power source.

Physical Products

The physical products associated with home water system controls and management are outlined in the following section. You will need to become familiar with each of the terms used in this section.

Water Alarms

Water alarms are used in any location in the home where leaks or intrusion of rainwater can cause damage if undetected. A basement or utility room area can be monitored for leaks with sensors capable of detecting water levels of 1/32". Wireless water alarms are easily installed in any location. Using battery-powered sensors and transmitters, the unit can send warning signals to a receiving unit equipped with a loud audible alarm or a telephone dialer.

Irrigation Controllers

Controllers designed for home water system management irrigation are available in a variety of designs and configurations. Controller types include electromechanical, electronic, and hybrid designs.

Electromechanical Controllers

Electromechanical controllers use clocks for controlling an irrigation program by using gears, dials, and pins. They have few sophisticated electronic components and are driven by electric motors and gears. Turning the dials or setting switches programs the controller to select watering times, dates, and duration of the watering cycle.

Electromechanical controllers usually have three separate dials, as shown in Figure 7.6. They enable the control of three separate functions mentioned earlier: start time, watering days, and individual zone run times.

Figure 7.6 An electromechanical irrigation controller.

Programming an electromechanical controller consists of pushing certain pins in for on or pulling them out for off. The spinning dials cause the pins that have been pushed in to contact specific terminals, completing an electrical circuit. Although electromechanical controllers are reliable and easy to use, they offer fewer features than electronic or hybrid controllers.

Electronic and Hybrid Controllers

All electronic controllers offer more features than electromechanical controllers. Programming a solid-state controller usually requires entering commands through a keypad, and the results of the commands are visible on a display, as shown in Figure 7.7.

Figure 7.7 Electronic irrigation system controller.

Hybrid controllers combine the ease of use of the electromechanical controllers with the versatility of the solid-state controllers. Hybrid controllers have sophisticated electronic circuitry combined with an easy-to-use panel that features dials and switches, such as those shown in Figure 7.8. A display guides the user through programming and displays information about the watering cycle.

Figure 7.8 A hybrid irrigation controller.

Control Points

Control points are those locations where keypads, sensors, timers, and alarms are used to activate watering systems. For example, water alarms are used to activate sump pumps when rising water levels are detected, whereas keypads are often included with an irrigation system controller for entering manual commands and to program automatic control parameters. Timers are used to control the on and off times of pool and Jacuzzi heaters and irrigation systems. Sensors also are an integral part of water management control points. They provide feedback to controllers and alarms to activate an audible warning when the sensor contacts are bridged by water. They can also serve as the control point to activate sump pumps.

Distribution Panels

Although a majority of manufactured distribution panels are designed for structured wiring systems to route telephone, data, and coax signals, some types also provide additional options for security systems, water control and management systems, audio distribution, satellite distribution, and home office computer networking.

An example of a distribution panel that provides a full range of control and utility options is shown in Figure 7.9. By using the plug-and-play feature, the panel provides the capability to upgrade to the various services as they are added to an automated home design.

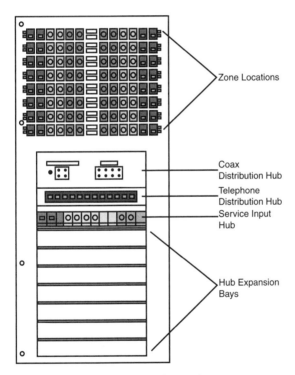

Figure 7.9 An expandable distribution panel.

Interface Locations

Water system interface locations are those areas where the installers, maintenance personnel, and home users have access to the controllers. Irrigation controllers are usually located on outside walls or garage areas to provide access to the low-voltage wiring that is run underground to each control valve.

Keypads can be located inside the home to provide easy access to manually control the watering system. Jacuzzi and pool filtration system interface controls and on/off switches are typically located outside in special all-weather waterproof outlet boxes.

Keypads

An important function in any home water system control and management system is the capability to remotely control any water system function via a keypad interface. The user might want to operate or disable the sprinklers at any time and from any location. This is particularly important during seasonal maintenance or when troubleshooting the sprinkler heads or valves.

The advanced X10 programmable keypad transmitter shown in Figure 7.10 provides eight buttons, which the user can individually configure to transmit any X10 code.

Figure 7.10 An X10 programmable keypad transmitter.

An X10 irrigation system can have different letter codes for each button, although an X10 home automation system specifies that water controlling equipment should be assigned house codes that begin with the letter *O* for optional equipment such as sprinklers, irrigation devices, fountains, birdbaths, and outside power outlets. Individual buttons can be configured to store and transmit up to 16 address/commands, giving it a possible total of 128. In addition, each button can be assigned different command codes and be configured to transmit standard code, extended code, or both.

Power Supplies

Most of the equipment associated with pools and spas runs from the 120v or 240v AC power outlets already supplied to households by the public utility companies. This includes circulation motors, filter pumps, underwater and surface area lights, motorized pool and spa covers, and heaters.

Home irrigation systems usually operate the external circuits to the in-ground valves using 24v DC. The lower voltage is produced from a transformer using the 120v primary power source in a power supply contained as an integral part of the main irrigation controller/timer. The lower voltage makes the installation easier. Because the voltage does not provide a hazard, the control wiring from the controller to the in-ground valves can be easily buried along with the PVC pipe routed to each control valve.

Sensors

Water system sensors provide the feedback to a central controller mentioned earlier to activate pumps and alarms or to override automatic irrigation programs. Rain sensors are used to temporarily suspend an automatic watering program when rainfall is detected by an external sensor. These units are mounted in a location exposed to normal rainfall but outside the watering spray of the sprinkler system. Various designs are available, but most have settings that allow some sort of measuring to take place. Rain causes the system to remain off during or after an event if sufficient rainfall is measured. The settings can be adjusted so that a light shower will not affect the system operation or eliminate a scheduled watering when rainfall is not sufficient to make up for a normal application.

Flood detector sensors are used in basements, utility rooms, or garage areas to monitor for water that might enter the area unexpectedly. These sensors are designed to check for emergencies such as overflowing washing machines, toilets, hot water heaters, aquariums, and air conditioner drain pipes and miscellaneous plumbing problems.

Sump Pumps

Sump pumps are designed to remove water from basements or potential flood areas. In addition, they are designed to operate in harsh environments and are usually installed in the deepest part of the basement. They also serve to eliminate water build-up in the humid soil around the foundation. If the basement begins to become damp because of excess water in and around the foundation or if water actually finds its way to the basement floor, the sump pump activates and begins pumping the water out of the basement and the ground below it into a runoff pipe.

Two types of sump pump designs are available. Their differences bear directly on where the pump is located—either in or out of the sump pit. The pedestal sump pump, shown in Figure 7.11, must be located so that the motor is placed above the water level. Notice how the motor is located at the top of the pedestal to protect it from water contamination.

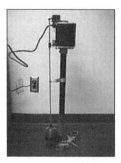

Figure 7.11 A pedestal sump pump.

The submersible sump pump can be located entirely below the water level in the sump hole without damage to the motor. An example of this type of sump pump is shown in Figure 7.12.

Figure 7.12 A submersible sump pump.

Sump Pump Sensors

Sump pumps operate using a simple float switch as a sensor that turns on the pump after the water raises the lever ball to a predetermined height. As shown in Figure 7.13, the switch then turns off the pump when the pump has removed enough water to sufficiently lower the float switch. The switch must be turned off before all the water is pumped out of the sump because serious damage will occur to the pump otherwise.

When installed in the sump, the float must be located so it is not hindered in its movement either up or down. This includes ensuring that no debris is allowed to enter the sump chamber. This system works well but does have one major drawback: During a severe emergency where electricity is cut off, it might not have the power to operate! A battery-operated backup sump can offer an additional level of protection.

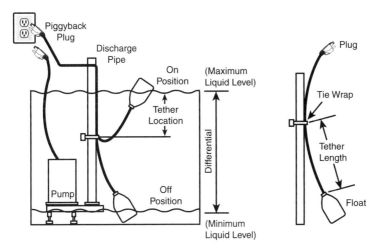

Figure 7.13 The operation of a sump pump.

If water is rushing in, the start and stop liquid level sensors need to be separated by a sufficient distance to keep the pump motor running long enough to dissipate the bubble head generated by the inrush current at startup. This can require continuous operation for 2–10 minutes or longer. You need to be aware of the details on the specific motor being used in the sump.

 Know where to locate the liquid sensors that are part of a protective sump pump system. You also need to remember the operational feature of sump pumps regarding the start and stop liquid level sensors mentioned in this section.

The liquid level sensors should be located in a calm area of the sump. If this is not possible, you might need to use a still well. A *still well* is a smaller well or pipe by the side of the sump chamber that is connected to the sump chamber by a small pipe or opening near the bottom. The liquid rises in this well to the same height as in the sump chamber and is practically undisturbed by the sump currents. In cases where the water in the sump is habitually agitated, you need to locate the level sensors in the still well.

You should always ensure that the pump stop level sensor is positioned at a high enough level to maintain sufficient submergence of the pump suction connection.

Solenoids

As discussed earlier, a solenoid is an electromechanical assembly that consists of a wire coil and a movable plunger that seats itself against the coil. When current is applied to the coil, an actuating magnetic field is created. Solenoids are frequently used as switches or controls for mechanical devices such as watering system valves.

Solenoid valves are electromechanical devices that use a solenoid to control valve actuation. Electrical current is supplied to the solenoid coil, and the resulting magnetic field acts on the plunger, whose resulting motion actuates the valve. Standard models are available in both AC and DC voltages.

Solenoid valves are useful in remote areas, rugged environments, and hazardous locations because they can be operated automatically.

Solenoids designed for irrigation work are designed as either 24v AC continuous solenoids, as shown in the Figure 7.14, or 12v/24v DC latching solenoids. A continuous solenoid must have current continually flowing through it to remain open. A latching type of solenoid remains in the latched state after the supply voltage is turned off.

Figure 7.14 Continuous solenoid.

Continuous solenoids are widely used in the design of electrically operated irrigation valves. When current is applied, the solenoid holds the valve in the open position; then, when voltage is removed, a spring plunger forces the valve into the closed position.

Sprinklers

Sprinklers are used in irrigation systems to disperse the flow of water over a designated area when water pressure is present in the pipe connected to the sprinkler. Several types of sprinkler heads for different locations are explained in the following sections.

Pop-up Sprinkler Head Features

Pop-up sprinkler heads are the most popular type of sprinkler for home lawn and garden irrigation systems. They are installed below the ground, and the sprinkler head remains out of sight while inactive. When the sprinkler system is turned on, a small portion of the head emerges above the surface to disperse water to the irrigation area. Spray-head sprinkler pop-ups are designed to spray a small fixed area, whereas rotor types are used for large, unobstructed areas.

Rotor Sprinkler Head Features

Rotor sprinkler heads are a pop-up type designed to disperse water over a large, circular area. Small rotors can cover radii of up to 50 feet, and large rotors can cover radii up to 100 feet. The two types of rotor sprinklers are called impact and gear driven. They differ only in the mechanism used to move the head in a circular motion.

Impact rotors, shown in the Figure 7.15, move in a circular pattern and slowly water the entire area within that circle.

Figure 7.15   Impact rotor.

Impact rotors are also available as part-circle sprinklers that can be used in corners or along walks or streets. The impact rotor is typically cheaper than a gear-driven rotor and provides the most uniform coverage of all sprinklers. However, the largest problem with impact sprinklers is their high maintenance requirement. As an impact sprinkler is activated, it rises out of its assembly to approximately 4" above the turf. During the time the sprinkler is in operation, this open cavity in the sprinkler case becomes an open catch for trash, mud, clippings, insects, and all types of yard debris. This debris is washed into the mechanism during the normal operation of the head, so periodic maintenance is required to keep sprinkler canisters clean and keep dirt and debris from causing damage to the mechanism.

Gear-driven rotors use the water pressure to power a small turbine (water wheel or fan) in the base of the unit, which drives a series of gears that cause the head to rotate. The gear drive mechanism is sealed from dirt and debris and operates with less noise than the older impact sprinkler heads. A typical gear-driven rotor is shown in Figure 7.16.

Figure 7.16   Gear-driven rotor.

Spray Sprinkler Heads

Pop-up spray heads, shown in Figure 7.17, spray water in specific circular patterns and can be easily changed to fit specific areas.

Figure 7.17   Pop-up spray head.

Spacing between sprinklers varies depending on the nozzle installed in the head. To operate efficiently, spray heads should not be spaced farther than 15 feet apart and should be supplied with 20–30 PSI of water pressure. Ideal for smaller, fragmented, hard-to-reach areas, these heads discharge 2–3 times the water of a rotor sprinkler head.

 Remember the preceding design features of sprinkler heads. The most popular type of pop-up sprinkler head is the gear-driven rotor because of its improved reliability over the older impact rotor. Spray heads deliver a lot of water and, although individual gear-driven rotor units are several times more costly than spray heads, their wider spacing capabilities means fewer heads are needed to cover a given area.

Standard Configuration and Settings

Proper watering system management requires separate configurations and settings for the controllers for daily, weekly, and seasonal watering needs. Each watering system has unique requirements based on the particular area and type of garden landscape to be irrigated. This section summarizes each of the seasonal as well as daily watering requirement.

Zone Programming

Irrigation systems for lawns and gardens require careful planning and consideration for the various needs of each landscaped area. Zone programming is the accepted method for configuring an irrigation system to meet the needs of different bed plantings, lawns, shrubs, and trees.

Watering zones can be planned by using a grid sheet similar to the type of plan used by a professional landscape company. You just add permanent features and structures such as driveways, sidewalks, fences, garages, pools, patios, lawns, trees, flowerbeds, gardens, and shrubs.

Next, you draw in the positions of the irrigation system supply lines, main valves, and tributary supply lines from the valves to each sprinkler head.

When all the areas have been identified according to their watering requirements, the lawn and garden areas can be divided into separate zones according to their needs. Try keeping the zones as rectangular as possible to make them easier to work with, as shown in Figure 7.18.

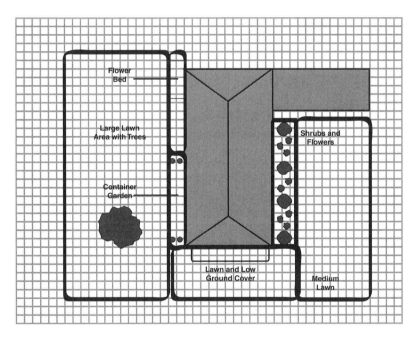

Figure 7.18 A division of watering requirements.

Time-of-Day Programming

Irrigation system controllers provide the capability to program the days of the week and start times for the system to begin a sprinkling cycle. The controller accepts instructions for the length of time for each zone to run. Each brand or model of controller might ask for the same basic information to be entered; however, the steps and methods vary from unit to unit. Most controllers accept from four to six different programs to allow for variations during drought or unusually wet trends.

Most landscape professionals advise that early morning is the best time to water with a typical daily cycle of three days per week. Water left on plant leaves or turf overnight can invite fungus problems, whereas watering during the daylight hours results in loss of water due to evaporation. Early morning watering ensures that the sun has ample time to dry out the surface of plants and soil to reduce disease problems. Ideally, the last zone of the sprinkler system should go off at dawn.

Some controllers have a water budget setting that allows an incremental reduction of the zone watering times. The reduction can be set temporarily for cool or cloudy days.

Seasonal Presets

Seasonal presets are used to adjust the watering cycles for each of the seasons when sunrise and sunset times vary. For example, watering times for early spring might not be appropriate for mid-summer days when the sun rises much earlier in the day. As indicated earlier, most controllers can store a minimum of four programs to allow seasonal programming to match the four seasonal watering needs.

In colder climates, most systems are turned off and the pipes are drained to keep them from freezing. If the drain valve is below the level of the underground pipes, a gravity drain will suffice. Irrigation system plumbing that is not level or drain valves that are not the lowest point in the system require the pipes to be drained with compressed air to remove all the standing water prior to the winter season shutdown. When the system must be temporarily rendered inactive, the controller retains the same programs for operating the system. When the controller is restored to the active mode after a winter shutdown, the seasonal programs are again operational.

Device Connectivity

Water irrigation and management systems depend on the proper cabling to connect components such as the timer/controller, valves, and keypads. Although local codes are the final authority for water system connectivity for both cabling and plumbing, some national published standards include both communications and low-voltage wiring guidelines discussed in the following paragraphs.

Water System Communications Cable

Communications cable for all new construction home wiring has been defined in recent published cabling standards. On December 13, 1999, the Category 5E Unshielded Twisted Pair (UTP) specification

(ANSI/TIA/EIA-568A-addendum 5) was officially ratified as a standard for residential communications wiring. This specification provides extended parameters for Category 5 wiring in home structured wiring and used for automated control of all internal residential telecommunications devices, including X10 water management systems.

Communications cable standards do not apply to the low-voltage structured wiring used to connect underground valves to the controller. This type of wiring is not required to meet Category 5E or any other Category X specification. Valve wiring is dictated by the valve manufacturer and the distance from the controller to the valve.

Category 3 UTP cable is now required by an FCC federal mandate to be the minimum category of UTP cable used for all residential structured wiring. The FCC ruling number 99-405 became effective July 8, 2000. The ruling states in part that, "for new installations and modifications to existing installations, copper conductors shall be, at a minimum, solid, 24-gauge or larger, twisted pairs that comply with the electrical specifications for Category 3, as defined in the ANSI EIA/TIA Building Wiring Standards."

Since the publication date of the ruling, the superior grade Category 5E cable has largely replaced Category 3 cable for new residential construction at approximately the same cost. The ruling is not yet well known by building contractors or home technology installers; however, it is a trend that all home technology integrators should be aware of for new water system installations and residential structured low-voltage wiring. It is no longer acceptable to use non-category-rated wiring for low-voltage communications cabling in residential wiring. Low-voltage structured wiring standards for communications cable are covered in more detail in Chapter 10, "10.0—Structured Wiring—Low-voltage."

Water System Low-Voltage Wiring

Home water systems use low-voltage cable to connect the controller to the various valves in each zone. If the length of the wire between the controller and the valve is less than 200 feet, #18-gauge wire is acceptable. This is the size of wire provided in most multiwire (or multiconductor) irrigation cable. Limited-energy, low-voltage circuits permit relatively inexpensive wire to be buried in the ground with the water PVC supply line.

Low-voltage wiring is a generic term used to describe all residential cabling that is not associated with the 240v/120v high-voltage alternating current (AC) primary power wiring used in all U.S. residential construction. Although no exact definition of low voltage exists in residential cabling standards, it is generally accepted in the industry to include those circuits that are connected to low-energy current limited sources described in Article 725 of the National Electrical Code (NEC).

As an example of limited-energy water system circuits, Article 725-2 of the NEC provides definitions of Class 1, Class 2, and Class 3 circuits. They are differentiated from each other by power limitations and maximum current permitted. Class 1 circuits might or might not be supplied by power-limited sources, whereas Class 2 circuits are limited to voltage and current values that do not usually present a shock or fire hazard. Finally, Class 3 circuits are allowed to have higher-voltage and current values than Class 2 circuits. The voltage and current levels of Class 3 circuits can present a shock hazard but generally do not present a fire hazard.

Water system low-voltage circuits are considered to be Class 2 circuits and generally operate at 30v or less at 100 VA. Class 2 power limitations with maximum ampere ratings are as follows:

➤ 0v–20v at 100 VA (5 amperes maximum)

➤ 21v–30v at 100 VA (3.3 amperes maximum)

➤ 31v–150v at 0.5 VA (5 milliamperes maximum)

Most home water management systems use 24v DC at less than 100 VA for valve control.

You will need to know that limited-energy circuits as defined in NEC Article 752-2 are classified according to the volt ampere (VA) rating. Volt amperes is the product of the applied voltage times the maximum current supplied by the load. As an example, an irrigation valve solenoid might require 24v DC at 3 amperes, which is equal to $(24 \times 3) = 72$ VA.

Termination Points

Termination points for home watering systems and controls include the irrigation controller wire connection block, the control valves, keypads, and wireless handheld remote controllers.

Industry Standards

The most important standard you will need to know about when preparing for the Infrastructure and Integration examination is ANSI/TIA/EIA 570A and its three addendums. It sets guidelines for residential single-family and multifamily telecommunications equipment and wiring. Among other aspects, the standard covers specifications for the network interface and demarcation point, auxiliary disconnects, a distribution device and cross-connects, cables and cable installation, outlets and connectors, and equipment cords.

Local Irrigation and Water Conservation Standards

With the number of residential irrigation systems on the rise, particularly in the warmer climates of the southern areas of the United States, local municipal governments are establishing water conservation laws. During severe drought periods, residential irrigation water use can cause dangerously low levels in local water utility capacity.

In some communities, a permit must be issued subject to a detailed plan for any irrigation system prior to the installation. The plans often require detailed specifications for all components and an estimate of water usage. Final inspections are usually required prior to the system being approved for use by the homeowner.

Installation Plans and Procedures

Installing a residential water system requires several key planning steps. Although local irrigation system codes vary, the following are typical of most installation plans and guidelines:

➤ *Install the control wire for the irrigation system valves while the trenches are open*—The wire should be at least 2" away from the pipe and either next to or under the pipe. Never place the wire above the pipe.

➤ *The preferred location for the control wire is 2"–4" below the pipe*—This way, the pipe protects the wire from damage and you are less likely to accidentally cut the wires when making repairs to the pipe. If you are using numerous wires rather than a multiwire cable, handling the wires is easier if you tape them together into a bundle approximately every 10 feet.

➤ *The size of each wire is determined by a chart provided by the valve manufacturer*—Every valve model is different, but we can make some generalizations. For residential systems in which the wire length between controller and valve (not the distance from controller to valve, but the length of the wire between the controller and the valve) is less than 200 feet, #18-gauge wire is recommended.

➤ *Schedule 40 PVC pipe is typically required by most local codes for underground installation*—All constant pressurized pipes should be buried at least 18" down (and 12" for nonpressurized lines).

- *Pop-up sprinklers in turf areas should have a minimum 4" pop-up height.*

- *Some local municipal codes require a detailed landscape plan to be submitted with the application for a building permit*—The plan must show the areas to be watered and the botanical names for all plants included in the landscape plan.

> Constant pressure irrigation pipes are those pipes that are connected to the main water supply prior to the control valves. The so-called (but somewhat misnamed) nonpressurized lines are the supply pipes between the control valves and the sprinkler heads. These lines are pressurized only when the control valves reopen.

Water System Troubleshooting and Maintenance Issues

The following are common maintenance and troubleshooting issues for residential watering systems:

- *The three most common physical problems in an irrigation system are broken components*—This includes risers, improperly designed or spaced heads, and dissimilar heads or nozzles.

- *Check the rotation and direction of the spray*—Adjust the radius and arc to avoid spraying sidewalks and buildings. Physical problems with the system can result in a lack of application uniformity, leading to the development of wet and dry spots.

- *As the system ages and the landscape matures, sprinkler heads sink*—They can also be pushed off vertically, causing them to stop turning or become clogged. All these physical problems affect the spray pattern.

- *Clean the sprinkler's trash filter screen if it has one*—Check the wiper seal at the base of the sprinkler heads. If it is worn, water can squirt out of the base. Make corrections to ensure proper operation and water distribution.

- *Use the sprinkler system checklist*—Although a hose and hose-end sprinkler system lacks many of these components, the checklist is still valuable in evaluating the proper use of this system and the effectiveness of the hose-end sprinkler.

Exam Prep Questions

Question 1

> Which hours of the day are recommended by most professionals for automatic settings for watering systems?
>
> ○ A. You should set water start times after midnight, with a completion of the last zone before noon.
> ○ B. You should water lawns only between sunrise and sunset for all zones.
> ○ C. The water start time should begin in the early morning, with the last zone finishing at dawn.
> ○ D. Watering start times should always be set to start at sunset and finish before midnight to reduce loss to evaporation.

The correct answer is C. Watering in the early morning hours with the last station ending at dawn reduces the risk of fungus. Answer A is incorrect because the cycle should ideally end at dawn to allow sufficient drying time. Answer B is incorrect because watering only during daylight hours is not recommended due to excessive loss through evaporation. Although answer D indicates that the nighttime hours have the potential for preventing water loss through daytime evaporation, a watering cycle ending at midnight is too soon for allowing sufficient drying time to occur during daytime hours with the attendant risk of fungus; therefore, answer D is incorrect.

Question 2

> What is the normal recommended depth for pressurized irrigation pipes?
>
> ○ A. 4"
> ○ B. 3 feet
> ○ C. 18"
> ○ D. 2 feet

Answer C is the correct answer. Although local building codes vary, the typical depth required for pressurized irrigation schedule 40 PVC pipe is a minimum of 18". Answer A is incorrect because 4" is too shallow for pressurized irrigation pipes in most climates. Answer B is incorrect because 3 feet is in excess of the depth needed to protect pipes from damage. Answer D is also incorrect because depths greater than 18" are not recommended because they make repair difficult and are deeper than the optimum value of 18".

Question 3

Which radius can normally be covered by a larger rotor sprinkler?
- A. 20 feet
- B. 60 feet
- C. 80 feet
- D. 100 feet

Answer D is correct. Large rotors are designed to cover radii up to 100 feet. Small rotors normally cover smaller radii of up to 50 feet. Answer A is incorrect because 20 feet is far less than the normal radius covered by a rotor sprinkler. Answer B is also incorrect because rotors are designed to cover radii larger than 60 feet. Answer C is also less than the optimum radius of 100 feet for rotors.

Question 4

Which water temperature range is recommended for spas for healthy adults?
- A. 100° F–104° F
- B. 95° F–99° F
- C. 106° F–110° F
- D. 115° F–119° F

Answer A is correct. Water temperatures between 100° F (38° C) and 104° F (40° C) are considered safe for a healthy adult, but under no circumstances should the water temperature be permitted to exceed 104° F (40° C). Answer B is incorrect because the values listed are lower than the recommended values for spa temperatures. Answer C is also incorrect because 106°–110° is too hot for most people to safely use a spa. Answer D is incorrect because the values are in excess of the recommended values of 100° F and 104° F.

Question 5

> Which voltage is typically used for electrically operated irrigation system valves?
> - A. 120v
> - B. 240v
> - C. 12v
> - D. 24v

Answer D is the correct answer. Irrigation system valves are usually designed for low-voltage 24v operation. 120v AC is used as the primary power source for the controller, and a transformer in the controller lowers the voltage level to 24v for in-ground valves. Answer A is incorrect because 120v is the normal power-line voltage for home AC wiring and is not used with low-voltage irrigation valves for practical safety reasons. Answer B is incorrect because 240v service is used only for appliances such as electric ovens, stoves, and clothes dryers. Answer C is incorrect because 12v is occasionally used for low-voltage exterior lighting but is not used in irrigation system components.

Question 6

> Which devices are required for isolating the irrigation system from the main municipal water supply?
> - A. Sump pumps
> - B. Zone controllers
> - C. Backflow and double-check valves
> - D. Booster pumps

Answer C is correct. Backflow and double-check valves are required for isolating the irrigation system from the main municipal water supply. Answer A is incorrect because sump pumps are designed to remove water from basements or potential flood areas and have no association with isolating the irrigation system from the main supply line. Answer B is incorrect because zone controllers have no functional interface with the main water supply line. Answer D is incorrect because a booster pump is used with irrigation systems to increase the water pressure when the main water line pressure is insufficient to provide the correct amount of water pressure for the irrigation system sprinkler heads.

Question 7

> When installing control wiring with irrigation system underground valves and supply line pipes, what is the proper procedure for locating the wiring with respect to the pipes?
> - A. The preferred location is 2"–4" above the pipe.
> - B. The preferred location for the control wire is 2"–4" below the pipe.
> - C. The preferred location is to place the wire directly on top of the pipe.
> - D. The preferred location is to place the wire 6"–8" above the pipe.

Answer B is correct. The preferred location for the control wire is 2"–4" below the pipe. The distance of 4" from the pipe protects the wire from damage and makes it less likely to accidentally be cut when repairs are made to the pipe. Answer A is incorrect because placing the control wires anywhere above the pipe can cause damage to the wires when repairs are required. Answer C is incorrect because placing the wire directly on the pipe can cause damage to the wires when repairing a broken or malfunctioning valve or pipe. Answer D is incorrect for the same reason as answer A—locating the control wires above the pipe raises the risk of future damage when repairs are made to the pipe.

Question 8

> What is the minimum height for pop-up sprinklers installed in turf areas?
> - A. 4"
> - B. 3"
> - C. 2"
> - D. 1"

Answer A is correct. Pop-up sprinklers in turf areas should have a 4" minimum pop-up height. Answer B is incorrect because most sprinkler heads made for turf areas have a minimum pop-up height of 4". A 3" pop-up height is below the minimum recommended height of 4". Answers C and D are incorrect for the same reason.

Question 9

What is the primary purpose of a water alarm?
- ○ A. It is a sensor that alerts the user when the lawn becomes excessively dry.
- ○ B. It is a device used to alert the user of leaks or intrusion of rainwater.
- ○ C. It is a monitoring system that disables an irrigation system if it is raining.
- ○ D. It is a monitoring system that warns residents if the temperature drops below 32°.

Answer B is the correct answer. Water alarms are used in any location in the home where leaks or intrusion of rainwater can cause damage if undetected. A basement or utility room area can be monitored for leaks with sensors capable of detecting water levels of 1/32". Answer A is incorrect because water alarms are not usually associated with soil water measurement. Answer C is incorrect because a water alarm is not the normal terminology used for rain sensors that automatically shut off an irrigation system when it is raining. Such sensors operate automatically and are not designed to activate any alerting signal or alarm. Answer D is incorrect because a water alarm is not a low-temperature sensor.

Question 10

What is the main advantage offered by a zoned water management system?
- ○ A. A zoned system is designed to provide different watering cycles based on the time of day.
- ○ B. A zoned system provides the capability to irrigate lawns when the humidity reaches a minimum value.
- ○ C. A zoned system provides the capability for different plants or lawns with different water needs to be irrigated separately.
- ○ D. A zoned water system is designed to water different areas based on the soil temperature.

Answer C is correct. Zoned water management systems provide the degree of automation needed to manage correct programmed water settings for selected areas. Irrigation systems are zoned so that plants with different water needs are irrigated separately. Answer A is incorrect because zoned systems are not designed for the primary task of watering based on the time of day. Most standard irrigation system controllers are designed with time-of-day settings but are not necessarily programmed for zone watering needs. Answer B is incorrect because a zoned system is not designed to water different areas based on any environmental condition. Answer D is also incorrect because zoned systems do not have sensors to monitor soil temperature.

Need to Know More?

 http://jessstryker.com/instal14.htm

 http://www.irrigationtutorials.com/sprinkler00.htm

 Gerhart, James. *Home Automation and Wiring*. Berkeley, CA: McGraw-Hill, 1999. (This book will help you plan a residential water management and sprinkler system.)

8.0—Home Access Controls and Management

Terms you'll need to understand:

- ✓ Code grabber
- ✓ Rolling code
- ✓ Remote control and remote access
- ✓ Condition monitoring
- ✓ Plenum cable
- ✓ Sensor
- ✓ Time-of-day settings
- ✓ ANSI/BHMA AI56.23
- ✓ Consumer Product Safety Commission
- ✓ Keychain keypad
- ✓ Relay
- ✓ Control box
- ✓ Solenoid
- ✓ Electromagnetic lock
- ✓ Card reader
- ✓ Programmable remote control
- ✓ 110-type connection block

Techniques you'll need to master:

- ✓ Planning a garage door safety system
- ✓ Understanding the concept of rolling codes
- ✓ Resetting the codes in a programmable garage door opener
- ✓ Memorizing the rated holding strengths for electromagnetic locks
- ✓ Understanding the role of BHMA in regard to home access control systems

Home access control and management systems are an extension of many features found in home security systems. Access control adds another dimension to standard home security monitoring and reporting functions. A modern integrated home can require electrically operated gates where the identity of the individuals entering an area needs to be authenticated using magnetic card readers or an access code. A homeowner can have a requirement for establishing a time-of-day setting for delivery personnel to have access to the gated community security entrance. Electrically operated door locks can be integrated into a home to provide an added access control convenience as well as an added residential security feature.

These are a few examples of the types of devices you will learn about in this chapter. Keep in mind the importance of knowing the terms, standards, and component names you might encounter on the HTI+ exam described in the following pages. Study the exam alerts and complete the practice exam questions at the end of the chapter.

Home Access Design Considerations

Design considerations for home access controls and management systems normally include features for controlling garage doors, gates, and residential access systems from a remote location. In addition, status monitoring features can be included in the home design that provide the residential user with information on when doors or gates were opened, closed, locked, and unlocked.

Remote Control

Remote control is a design feature used to perform monitoring, controlling, and supervisory functions of garage doors, gates, and residential entrance doorways at a distance. Remote-control operation is a convenience feature allowing home residents to open garage doors and gates from a vehicle using a wireless handheld controller.

Garage Door Remote Control

Motorized remote-controlled garage door lifters (also known as *garage door operators*) have become a standard item in most new residential construction. Unfortunately, this also has contributed to a weakness in home security. Criminals have recently found methods to enter garages using devices called *code grabbers.* When a garage door remote transmitter is in operation, the code grabber device can read and record the fixed radio code of the user's transmitter. Fortunately, a rolling code radio system has been designed for remote-control transmitters. This feature automatically changes the signal code each time the remote-control transmitter is used.

Garage Door Remote Control Safety

The Consumer Product Safety Commission (CPSC) requires that all garage door operators manufactured or imported after January 1, 1993, for sale in the United States be outfitted with an external entrapment protection system. This system can be an infrared sensor, a door edge sensor, or any other device that provides equivalent protection. If an infrared system is used, it should be installed at a height of 5"–6" above the floor. An example of a garage door safety system is shown in Figure 8.1. As a result of the CPSC action, every residential garage door operator manufactured since January 1993, has the infrared noncontact safety sensors to help prevent an automatic garage door from closing on a child or pet.

Figure 8.1 A garage door safety beam system.

Remote Access

Remote access door and gate systems, such as the one shown in Figure 8.2, permit residents to admit or refuse admittance to selected individuals from some distance away. Alarm systems for gates and doors can be remotely armed, accessed, and operated by authenticated users.

Figure 8.2 A door/gate remote access system.

This remote access controller is suitable for use with any size system—residential or commercial. Each controller operates up to four individual ports and can be expanded through networking to include as many as 32 individual controller units. Secure keyless transmitter entry is supported through the use of a built-in radio receiver.

Although the differences are minor, the design considerations for residential remote-control systems are different from remote access systems. Remote control is a design feature used by anyone to perform monitoring, controlling, and supervisory functions of garage doors, gates, and residential entrance doorways at a distance. Remote access is a design feature that requires some type of authentication of the identity of a user attempting to gain access to a locked or password-protected home security or computer system from a distant location.

Home Access Condition Monitoring

Design considerations for monitoring specific conditions at each door or gate have been incorporated into most residential access control systems. As an example, a Microsoft computer monitoring application can be installed that is capable of networking up to seven additional controllers, as shown in Figure 8.3.

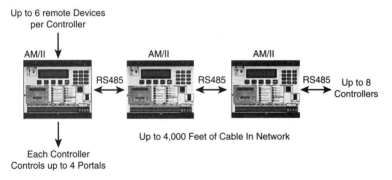

Figure 8.3 A linear network controller configuration.

Locked Condition Monitoring

In addition to monitoring the locked status of a door or gate, the condition monitoring system can also provide details as to how long and during what time periods the door or gate has remained locked.

Unlocked Condition Monitoring

The condition monitoring system can record and signal each time a specific gate or door is unlocked (granting access) and what type of access was granted, and it can identify who was granted access.

Opened and Closed Condition Monitoring

Because data between locked/unlocked conditions on a specific gate or door do not necessarily correlate to that door or gate actually being opened (providing access), the recorded data differentiates between these conditions.

If a door or gate is left open for a specified time, that information is also provided by the system. Signaling and reporting is continuous during the elapsed time between the opening and closing of any given door or gate. The condition monitoring system includes an event log detailing the times and dates of various events as they occur.

Equipment Component Location Considerations

In addition to design considerations, the location of home access control devices and components is also important. Keypad, relay, and sensor location considerations and functionality are discussed in this section.

Keypads

Keypads can be located in any area of a home that is accessible, affords a measure of security, and is convenient for the resident to operate external gates and doors. Keypads are used for programming, controlling, and operating various residential access control and management devices. Regardless of the controller's format (computer keyboard, wall-mounted keypad, or portable handheld unit), it needs to be located where it can be used effectively.

A handheld keypad, such as the example shown in Figure 8.4, is portable and permits the home user to carry it where it is needed.

Figure 8.4 A handheld keypad.

Relays

Relays are electromechanical devices used to activate electric gate and door locks. Single pole double throw (SPDT) dry-contact relays provide either fail-safe or fail-secure electronic locking functions. In dry-contact configurations, they control gate operators, garage door openers, or release door strikes. Their latching operations are often time-adjustable from 1 to 99 seconds, and they provide manual latching.

Activating relays are frequently built in to the keypads, door locks or gate operators. The relay kit shown in Figure 8.5 is designed for use as a gate controller and provides a view of the relay and associated components. The accompanying keypad is used to operate it.

 Relays are used in many door and gate home access systems to turn higher-voltage sources on and off, such as gate motors and high-current electromagnetic doors. The primary purpose of a relay is to control high-voltage access components from a low-voltage device such as a keypad. Relays are often included as components in a solenoid switch.

Figure 8.5 A gate controller relay and components.

Sensors

Various sensors can be used to detect the opened, closed, locked, or unlocked condition of an automated door or gate. They can also be configured to initiate an opened, closed, locked, or unlocked condition at a specified door or gate.

A simple sensor and a matching set of contacts for a door or window are shown in Figure 8.6. Its transmitter sends a signal to either a control panel or an emergency dialer when the magnetic switch contacts are broken as the door is opened.

Figure 8.6 A door sensor with magnetic switch contacts.

Physical Products

The key products that make up a home access control and management system include the various types of keypads mentioned earlier. Control boxes, power supplies, and solenoids are found in almost all types of access systems such as gates, doors, windows, and doors. In addition to these components, distribution panels are used for maintaining the structured wiring design concept for installed home access control systems. You will need to review the physical features of these products, as discussed in the next few paragraphs.

Keypads

Keypads are designed for wall mounting or portable handheld use. They selectively allow residential visitors to enter through gates and doors by requiring them to enter a code.

Multiple-keypad systems enable the programming of an entire network of keypads in a home.

More sophisticated access control systems enable the programming of additional access criteria for individual visitors based on the time of day, day of the week, or special holidays. Home security management can benefit from the use of networked keypads because individual events are stored on a computer hard drive for later retrieval. Often, events can be printed on a printer as they occur.

Wall-mounted Keypads

An important feature in home access control and management systems is the capability to prevent the unauthorized entrance through residential doors and gates. Most experienced burglars can pick through a conventional key lock within seconds. Access to a residence can be greatly strengthened with the use of mounted keypads, such as the one shown in Figure 8.7.

Figure 8.7 A weatherproof keypad control module.

Portable Keypads

Portable keypads small enough to fit on a key chain, as shown in Figure 8.8, can be used to remotely operate the homeowner's motorized garage doors and gate systems.

Figure 8.8   A learning remote keychain keypad.

This type of remote keypad can store the coded signals required for opening up to four gates, garage doors, or X10 system components located in the home. Because this keypad can learn the codes of various manufacturers' equipment, there is no need to carry several controllers when away from home. Any buttons not needed for garage doors can be used to turn the lights on or off from outside.

As shown, the learning keypad is the same size as the familiar keychain keypad that operates car alarm or entry systems. When pressed, its multiple buttons send different frequencies and coded signals using the same coding systems of typical garage door transmitters and receivers.

Control Boxes

Control boxes provide the intelligent adjustments required in a residential access control system. The box itself can be a simple intrusion timer unit, smart receiver, or powerful computer. When peripheral devices are being controlled through the transmission of command signals, the sending/receiving device can be thought of as a controller. If it is packaged as a separate component, the controller is called a *control box*.

A control box/receiver can be mounted into a standard single-gang electrical box, as shown in Figure 8.9, to control the entry through a door or gate. Also shown is a palm-sized transmitter that provides remote point-and-click access control.

Figure 8.9 A control box receiver and remote transmitter.

Because the system is designed for indoor or outdoor use, the receiver's relay output can be connected to an electric strike, a magnetic lock, or a garage door opener. A plug-in transformer provides the controller's 12v AC or DC input power using an RJ-45 connection.

Each transmitter is programmed at the factory with its own unique code and sports a range of 8–50 feet, depending on the environmental conditions. In addition, these codes can be added or deleted from any access control system.

Access Control Power Supplies

Access control power supplies designed for use with door/gate locking systems normally operate using low-voltage AC or DC, at either 12 volts (v) or 24v. The easiest method of supplying power to an electric locking system is to use a plug-in transformer such as the one shown in Figure 8.10, which is suitable for a small system consisting of one or two doors.

Figure 8.10 An AC plug-in low-voltage power supply.

For a larger system, a multiterminal power supply such as the one shown in Figure 8.11 is recommended. This installation would allow all required door and gate locks to be powered from a single source.

This multioutput access control power supply/charger is specifically designed for use with access control systems and accessories by converting a 115VAC/60Hz input into five, individually regulated 12v DC or 24v DC power-limited outputs. Outputs are suitable for powering access control hardware devices, such as magnetic locks, electric strikes, or magnetic door holders.

Figure 8.11 A multiterminal low-voltage power supply.

Solenoid-Operated Locks

Electrically operated deadbolt locks offer increased levels of security for the home. Adaptable to any security system, electric deadbolts perform well as auxiliary locks on doors where access control is desired.

Electric deadbolts such as the one shown in Figure 8.12 are solenoid actuated. This type of solenoid operates from a low-voltage DC power supply contained in the deadbolt assembly. The low-voltage power supply is wired on the input side to the 120v AC home power source. They can be used with swinging, sliding, power-operated, and vertical lift doors. In addition, accessory mounting tabs can be used to mount them on fence gates.

Figure 8.12 A solenoid-actuated electric deadbolt.

These deadbolts can be set to either lock or unlock when energized and are listed by Underwriters Laboratories (UL) as burglary-protection devices. Controlled by remote switches or card-access devices, they deadlock automatically when the door is closed.

 Solenoids are the main component of home access devices used to actuate electrically operated deadbolt locks. They are often listed as a component used to operate relays for opening and closing doors and gates.

Electromagnetic Locks

Electromagnetic locks consist of two basic parts: the *electromagnet* and the *armature*, or strike plate. The electromagnet portion is normally a long E-shaped bar made of special steel. A coil of insulated copper wire is placed inside the bar and held in place by an epoxy compound, which is usually forced in place under high heat and pressure. It cures into a hard, strong sealant that is resistant to chemicals, heat, and vandalism. The entire assembly is then machined: The top surface is ground flat to produce good mating contact with the armature, and holes are tapped for later attachment of the electromagnet to a mounting plate and housing. The exposed portion of the steel bar is plated with a thin coating of material to resist corrosion.

Lead wires attached to each end of the coil are brought out of the magnet assembly. When DC voltage is applied through these leads to the coil, an intense magnetic field is created, causing the "E" frame to become magnetized.

The armature is simply a bar of steel, its size calculated to provide a certain mass to interact with the magnetic field produced by the electromagnetic lock. It is also machined—its face is ground to provide a good mating surface and holes are drilled to accept mounting hardware. The entire armature is plated to resist corrosion. Highest security is attained when mounting an electromagnetic lock to a steel door and frame.

Distribution Panels

When access control system cabling is located behind walls, a distribution panel is the logical location from which to make the necessary connections or changes. This typically necessitates the location of applicable power supplies (both primary and backup) for electric door and gate locks in the wiring closet.

An example of a universal distribution panel is shown in Figure 8.13.

Figure 8.13 A universal distribution panel.

This panel is designed to meet the requirements for residential structured wiring and enables the arrangement of terminals to be customized depending on the current situation. This includes any wiring arrangements required for multiterminal power supplies that operate electric gate and door locks.

Gates for Home Access Control

Residential gates can be classified into two main types: sliding and swinging. Both types can be opened or closed by the use of a motorized operator. The size of the gate operator is determined by the width of the pathway and the weight of the selected gate. Common motor sizes include 1/2, 3/4, and 1 horsepower, and the required voltages to operate them range from 120v AC, single-phase to 360v, three-phase. Sliding gates, such as the one shown in Figure 8.14, are used where high levels of operational safety and security are major concerns.

Figure 8.14 A sliding gate.

The swinging gate, shown in Figure 8.15, is equipped with hinges that are fully adjustable, allowing each leaf to swing through 180°.

An articulated arm, which runs from the operator to the gate, is used to pull or push the gate open and closed. Manufacturers sometimes use a piston/cylinder configuration for the arm.

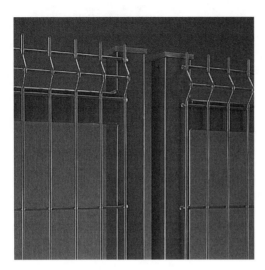

Figure 8.15 A swinging gate.

Standard Configuration and Settings

The proper configuration and setting of each type of access control and management system must be established by the homeowner to ensure all access systems satisfy the design objectives. This includes setting requirements for card readers, programming the software settings for entry keypads, and installing the proper sensor systems. Ensuring that garage door sensors are placed in compliance with federal law and establishing a timetable for time-of-day access systems where applicable are also necessary.

User Access

Apart from a system whereby all residents of a home have personal key access to gate and door locks, other systems offer increased security controls. These systems are designed to permit expected visitors to enter a residential property without the homeowner having to personally check the door/gate, visually scan a TV monitor, or physically press an admission button. The required checking can be done using a card reader that can be accessed from an auto like the one shown in Figure 8.16.

Figure 8.16 A card reader for user access.

Programming

Sophisticated user access systems are designed to be simple to use in spite of the fact that they provide high levels of security. One of the reasons for this is that much of the programming for access systems can be performed by off-the-shelf computer software applications.

This type of software can determine not only whether a specific card is eligible for entry, but also information as to which entrance should be used, valid times for the use of specific entrances, and dates beyond which entry is invalid under any circumstances. A similar system can be used to validate keys, keypads, fingerprints, or visual ID data.

Sensors

Automated residential systems must be capable of not only controlling access at various doors and gates, but also indicating their current condition, whether opened, closed, locked, or unlocked.

The door/window sensor shown in Figure 8.17 can be used to protect doors and windows. It alerts residents when someone attempts to gain entry through illegal means. Its magnet and reed-switch mechanism detects any intrusion and signals an alarm during any unauthorized entry. This type of sensor can indicate the open/closed condition of a door or window, but it cannot determine whether the door is locked or unlocked.

Figure 8.17 A recessed door sensor.

Even though up to 70 varieties of door sensors are available, a sensor alone cannot offer true control other than detection and warning. Its signal, however, can be used by other access control system components to provide automated responses. Residential door sensors are usually recessed in the door.

Garage Door Sensors

Because a garage door is the largest moving object in a residence, it can be dangerous. Most garage door openers include an internal reversing mechanism that switches the door into reverse (open mode) on contact with any obstruction during a close operation.

A federal law requires that all residential garage door openers sold in the United States since 1993 must include this type of additional protection against entrapment and comply with the latest UL 325 standards, "Federal Laws for Reversing Mechanisms and Sensing Edges." This protection usually takes the form of photoelectric sensors or sensing edges. The law also requires that, if these sensors become inoperative, the garage door opener also becomes inoperative.

Reverse signaling apparatus such as photoelectric sensors are located about 5"–6" off the floor on both sides of a garage door, as shown in Figure 8.18.

Figure 8.18 A garage door photoelectric sensor.

Garage door operator motors are now required by federal law to include sensors that switch the operation into reverse. The sensors must be installed on each side of the door opening approximately 5"–6" off the floor.

Time-of-Day Settings

Most automated user access control systems base decisions about valid or invalid entry requests, also called *transactions*, on the time of day. This is normal because any request for entry that does not fit the predefined time profile or time schedule of an identified user is subject to suspicion. Such a situation might occur for a daily delivery that arrives later than normally expected. The user, in this case a recognized delivery agent, has been identified but is seeking entry at an unauthorized time. The residential homeowner will undoubtedly need to intervene to accept delivery.

Device Connectivity

Connectivity for home access control and management systems is achieved with communications cable according to the standards for various types of locations defined in this section. Also, discussed next are wireless alternatives for connecting home access components. Phone-line connectivity, low-voltage wiring, and termination points for home access control systems are also covered in the following paragraphs.

Communications Cable

Cable ratings are established by Article 800 of the National Electrical Code (NEC) and also by Underwriter Laboratories. Table 8.1 lists the naming conventions and cable markings for communications cable contained in NEC Article 800-50, "Listings, Markings, and Installation of Communications Wires and Cables."

Table 8.1 NEC Markings

NEC Cable Marking	Cable Type
MPP	Multipurpose plenum [cable]
CMP	Communications plenum
MPR	Multipurpose riser
CMR	Communications riser
MPG	Multipurpose general purpose
CMG	Communications general purpose
MP	Multipurpose
CM	Communications general purpose
CMX	Communications, limited use
CMUC	Communications, under-carpet wire and cable

Communicationsspread cable has two rating systems. One rating method describes how the cable will react to fire; the other rating defines the cable data handling or performance capabilities.

Communications Cable Fire Ratings

Residential fires can spread quickly throughout a structure along the paths of cables that do not contain fire-resistant material. The NEC ANSI/NFPA 70 standard assigns fire ratings to communications transmission cables and discusses suitable applications for each cable type.

Table 8.2 lists the various fire ratings and their labeling marks and indicates their suggested applications.

Table 8.2 Cable Fire Ratings and Applications

Cable Classification	Residential	General Purpose	Riser	Plenum
Signaling circuits	CL2X	CL2	CL2R	CL2P
Optical fiber cables		OFN	OFNR	OFNP
Communications circuits	CMX	CM	CMR	CMP
Antenna TV and radio distribution systems	CATVX	CATV	CATVR	CATVP

Communications Plenum Cable

A *plenum* is any air space between walls, under structural floors, or above dropped ceilings that is used for environmental air, such as an air-conditioning duct. The Plenum test, UL 910, is the most stringent NEC test and is the only test that measures smoke generation as well as flammability. Plenum-rated cable can be used anywhere that residential, general-purpose, or riser cable is used.

Communications Riser Cable

Copper communications riser cable (CMR) is suitable for installation in vertical riser shafts designed to run from floor to floor. It cannot be used in any environmental air spaces, such as air-conditioning ducts, unless the local code allows it.

A riser connects one floor to another, usually in a commercial building, and penetrates fire-rated walls or floors.

Communications Cable Limited Use

Copper communications cable limited use (CMX) is specified for use in modems and residential indoor/outdoor LANs with transmission rates up to 100MHz. It is also suitable for power-limited circuits and communications for non-riser applications such as remote controls, signaling, security systems, single line telephones, intercoms, and PA systems.

Multipurpose Plenum Cable

Copper multipurpose plenum (MPP) cable is suitable for residential or commercial installation in the space between a ceiling and the floor above it. It is coated to prevent the production of toxic fumes in case of fire.

Multipurpose Riser Cable

Copper multipurpose riser cable (MPR) is suitable for installation in vertical riser shafts designed to run from floor to floor. It is coated to prevent the production of toxic fumes in case of fire, but its data carrying capabilities are not as good as CMR cable.

Power-Limited Plenum Cable

NEC Article 725 covers CL2P, or power-limited plenum cable. This cable can be used with remote control, signaling, and power-limited circuits that are not an integral part of a device or appliance. Type CL2P (class 2 plenum) cables are listed as being suitable for use in ducts, plenums, and other spaces used for environmental air. As such, they are listed as having adequate fire-resistant and low smoke-producing characteristics.

CL2P is the UL designation you will need to remember for the NEC cable rating for class 2 low-voltage circuits. This type of rating is suitable for installation in plenums.

Wireless Communication

Wireless communications devices are often used for connecting residential access system components. This type of connectivity is an economical solution that eliminates the need for new wiring between control devices, residential intercoms, and electrically operated security gates and doors.

Wireless Intercom Systems

Intercom connectivity to all parts of a residence can be accomplished with wireless intercom systems. Most wireless intercom systems support voice and paging communications between all areas, inside or outside the home. For example, master units such as the one shown in Figure 8.19 are normally located in the kitchen area and are designed to control an entire whole-home intercom system.

Figure 8.19 A master wireless intercom unit.

From the master unit, you can monitor any room station in the residence. The master unit's built-in AM/FM digital radio can also provide music to any remote room station.

Phone Line

Existing residential telephone lines provide an economical resource for adding access control systems to a home. An example of such use is the advanced telephone entry system for securing gated residential entrances, shown in Figure 8.20.

Figure 8.20   A telephone gate or door access control system.

The level of security to any residence can be increased with the use of this type of telephone-based access control system. The small, self-contained unit provides connectivity between the existing phone system and a gated entry to the home.

Low-voltage Wiring

Wire types are also categorized according to the location where they are used, such as inside walls, plenums, and hazardous areas (as mentioned earlier in the section "Communications Cable").

Additional data on low-voltage cabling for residential construction included in the HTI+ examination is covered in more detail in Chapter 10, "10.0—Structured Wiring—Low-voltage."

Termination Points

Termination points in a low-voltage structured wiring system involve the outlets in each room, push-down blocks, and the central distribution panel.

Low-voltage cables might terminate in a wiring closet on some types of terminal blocks, such as the 110-type wiring termination block discussed earlier in Chapter 2, "2.0—Audio and Video Fundamentals."

The advantage of this type of termination system is that the wiring can be quickly rerouted to any room, door, lock, or gate operator as conditions at the owner's residence evolve. This arrangement simplifies the amount of rework necessary to accommodate situations in which equipment and the associated access controllers must change locations. All the rewire activity occurs at the block in the wiring closet.

Industry Standards for Home Access Control

As described in earlier chapters, numerous standards organizations are involved in writing and publishing standards related to the home technology industry. The following organizations are the major groups you will need to remember relative to home access control and management systems.

The Builders Hardware Manufacturers Association

The Builders Hardware Manufacturers Association (BHMA), in association with the American National Standards Institute (ANSI), has published a wide array standards related to home access control systems.

ANSI/BHMA Electromagnetic Lock Standards

The amount of pressure an electromagnetic lock can withstand is determined by its integrity rating. The BHMA, together with ANSI, has determined three integrity grades for electromagnetic locks. ANSI/BHMA Al56.23, "American National Standard for Electromagnetic Locks," provides full requirements for electromagnetic locks including operational, strength, and finish tests. The locks are ranked in three distinct pounds of force (LBF) holding strength as follows:

➤ *Grade 1*—1500 lbs. minimum

➤ *Grade 2*—1000 lbs. minimum

➤ *Grade 3*—500 lbs. minimum

Highest security is attained when mounting an electromagnetic lock to a steel door and frame. Remember the three grades of locking strength for electromagnetic locks.

The National Electrical Code

The National Electrical Code covers the requirements for electric conductors and equipment installed within or on public and private buildings or other structures. It also covers conductors that connect the installations to a supply of electricity and other outside conductors and equipment on the premises, including optical fiber cable. It covers buildings used by the electric utility, such as office buildings, warehouses, garages, machine shops, and recreational buildings that are not an integral part of a generating plant, substation, or control center.

Several articles of the NEC applicable to low voltage wiring are covered in more detail in Chapter 10.

Underwriters Laboratories, Inc.

Underwriters Laboratories, Inc. (UL) is an independent, nonprofit product safety testing and certification organization. It has tested products for public safety for more than a century. Each year, more than 17 billion UL marks are applied to products worldwide. UL lists several types of cables approved for connecting home access control and management systems.

The U.S. Consumer Product Safety Commission

The U.S. Consumer Product Safety Commission (CPSC) is an independent federal regulatory agency created by Congress in 1972 under the Consumer Product Safety Act. The agency's public charter is to "protect the public against unreasonable risks of injuries and deaths associated with consumer products."

It has jurisdiction over about 15,000 types of consumer products, from coffeemakers and garage door openers to toys, lawn mowers, and fireworks.

However, some types of consumer products are covered by other federal agencies. For example, cars, trucks, and motorcycles are within the jurisdiction of the Department of Transportation, whereas food, drugs, and cosmetics are covered by the Food and Drug Administration.

Installation Plans and Procedures

The sensors for garage doors require alignment during the installation. The procedure outlined here is typical of almost all types of garage door safety sensor systems.

Reverse Garage Door Sensors

Garage door sensor alignment can be achieved by performing the following steps:

1. Remove the nut holding one of the sensors in place and add a lock washer under the nut.

2. Replace the sensor and tighten the nut with the socket driver. A lock washer will keep the sensor in position.

3. Repeat the process with the other sensor.

4. Pull a length of string between the two sensor devices to ensure that they are in alignment, as shown in Figure 8.21.

Figure 8.21 Checking sensor alignment.

5. Make adjustments to one or both devices if needed, and tighten the nuts.

6. Test the operation of the sensors by initiating a close operation.

7. Wave a hand or other obstruction in front of one of the sensors. The door should stop and move in reverse.

Troubleshooting and Maintenance Plans

As with other automated systems the owner might have installed in the home, periodic adjustments and regular maintenance routines are required by home access control and management hardware/software. This is true in spite of the fact that the associated equipment might have been operated exactly as recommended by the manufacturer. Maintenance is also required periodically on wireless garage door controllers.

Wireless Garage Door Openers

Some programmable garage door openers can be reset to the factory settings. To reset the code, press both the Access Learn and Radio Learn buttons together for 10 seconds. This automatically sets the access code to 1-2-3-4 and the radio code to 0-0-0-0, which are the default settings. The codes can then be set for any desired secure code.

Battery replacement does not cause loss of the settings for most programmable remote openers.

Exam Prep Questions

Question 1

Which organization established national safety standards for garage door operators manufactured or imported after January 1, 1993?
- A. The National Fire Protection Association
- B. The Federal Communications Commission
- C. The Institute for Electrical and Electronic Engineers
- D. The Consumer Product Safety Commission

The correct answer is D. The Consumer Product Safety Commission (CPSC) requires that all garage door operators manufactured or imported after January 1, 1993, for sale in the United States be outfitted with an external entrapment protection system. Answer A in incorrect because the National Fire Protection Association is a nonprofit organization that works toward reducing the risk of fire and other hazards on the quality of life by providing and advocating fire codes and standards. It is not directly involved in garage door safety standards. Answer B is incorrect because the FCC is a federal agency charted to regulate telecommunications and wireless infrastructures in the United States and is not involved in garage door safety regulation. Answer C is incorrect because the IEEE is a nonprofit, technical professional association that publishes standards for a variety of technologies, including the well-known IEEE 802.X standards for computer networking.

Question 2

Which home access design feature is used to perform monitoring, controlling, and supervisory functions of garage doors, gates, and residential entrance doorways at a distance?
- A. Remote access
- B. Access operation
- C. Access control
- D. Remote control

The correct answer is D. Distinct differences exist between remote access and remote control, as mentioned earlier. Remote control is a design feature used by anyone to perform monitoring, controlling, and supervisory functions of garage doors, gates, and residential entrance doorways at a distance.

Answer A is incorrect because remote access is a design feature that requires some type of authentication of the identity of a user attempting to gain access to a locked or password-protected home security or computer system from a distant location. Answer B is incorrect because access operation is the process involved for authenticating the identities of individuals who are entering a protected area and is not directly associated with remote control. Answer C is also incorrect because access control is not the correct term for systems that are controlled and monitored from a remote location.

Question 3

Which amount of pressure can a grade 1 electromagnetic lock withstand?
- A. 500 lbs. minimum
- B. 1000 lbs. minimum
- C. 1500 lbs. minimum
- D. 2000 lbs. minimum

The correct answer is C. The locks are ranked in three distinct pounds of force (LBF) holding strength as follows: A grade 1 electromagnetic lock can withstand 1500 lbs. of holding strength. A grade 2 electromagnetic lock can withstand 1000 lbs. of holding strength, making answer B incorrect. A grade 3 electromagnetic lock can withstand a holding strength of 500 lbs., making answer A incorrect.

Question 4

Which type of copper communications riser cable is suitable for installation in vertical riser shafts that are designed to run from floor to floor?
- A. CMR
- B. CMX
- C. CM
- D. CL2

The correct answer is A. Copper communications riser cable (CMR) is suitable for installation in vertical riser shafts that are designed to run from floor to floor. Answer B is incorrect because CMX is a type of residential communications cable and is not vertical riser cable. Answer C is incorrect because CM cable is a general-purpose cable and is not rated for installation in riser ducts. Answer D is incorrect because CL2 is a general-purpose signaling cable and is not acceptable for installation in riser ducts between floors.

Question 5

When installing a reverse signaling apparatus such as photoelectric sensors for a garage door, what is the proper clearance above the floor for the sensors?

○ A. 2"–3"
○ B. 4"–5"
○ C. 5"–6"
○ D. 8"–10"

The correct answer is C. Reverse signaling apparatus such as photoelectric sensors for garage doors should be located about 5"–6" off the floor. Answer A is incorrect because 2" is lower than the recommended clearance for garage door sensors. Answer B is also incorrect because 4" is not the correct value for photoelectric cell clearance above the garage floor. Similarly, answer D is incorrect because 5" is above the required height for garage door sensors.

Question 6

Which type of residential access control system provides details as to how long and during what time periods a door or gate has remained locked?

○ A. Remote operation access
○ B. Condition monitoring
○ C. Limited access
○ D. Locked status control

The correct answer is B, condition monitoring. In addition to monitoring the locked status of a door or gate, a condition monitoring system can provide details as to how long and during what time periods the door or gate has remained locked. Answer A is incorrect because remote operation access is a name for a system that supports operation or monitoring of an access system from a remote location. Answer C is incorrect because limited access deals with a system that prohibits access based on a set of fixed criteria and does not provide or support condition monitoring. Answer D is incorrect because locked status control does not have any relationship to the definition of condition monitoring.

Question 7

Which voltages are normally provided from power supplies designed to operate electrical locks for doors and gates?

○ A. 220v–240v AC
○ B. 60v–90v AC only
○ C. 12v–24v AC or DC
○ D. 120v–140v DC

The correct answer is C. Access control power supplies designed for use with door/gate locking systems normally operate using low-voltage AC or DC, at either 12v or 24v. Answer A is incorrect because 220v–240v AC is used for kitchen appliances and large motors. This voltage range is not applicable to the design of power supplies for locking systems. Answer B is incorrect because 60v–90v AC only is not a typical voltage used for any electromechanical locking system. Answer D is incorrect because 120v–140v DC is not a voltage range or type used for any home lock system power supply.

Question 8

Which type of electromechanical device is used in access control systems to turn higher-voltage sources, such as gate motors and high-current electromagnetic doors, on and off?

○ A. Sensor
○ B. Relay
○ C. Receiver
○ D. Deadbolt

The correct answer is B. Relays are electromechanical devices used to activate electric gate and door locks. They are controlled from a low-voltage source such as a keypad. The relay contains contacts that switch a high-voltage source on or off, thereby isolating the keypad from the high-voltage circuit. Answer A is incorrect because a sensor is any unit designed to monitor or detect a voltage, door position, or environmental condition and is not used to turn on or off high-voltage sources. Answer C is incorrect because a receiver is a radio component that receives a signal from the transmitter in a wireless garage door opener system. It is not used to directly control high-voltage circuit switching. Answer D is incorrect because a deadbolt is a device used to mechanically secure a door by inserting a metal cylinder into a latching mechanism in a door frame.

Question 9

What is the main electromechanical component of home access devices used to actuate electrically operated deadbolt locks?

- A. Solenoid
- B. Photocell
- C. Power supply
- D. Sensor

The correct answer is A. A solenoid is used to electrically operate deadbolt locks by a magnetic coil that pulls the deadbolt into the open position when activated by a card key or any other device used to identify the person who wants to enter a secure area. Answer B is incorrect because a photocell is a component used for monitoring garage door openings for an obstruction. They are mounted near the floor of a garage area near the doors to reverse the overhead door if an obstruction is present. Answer C is incorrect because a power supply supplies the correct voltage to an access control system, such as an automatic gate. Answer D is incorrect because a sensor is a unit designed to monitor or detect a voltage, door position, or environmental condition and is not used to operate a deadbolt.

Question 10

Which component of a home access control system is used to detect a locked or unlocked condition of an automated door?

- A. Relay
- B. Deadbolt
- C. Sensor
- D. Solenoid

Answer C is correct. Sensors are small magnetic switches installed in door frames. They can be used to transmit an open, locked, or closed condition of a door to a control panel. Answer A is incorrect because relays are electromechanical devices used to activate electric gate and door locks. They are controlled from a low-voltage source such as a keypad. Answer B is incorrect because a deadbolt is a device used to mechanically secure a door by inserting a metal cylinder into a latching mechanism in a door frame. Answer D is incorrect because a solenoid is an electromechanical device used to operate deadbolts in high-security access systems.

Need to Know More?

 http://www.smarthome.com/howto05.html

 http://www.smarthome.com/5181B.html

 http://www.nokey.com/853111/inwiracconsy.html

 http://www.mdsbattery.com/shop/productprofile.asp?ProductGroupID=539

 http://www.geniecompany.com/Products/intellicode.htm

9.0—Miscellaneous Automated Home Features

Terms you'll need to understand:

- ✓ Power converters
- ✓ Infrared sensors
- ✓ Piezoelectric
- ✓ Stair lifts
- ✓ Photodiodes
- ✓ PIR
- ✓ ANSI/TIA/EIA-570-A-99
- ✓ Dumbwaiter
- ✓ Electronic ignition
- ✓ IR Xpander
- ✓ Solenoids
- ✓ Relays
- ✓ X10 fan controller
- ✓ Occupancy sensor
- ✓ Entrance-protection system

Techniques you'll need to master:

- ✓ Identifying the benefits of time-of-day-settings
- ✓ Connecting a programmable keypad to a computer
- ✓ Identifying the safety features of fireplace igniters
- ✓ Describing the functions performed by portals
- ✓ Making a list of the common home automation products

This chapter introduces some of the automated home features that do not necessarily fall into the other technology areas in the residential system HTI+ objectives. In this chapter, which covers the remaining objectives in HTI+ exam 1, you will discover the design features and mechanical construction of A/V cabinetry, portals, distribution systems, and automated lift systems. Also covered here are environmental systems such as window shade controls, fireplace igniters, automated fans, and skylights.

The chapter also describes the types of components, cable, and wire used for implementing and connecting each of the control systems for the automated home. Some of the components will be familiar to you, such as keypads, relays, and solenoids, because they are used in several automated home system designs; however, they are included here because they have special design features that are important.

Automated Home Design Features

In the following section, you will discover how motorized audio/video (AV) cabinetry and window coverings help provide a setting for the operation of home theater equipment. Additional automated home design features you will need to remember include automated lift systems, fireplace and fan automation, and skylight controller options.

Automated A/V Cabinetry

Various automated systems are available that provide motorizing and remote control conversion kits for A/V cabinetry. In consultation with the cabinetmaker, the homeowner can easily install a motorized door mechanism that quietly glides an impressive A/V cabinet open with the touch of an RF remote controller. Using screws to attach the top and bottom rails, the user simply hangs the doors and plugs it in. Easy integration with X10 or other home automation systems is achieved using the included inputs for contact closure control.

For example, the motorized door mechanism that opens the A/V cabinet, shown in Figure 9.1, features an automatic reverse switch that stops and reverses the doors if any obstruction is detected during travel.

The operation of the doors is controlled using a multibutton radio control transmitter together with solid-state microprocessor logic.

The remote transmitter can also provide any required lighting control, including dimming. The cabinet valence can hang directly on the mechanism through the use of custom clips, and the unit is shipped with the motor box already assembled.

Figure 9.1 The A/V cabinet in closed position.

Another style of lift is designed to conceal a TV in the ceiling. This type of design is specific to a plasma screen television where the unit takes up only 11 1/2". This is because it lays flat when not in use, as shown in Figure 9.2.

Automated Lift Systems

Automated lift systems include a number of disciplines that serve to improve the quality of residential living. Among the home enhancements grouped under this banner are products that provide customized shading, vehicle parking and storage, and various elevations for moving disabled people or various material from floor to floor.

Figure 9.2 The plasma screen ceiling television lift.

Automated Window Shading Designs

Shading products that feature roller shades and sunscreens are widely available; an example is shown in Figure 9.3.

Figure 9.3 A shading system featuring automated rollers.

Shades using either manual chain lift or motorized lift systems can be custom manufactured to the owner's specifications and are supplied with complete installation hardware. Both 12 volt (v) DC and 120v AC motor options are available depending on the lifting capacity required. They are suitable for both residential and commercial applications. Sunscreens are excellent for sun and glare control and are made from vinyl-coated polyester or fiberglass yarns. They are capable of protecting the resident's interior furnishings from harmful ultraviolet (UV) rays and can provide noticeable energy savings in warm climates.

As shown in Figure 9.4, various controllers can be incorporated into a home for use in light management.

Figure 9.4 A collection of shading controls and switches.

Group controls enable someone to raise and lower blinds individually or as a group. Remote controls offer the capability to use infrared or radio control to operate blinds with or without a wall switch. Timer controls allow blinds to raise and lower according to preset schedules, whereas comfort controls work with light and temperature sensors and automatically raise or lower the motorized blinds based on sun conditions.

Automatic window shading designs are classified according to four distinct groups based on the type of control system used. The types and their functions are group controls, manual remote control, timer controls, and controls using light and temperature sensors. Review the functional performance and names associated with each type of shading.

Automated Lift Systems

Although custom home elevators were once considered commercial equipment and chosen strictly for mobility, residential elevators now provide convenient, sophisticated, and affordable design solutions for many automated home systems.

They provide increased flexibility in residential floor planning while making the most of the available space. Shown in Figure 9.5, residential elevators provide dependable convenience and increase the market value of a home.

Figure 9.5 A convenient residential elevator.

Stair Lift Designs

Homeowners wishing to maintain a certain amount of independence in spite of the physical challenge of mobility can opt for a custom stairway chair lift, often called a *stair lift*. Removing the obvious concerns about the physical challenge of stair navigation, this device enables the user full access and a lifetime of independence in any home, at any time. An unoccupied chairlift is shown in Figure 9.6.

Figure 9.6 An unoccupied residential chairlift.

Dumbwaiters

The task of moving many items between the lower and upper levels of a home can be simplified with the use of a modern, multifunction dumbwaiter, as shown in Figure 9.7.

Figure 9.7 A modern dumbwaiter.

Affordable and efficient, a modern dumbwaiter can easily move heavy items for storage, including groceries, trash, firewood/ashes, and laundry.

A dumbwaiter can easily carry loads up to 200 lbs. and can travel approximately 30 feet per minute. This means it can help eliminate the strain of lugging packages or the frustration of spilling or breaking items that would otherwise have to be carried to upper levels.

Automatic Fireplace Igniters

The mechanism used to create the spark that lights a gas fireplace is called the *igniter*. Three general classes of fireplace igniters are available: pilot, piezoelectric (push button or rotary), and electronic.

A fireplace equipped with a pilot light ignition uses a continuous, small flame to ignite the gas at the main burner.

Piezoelectric technology is based on physical contact and elastic coupling. It's a property of certain classes of crystalline materials. Because of this property, a high-voltage spark can be triggered by the impact of a spring-driven push-button hammer on a crystalline material.

A twisting rotor or a squeeze-action lever can also produce the necessary spark without the need for external electrical power. This makes piezoelectric gas igniters suitable for applications where batteries or commercial power sources are not available. However, piezoelectric igniters are also available that are capable of performing with systems supplied with commercial power.

Fireplaces equipped with electronic ignitions eliminate the need for a continuous pilot because they light the main burner by electronic spark or hot surface interaction. These fireplaces offer increased levels of energy efficiency and convenience.

Electronic ignition, also referred to as *intermittent* ignition, can be achieved using either battery power or 120v AC sources.

Automated Home Fans

Automated home ceiling fans can be operated with X10 controllers. Automated fan controllers operate similarly to a dimmer switch but are designed specifically for ceiling fans. Unlike dimmer switches, X10 controller modules can handle 500v inductive loads and offer 16 steps in speed. The dimmer control on the X10 remote can be used to slow the fan, and the brighten control can be used to speed it up. Most X10 fan controllers respond to All Lights ON / All OFF commands.

Other design features of automated home fans include the volume of airflow in cubic feet per minute (CFM) and static pressure (SP), in inches of water, which represents a fan's potential energy and required horsepower.

Automated Skylights

Automated skylights can be operated using a standard handheld remote. The homeowner using this type of controller can operate the system from several locations in the home. Many new residences are being prewired for skylight and window automation, but this automated design feature should be considered prior to installing the drywall. In older homes, a qualified electrician can usually fish the required wiring behind the walls. However, a sophisticated whole-house system can require a dedicated distribution panel.

Location Considerations

The automated home has three essential components located in many places identified in the physical components section. They are keypads, relays, and solenoids. You will need to be prepared to identify the location within each assembly as well as know the functions performed by each automated system covered earlier.

Keypads

Keypads are used as the control interface for lifts, fireplace igniters, automated fans, and various skylight designs. Keypads are either permanently mounted in wall outlet boxes or included with wireless remote control units.

Relays

Relays are electrical components used primarily to isolate high-current or high-voltage devices from the low-voltage control circuits. They are used in many locations, such as lifts, fireplace igniters, fans, and skylights.

Sensors

Sensors provide the feedback to control units to limit door travel and operate the start and stop limits for lift systems. They are also a key component of fireplace igniters, automated fans, and skylights.

The three key components of most automated home systems are sensors, keypads, and relays. They are frequently used in the physical products discussed in the next section including lifts, fireplace igniters, outlet boxes, fans, and skylights.

Physical Products

Physical products and components for automated home features include various control devices, such as keypads and control boxes with integrated X10 control systems. Also included are solenoid-operated control valves for lift systems and fireplace safety devices, power supplies, and home Internet access portals.

Keypads

Keypads enable the home user to control audio/video entertainment systems as well as a number of other appliances, including automated fireplaces. The trend in the industry is to develop universal wireless remote keypads that integrate several functions into a single remote keypad. Keypads can also be permanently mounted on a wall box when required.

Audio/Video Controller Keypads

Home theater control systems permit the user to control the TV, VCR, DVD player, or satellite receiver from a single remote keypad. This type of remote can be operated from any location within the residence. For example, the universal remote keypad and its companion RF receiver shown in Figure 9.8 can operate from any room located within 100 feet of the original A/V equipment location. This range can be even longer depending on the layout of the residence.

Figure 9.8 A remote A/V keypad and RF receiver.

Although the small RF receiver is located in the room where the A/V equipment is set up, remote operation of the home entertainment system is possible from anywhere in the home, including the outside patio.

Fireplace Controller Keypads

Simple electric fireplace control can be achieved through the use of the keypad shown in Figure 9.9, which can be used to turn an electric fireplace on or off remotely.

Figure 9.9 Keypad remote for electric fireplace.

A variety of remote keypads are available for gas fireplaces. In addition to simple on/off functions, some models have other functions. Figure 9.10 shows a gas fireplace remote that not only turns the fireplace on or off, but also offers a digital temperature display.

Figure 9.10 Keypad remote with temperature readout for gas fireplace.

Control Boxes

Used in conjunction with learning remotes are control boxes such as the X10 IR command center shown in Figure 9.11.

Figure 9.11 The X10 IR command center control box.

Control is provided for up to 16 X10 devices by directing a keypad remote toward this box. In this manner, the resident can use the same remote to control the TV, VCR, and stereo, as well as X10-controlled lights, appliances, draperies, and thermostats. Manual control of up to 8 X10 devices is provided by the keypad on top of the control box, including on/off, dim/bright, and all lights on/all off.

The command center is specifically designed to receive the IR signals from a keypad remote and transmit those signals to the various X10-compatible receivers in the home. The various home automation codes required for specific devices are stored under the AUX or CD function of the remote keypad. The command center can be used with a variety of infrared range extenders, including the IR Xpander.

It can also be tucked away in the A/V shelves in the living room or a bedroom. When a remote keypad is pointed toward the X10 infrared command center, the IR signals are converted to X10 commands. These commands are then relayed to the IR Xpander, shown in Figure 9.12, mounted in the A/V cabinet.

With a possible capacity of 500 commands, the Xpander is responsible for sending the appropriate commands to the individual pieces of A/V equipment and to other electronic devices plugged in to the whole-house network. In addition to the capability to initiate events with IR signals, this unit can also learn new commands. It offers four zoned output ports, each of which can drive up to four standard IR emitters.

Figure 9.12 The IR Xpander.

The IR Xpander is also capable of programmed computer control by allowing its 500 IR commands to be recorded and played back via software scheduling. The computer software can also respond to incoming IR information received by the IR Xpander, which is connected to a controller through the AUX bus. The controller is interfaced to a PC serial port when operating in the serial mode.

Power Supplies

Depending on the specific design application, power supplies engineered for audio/video purposes can be incorporated within a specific piece of equipment or serve as standalone units. For the purposes of this section, only standalone units are discussed.

External power supplies designed for audio/video applications range from small power transformers that operate devices requiring limited power resources to large power amplifiers needed to drive concert PA systems and sophisticated video cameras and consoles.

AC-to-DC Power Converters

Power supplies that operate to convert household AC power into DC voltages ranging from 3v to 12v are usually rated at 1000–1800 milliamperes (mA) and are often supplied by manufacturers as accessories for battery-powered devices.

Most users are accustomed to running such devices from an AC outlet—when one is available—to conserve battery power. This is especially true for battery-operated tape decks and camcorders, where weak batteries can lead to unsatisfactory recording results.

Although many manufacturers produce power transformers designed to operate only their own specific products or only at specific voltages, as shown in Figure 9.13, other units are available for multipurpose use.

Figure 9.13 A 4.5v DC power transformer.

The power transformer shown in Figure 9.14 is adjustable in increments of 3v, 4.5v, 6v, 9v, and 12v DC. This makes it suitable for use with a variety of devices rather than for only one specific piece of equipment.

Figure 9.14 An adjustable power transformer.

Solenoids

Electric solenoids are used for numerous control functions for home automation. Two typical uses are found in the design of home lift systems and fireplace safety pilot shutoff valves.

Elevator Solenoid Applications

Because a solenoid is a form of an electromagnet, it can be shaped into a tube and used to move a piece of metal linearly. This property can be used in many ways, especially with valves, which is why a solenoid valve is suitable for use in elevators and other lift systems.

The two major elevator designs are hydraulic elevators and roped elevators. A *hydraulic* elevator system lifts the enclosure by using a hydraulic ram. This is a fluid-driven piston mounted inside a cylinder. The cylinder is connected to a fluid-pumping hydraulic system that uses oil or some other incompressible fluid. The hydraulic system has three parts:

➤ A fluid reservoir

➤ A pump

➤ A solenoid valve

The pump is powered by an electric motor, and the valve is located between the cylinder and the fluid. Fluid from the reservoir is forced into the pipe leading to the cylinder by the pump. Opening the valve causes the pressurized fluid to take the path of least resistance, returning to the fluid reservoir. Closing the solenoid valve forces the pressurized fluid into the cylinder because it has nowhere else to go. The piston lifts the elevator enclosure due to the fluid being forced into the cylinder, as shown in Figure 9.15.

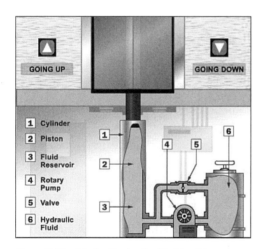

Figure 9.15 A hydraulic elevator lifted by piston.

Fireplace Solenoids

Because factory-built gas fireplaces are tested for safety and listed by Underwriters Laboratories, Inc. (UL) and the American Gas Association, today's fireplaces use the same types of controls and safety features as furnaces. For example, all gas fireplaces are required to have a safety pilot system and a safety combination valve, which prohibits the flow of gas to a burner until a pilot light is lit.

Gas fireplaces often use a U-shaped electromagnet or solenoid as part of the safety pilot shutoff valve, which shuts off gas flow on the basis of the presence or absence of a flame.

> Fireplaces are tested for safety to the same standards required for furnaces, and the organizations responsible for fireplace safety standards are Underwriters Laboratories and the American Gas Association.

Distribution Panels

Although a majority of distribution panels are designed for providing a central location for residential structured wiring, some units are used for special automated home functions such as skylights.

Skylight Distribution Panels

A distribution box designed for an automated skylight or window shade can be as small as a single-purpose unit, which can be located along an attic rafter (see Figure 9.16).

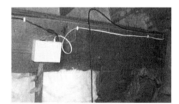

Figure 9.16 A distribution box for automated skylight.

The wiring for this distribution box is simplified by running low-voltage wiring from the skylight's motor through the wall and into the attic. The distribution box provides connections for the two motor leads, the cable running to the remote eye and the power cord for the low-voltage DC power system. Figure 9.17 shows the skylight window with the shade in the open position.

Figure 9.17 An open automated skylight.

Lift Systems

To a disabled driver, the term *lift system* might refer to a system that lifts a wheelchair and its occupant from ground level directly into a vehicle's operator position. A mechanic might envision a system that raises a vehicle in the garage for maintenance work, and a disabled veteran might contemplate a residential elevator to help in getting from floor to floor. Let's focus on the type of lift system products an HTI+ technician might encounter.

Automobile Lifts

Home mechanics and automobile collectors use various lifting systems to either work on their own cars or provide the additional space required for storing vintage vehicles.

For example, the two-post model shown in Figure 9.18 is the type used for most home auto lift systems.

Figure 9.18 A two-post auto lift.

It provides an ample amount of unobstructed floor space, which enables the user to easily move transmission jacks or other service equipment around or underneath a car, van, or light truck. All the required mounting hardware and various truck adapters are included.

Portals

Home portals enable the sharing of a high-speed Internet connection throughout the home. A *portal* is a term used to describe a type of network switch that uses the telephone outlets in the home to provide additional ports for high-speed Internet access without adding new wiring. The portal device connects to a home computer, a phone jack, and the Ethernet port on the high-speed modem. This enables any phone jack in the home to share the high-speed modem connection. Any phone jack can then be used for connecting a laptop or desktop computer equipped with a network interface card to the high-speed modem.

NOTE | Portals are a type of gateway/router covered earlier in Chapter 1, "1.0—Computer Networking Fundamentals." Portals provide additional capability for home phone-line networking where each phone line in the home acts as an additional portal for sharing access to the gateway or high-speed modem.

Standard Configurations and Settings

Automated home configuration and settings topics discussed in the following sections include user access programming with a computer for remote keypads, sensors, and time-of-day settings.

User Access Programming

The programmable type of remote keypad can be used for many automated home features. Gone are the complicated program macros of other universal remotes. Instead, the owner uses a PC equipped with a USB port connection. A CD-ROM drive, a Windows 98SE/ME/2000/XP operating system, Internet access, and a minimum of 10MB of hard drive space are also required to configure the remote keypad.

The procedure is included in the setup software. The program prompts the user for answers to simple questions, permitting the automatic programming of activities into the remote's memory.

Sensors

Sensors perform various functions in residential home system equipment. The sensor technologies include infrared, thermal, oxygen depletion, and photodiodes. Typical configurations and settings for home automation sensors are explained in the following paragraphs.

Infrared Sensors

The doors on a modern home elevator system use infrared sensors to detect family members and guests entering or leaving the elevator. These infrared (or photoelectric) sensors detect not only people, but also objects in the door opening preventing the continued closing of the doors. Older systems use mechanical safety edges to stop or retract the doors when they make contact with a person or object, but modern systems use a large number of invisible light rays to detect people or objects in the doorway. This approach permits the reversing or stopping of the doors without them having to make actual physical contact.

For example, one company has created an entrance-protection system that features an invisible safety net of infrared beams. As shown in Figure 9.19, this system is designed to ensure the physical well-being of passengers.

Figure 9.19 An infrared elevator entrance protection system.

Fireplace Igniter Sensors

Fireplace igniters use sensors to determine when the gas supply valve should be turned on or off. In Figure 9.20, the sensor connects directly to the fireplace ignition module.

A safety feature normally included with all vent-free fireplace systems is an oxygen depletion sensor (ODS) connected to the automatic shutoff valve. If the oxygen level in the surrounding area falls below 18%, this sensor signals the controller to shut off the unit before dangerous levels of carbon monoxide are produced.

Figure 9.21 shows an ODS pilot under normal conditions. The tip of the thermocouple is engulfed in the flame. Consequently, the flame failure device is held open, allowing gas to flow and the fireplace to operate.

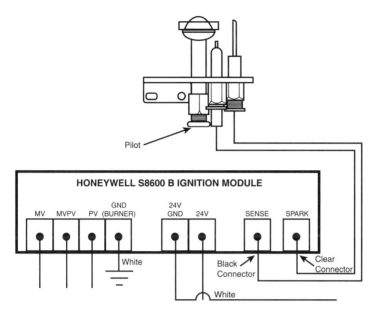

Figure 9.20 A partial diagram showing fireplace igniter sensor connection.

Figure 9.21 An ODS sensor under normal operation.

Fan Sensors

Sensors can be configured for fan operations in several interesting ways. For example, programmable thermostats can be used to run heating, cooling, or exhaust fans whenever their sensors detect that a room is occupied.

Other occupancy sensors can be configured for mounting on the wall or in the ceiling. They can operate using either passive infrared (PIR) or ambient light detection. They have adjustable delay times of 10–30 minutes.

Fans designed for operation in a bathroom can be fitted with sensors that automatically detect humidity. They can signal the fan's controller to turn on and remove the accumulated steam and potentially damaging moisture.

Light Sensors

Photodiodes are often used as skylight sensors designed to precisely monitor either work or ambient light levels. The measured light levels are converted to analog signals that are sent to an automated system controller that manages energy use for the entire residence. When tied to the heating ventilation air conditioning energy management system (HVAC/EMS), the sensors provide the information required to control area lighting by switching banks of lights on and off or by operating electronic dimming ballasts for fluorescent fixtures.

Time-of-day Settings

The capability to automatically turn residential equipment on or off at predetermined times is known as *time-of-day scheduling*. These schedules routinely allow for holidays, weekends, daylight savings time, and vacations. Specific override features can be employed that enable certain equipment operations outside of normal programmed hours, provided they do not interfere with normal settings. Notable utility savings can be achieved when residential devices are not operating unnecessarily.

Methods of Device Connectivity

The components of various automated home systems are usually connected to the termination points in the home using low-voltage cable. In this section, you will learn about the general classifications for low-voltage cabling as well as communications cabling used for device connectivity.

Communications Cable

Some types of communications cables are suitable for use within the vertical shafts of lifts and elevators. Figure 9.22 depicts a typical 15-pair communications cable designed for installation within the automated lift systems environment.

Figure 9.22 A communications cable for hoistway deployment.

Available in versions of 1, 2, 3, 4, 6, 9, 12, and 15 shielded pairs, these communications cables are composed of 20AWG, stranded insulated conductors.

The outer sheath of each cable is composed of a chrome and polyvinyl chloride (PVC) layer that provides a smooth, wear-resistant surface.

Low-voltage Wiring

Residential wiring designed to carry 30v or less does not present a hazard when compared to 120v/240v AC house wiring. Low-voltage wiring is used to carry audio and video signals, intercom communications, telephone hookups, and signaling for remote-controlled devices.

The standards for the residential use and installation of low-voltage cabling are contained in the National Electrical Code Article 725 and the ANSI/TIA/EIA 570-A standard. Low-voltage information you will need to know prior to taking the exam is described in Chapter 10, "10.0—Structured Wiring—Low-voltage."

Termination Points

Termination points in a low-voltage structured-wiring system involve the outlets in each room, push-down blocks, and the central distribution panel.

In an automated home design, the outlets serve as the termination points for a structured-wiring system. The standard for grade 2 residential structured wiring recommends a four-outlet connector in each room. Each outlet provides a connection point for two Category 5e UTP cables and two RG-6 coaxial cables. The push-down blocks provide the demarcation point for all telecommunication cables entering the home, and the central distribution panel provides the other termination location for all structured UTP and coaxial cable in an automated home.

Connection blocks are termination systems used to connect incoming phone lines on one side and telephone wires going to each room in the house on the other side. Wires are connected to each side of the block using a punch-down tool, which inserts the wire into a terminal on the block. As the wire is seated, the terminal cuts through the insulation to make an electrical connection and the spring-loaded blade of the tool trims the wire flush with the terminal. The two common families of punch-down blocks are 66 blocks, which are used for phone work, and 110 blocks, which are used for both data and voice circuits in a structured-wiring design.

Industry Standards

Several standards are used as guidelines for the installation and design of automated home features. As indicated in previous chapters, the National

Electrical Code (NEC), the Telecommunications Industry Association (TIA), and the Electronic Industries Alliance (EIA) are important standards organizations associated with the home automation industry.

National Electrical Code

As a guideline for electricians, electrical contractors, engineers, and inspectors, the National Electrical Code is the preferred guide. Although every state might differ slightly in its requirements for inspection and code compliance, in most areas the local wire inspector is the authority having jurisdiction. The authority having jurisdiction has the responsibility for making interpretations of the rules.

Building codes vary within municipal as well as U.S. state inspection and regulatory organizations. Local authorities have precedence over national NEC and ANSI standards. ANSI and NEC electrical wiring standards are used as guidelines by most contractors, but final inspections for residential occupancy are the responsibility of local authorities.

TIA/EIA Standards

TIA/EIA standards are certified by the American National Standards Institute (ANSI). Telecommunications Systems Bulletins (TSBs) are addenda to, or explanatory comments about, either an industry standard or an interim standard and are often integrated into the next standard revision.

Several standards related to miscellaneous residential automated features are as follows:

- *TIA/EIA 570-A Residential Telecommunications Cabling Standard (ANSI/TIA/EIA-570-A)*—This document standardizes requirements for residential telecommunications cabling. These requirements are based on the facilities necessary for existing and emerging telecommunications services.

- *EIA 600 CEBus*—This standard provides the necessary specifications for the Consumer Electronic Bus (CEBus), a local communications and control network designed specifically for the home. The CEBus network provides a standardized communication facility for exchange of control information as data among devices and services in the home. It is intended for such functions as remote control, status indication, remote instrumentation, energy management, security systems, and entertainment device coordination.

Underwriters Laboratories

Underwriters Laboratories, Inc. (UL) is an independent, nonprofit product safety testing and certification organization. Since the UL organization was founded in 1894, it has held the undisputed reputation as the leader in U.S. product safety and certification. Numerous safety features for doors, lifts, fans, and fireplaces discussed in this chapter are tested and approved for consumer use by the UL.

Exam Prep Questions

Question 1

> What are the typical voltage outputs for low-voltage power supplies that convert household AC power into DC voltages?
> - A. 6v–24v DC
> - B. 30v–60v DC
> - C. 25v–45v DC
> - D. 3v–12v DC

The correct answer is D. The standard low voltages for powering home automation systems are normally in the range of 3v–12v. Answer A is incorrect because 6v–24v is above the typical voltage range for most home, general-purpose DC power supplies. Answer B is incorrect because 30v–60v is not a voltage range used in household power supplies. Answer C is also incorrect because 25v–45v is not a typical range used in power supplies for home devices referenced in this chapter.

Question 2

> Which type of sensor is used to protect against doors closing when people are entering or leaving an elevator?
> - A. Piezoelectric
> - B. Infrared
> - C. Solenoid
> - D. Pressure

The correct answer is B. Infrared (or photoelectric) sensors detect not only people, but also objects in the door opening preventing the continued closing of the doors. Answer A is incorrect because a piezoelectric device is used in gas fireplace igniters. Answer C is incorrect because a solenoid is an electromechanical used for numerous control functions where an electrically activated piston action is needed to operate valves and lift systems. Answer D is incorrect because a pressure sensor is not used for the purpose of observing obstructions in elevator doors. It is used in security systems and pressure lines to detect a change in water or gas pressure.

Question 3

What is the typical load limit for home dumbwaiter lift systems?
- ○ A. 50 lbs.
- ○ B. 100 lbs.
- ○ C. 200 lbs.
- ○ D. 400 lbs.

The correct answer is C. A dumbwaiter can easily carry loads up to 200 lbs. and can travel approximately 30 feet per minute. Answer A is incorrect because most home dumbwaiter lift systems are designed for considerably more than 50 lbs. Answer B is also lower than the typical load limit for dumbwaiters, so it is incorrect. Answer D is incorrect because 400 lbs. is considerably more than the load limit for a dumbwaiter manufactured for home use.

Question 4

What is the main advantage of a fireplace electronic ignition system?
- ○ A. It eliminates the need for a continuous pilot.
- ○ B. It is less expensive than manual systems.
- ○ C. Electronic lighters rely on an impact of a spring-driven push-button hammer on a crystalline material.
- ○ D. Electronic lighters use a continuous pilot for safety purposes.

The correct answer is A. Electronic ignition systems eliminate the need for a continuous pilot and are therefore more energy efficient than continuous pilot fireplace ignition systems. Answer B is incorrect because fireplace electronic ignition systems are more expensive (not less expensive) than non-electronic systems. Answer C is incorrect because impact igniters are classified as piezoelectric igniters and are not the same as electronic lighters that require an electrical power source for operation. Answer D is incorrect because electronic igniters do not rely on a pilot light for proper ignition.

Question 5

Which type of sensor is often used with skylights to monitor either work light or ambient light levels?

○ A. Piezoelectric
○ B. Photodiode
○ C. ODS
○ D. Contact switch

The correct answer is B. Photodiodes are often used as skylight sensors designed to monitor either work or ambient light levels. The measured light levels are converted to analog signals that are sent to an automated system controller that manages energy use for the entire residence. Answer A is incorrect because piezoelectric is a type of fireplace igniter and is not used to measure light. Answer C is incorrect because an ODS is used to measure oxygen depletion for fireplace shut-off valves and has no operational relationship with a light sensor for skylights. Answer D is incorrect because contact switches are not designed to measure light levels; they are usually found in door closure monitoring sensors.

Question 6

Which electrical component of automated home systems is often used primarily to isolate high-current or high-voltage devices from the low-voltage control circuits?

○ A. A relay
○ B. A sensor
○ C. A photodiode
○ D. An RF receiver

The correct answer is A. Relays are electrical components used primarily to isolate high-current or high-voltage devices from the low-voltage control circuits. They are used in many locations, such as lifts, fireplace igniters, fans, and skylights. Answer B is incorrect because a sensor is a device used to monitor a condition such as a window glass break, water level, or temperature. Answer C is incorrect because photodiodes are used as skylight sensors that precisely monitor light levels. Answer D is incorrect because an RF receiver is a wireless device used with a handheld transmitter unit to control automatic garage door lifters.

Question 7

> Which document standardizes requirements for residential telecommunications cabling?
>
> ○ A. EIA 600
> ○ B. TSB
> ○ C. ANSI/TIA/EIA-570-A
> ○ D. ANSI/TIA/EIA 568-B

The correct answer is C. This document establishes the standard guidelines for selecting and installing residential telecommunications cabling. These requirements are based on the home facilities that are necessary for existing and emerging telecommunications services. Answer A is incorrect because EIA 600 is the standard for the CEBus home automation control protocol. Answer B is incorrect because TSBs are addenda to an industry standard or an interim standard and are often integrated into the next standard revision. They are not related to the published version of ANSI/TIA/EIA 570-A. Answer D is incorrect because the ANSI/TIA/EIA 568-B standard title is "Commercial Building Telecommunications Cabling Standard." It is not the standard that was developed for residential communications cabling.

Question 8

> Which type of TV system can be concealed when not in use by a ceiling lift system?
>
> ○ A. A plasma TV
> ○ B. A rear-projection TV
> ○ C. A CRT TV
> ○ D. A front-projection TV

The correct answer is A. This ceiling-concealed design is specific to a plasma screen television, where the unit takes up only 11 1/2" because it lays flat when not in use. Answer B is incorrect because a rear-projection TV system is too large for use with a ceiling lift system. Answer C is incorrect because a CRT TV is also physically larger than a plasma TV and will not work with concealed-ceiling lift systems. Answer D is incorrect because the large dimensions for this type of TV projection system make it unsuitable for ceiling lifts.

Question 9

> Which type of network device uses the telephone outlets in the home to provide additional connections for high-speed Internet access without adding new wiring?
>
> ○ A. NIC
> ○ B. Relay
> ○ C. Keypad
> ○ D. Portal

The correct answer is D. Portals are switches that enable additional computers, such as laptops, to use telephone outlets to connect to a high-speed modem using any telephone outlet in the home. Answer A is incorrect because NICs are used as expansion cards for PCs when connecting computers in a home network. Answer B is incorrect because a relay is an electrical component used primarily to isolate high-current or high-voltage devices from the low-voltage control circuits. Answer C is also incorrect because a keypad is a device used as a control interface for entering commands to operate home automation systems. Keypads contain buttons or keys for entering commands using an alphanumeric keyboard or special command push buttons.

Question 10

> How many distinct types of control systems are used in the design of residential automatic home window shading systems?
>
> ○ A. 6
> ○ B. 8
> ○ C. 4
> ○ D. 3

The correct answer is C. Automatic window shading designs are classified according to four distinct groups based on the type of control system used. The types and their functions are group controls, manual remote control, timer controls, and controls using light and temperature sensors. Answer Answers A, B, and D are therefore incorrect.

Need to Know More?

 Gerhart, James. *Home Automation and Wiring*. Berkeley, CA: McGraw-Hill, 1999. (This book will help you install or plan a structured-wiring system for a home.)

 http://www.homeautomator.com/

 http://www.home-automation.org/Example_Homes_and_Ideas/

PART II
Systems Infrastructure and Integration

10 10.0—Structured Wiring—Low-voltage

11 11.0—Structured Wiring—High-voltage

12 12.0—Systems Integration

10.0—Structured Wiring—Low-voltage

Terms you'll need to understand:

- ✓ Home-run
- ✓ Structured wiring
- ✓ Daisy chain
- ✓ Plenum cable
- ✓ BNC connector
- ✓ Distribution panel
- ✓ Filters
- ✓ Class 1 and 2 standards
- ✓ RJ-45
- ✓ RJ-11
- ✓ Bundled cable
- ✓ Multipurpose outlet
- ✓ ISO/IEC

Techniques you'll need to master:

- ✓ Preparing a sketch of a home-run wiring design
- ✓ Describing a grade 2 low-voltage wall outlet
- ✓ Identifying the type of connectors used for structured wiring
- ✓ Making a drawing of a line 1 and line 2 phone connection to an RJ-11 jack
- ✓ Identifying the correct type of cable for video distribution
- ✓ Describing the basic concept of a low-voltage structured-wiring design
- ✓ Describing how fiber-optic cable is used in a home audio/video system

Structured-wiring and low-voltage cabling concepts are topics included in the second (HT0-102) HTI+ examination, "Systems Infrastructure and Integration." This chapter covers the essential points you will need to know for this portion of the HTI+ exam.

Structured wiring for homes is a technology that has evolved from a 1999 residential wiring and cabling standard. This standard is the core document that establishes the basis for residential structured wiring. It has consolidated all the residential wiring for telecommunications, computer networking, security systems, audio/video systems, and control cabling into a coherent cabling design plan. The type of wire used for residential structured wiring is also established in selected portions of the National Electrical Code (NEC) and Underwriters Laboratories, Inc. (UL) standards.

Low-voltage wiring is a generic term used to describe all residential cabling that is not associated with the 240v/120v, high-voltage, alternating current (AC) primary power wiring used in all U.S. residential construction. Although no exact definition of low-voltage exists in residential cabling standards, it is generally accepted in the industry to include those circuits that are connected to low-energy, current-limited sources described in Article 725 of the National Electrical Code.

NOTE: The important standard that is referenced often in this chapter is ANSI/TIA/EIA 570, "A Residential Telecommunications Cabling" (September 1, 1999). Three addendums have subsequently been published covering alarm and security wiring, home theater wiring, and home automation control wiring. Key points in the 570-A standard are described later in this chapter.

Structured-wiring Design Considerations

The design concept for structured wiring has been implemented in the past by the commercial industry, but it is now being used extensively in residential designs. It employs a home-run wiring plan in which all circuits are terminated at a central location rather than using the daisy-chain wiring of the past. This section discusses the important design considerations, including wire types, types of conduit, and the features of home-run versus daisy-chain wiring designs.

Wire Types

Wire types used in low-voltage, power-limited circuits are derived from the standards documents mentioned previously. Wire types for telecommunications,

10.0—Structured Wiring—Low-voltage

security systems, audio/video, and computer networking are grouped into the following categories:

➤ Shielded twisted pair (STP)

➤ Unshielded twisted pair (UTP)

➤ Coaxial cable

➤ Fiber-optic cable

Wire types are also categorized according to the location where they are used, such as inside walls, plenums, and hazardous areas. Later, you will also learn about the classes of limited-energy circuits and related wiring specifications for residential designs included in the HTI+ structured-wiring and low-voltage examination elements.

Shielded Twisted Pair

The term *shielded twisted pair (STP)* usually refers to the 150-ohm twisted-pair cabling defined by the IBM Cabling System specifications for use with Token-Ring networks. The twisted pairs in 150-ohm STP are individually wrapped in a foil shield and enclosed in an overall outer braided wire shield. The shielding is designed to minimize electromagnetic interference (EMI) radiation and susceptibility to crosstalk. 150-ohm STP is not generally intended for use with Ethernet local area networks (LANs). In addition, this cable has not been used for residential structured wiring because unshielded twisted pair (UTP) has evolved as the preferred standard for home structured wiring and commercial networking cabling. UTP cabling is also lower in cost than STP cabling, is much easier to install, and uses lower-cost connectors.

The revised ANSI/TIA/EIA 568-B, "Commercial Building Telecommunications Cabling Standard October 1995," no longer recommends STP cable for commercial installations. Older 150-ohm STP cabling should not be confused with the newer 100-ohm STP cable recently used with Fast Ethernet LAN installations. This new cable is commonly known as *screened twisted pair (ScTP)* or *foil twisted pair (FTP)* cable.

Screened Twisted Pair Cable

A new type of 100-ohm twisted pair cabling called screened twisted pair (ScTP) or foil twisted pair (FTP) cabling has evolved from a new (interim) ANSI/TIA/EIA standard IS-729. The *IS* in the standard name indicates it is an interim standard. ScTP can be described as four-pair 100-ohm UTP with a single foil or braided screen surrounding all four pairs to minimize EMI radiation and susceptibility to outside noise. ScTP cable is also being developed to meet new, more rigorous criteria for future Category 5E, 6E, and 7F cable specifications.

ScTP cable conforms to the general specifications described in the following list:

- *Physical properties*—0.51mm (24 AWG) 100-ohm nominal impedance, four-pair enclosed by a foil shield
- *Color-coding*—ScTP cable is color-coded as follows:
    - Pair 1 = white/blue – blue
    - Pair 2 = white/orange – orange
    - Pair 3 = white/green – green
    - Pair 4 = white/brown – brown
- *Shielding drain wire*—A copper conductor drain wire of .040mm (26 AWG) or larger is included. (In a cable, the uninsulated wire is in intimate contact with a shield to provide for easier termination of such a shield to a ground point and is called the *drain wire*.)

Unshielded Twisted Pair

Unshielded twisted-pair cable is the most popular type of cable used in structured-wiring designs. UTP cabling is twisted-pair cabling that contains no shielding but has each pair of conductors twisted together to reduce the effects of EMI. For networking applications, the term *UTP* generally refers to the 100-ohm, Category 3, 4, and 5 cables specified in the TIA/EIA 568-A standard. Category 5E, Category 6/class E, and Category 7/class F standards have also been proposed to support higher-speed transmission. UTP cabling most commonly includes four twisted pairs of wires enclosed in a common sheath. The physical configuration for a four-pair UTP cable is shown in Figure 10.1.

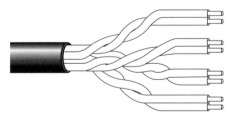

Figure 10.1 Twisted-pair cable design.

10.0—Structured Wiring—Low-voltage

Addendum 5 to the ANSI/TIA/EIA-568-A standard published in December 1999 specifies enhanced Category 5 (Category 5E) performance requirements. These requirements were recommended for new Category 5 cabling installations. A major new standard, TIA 568-B.1-2000 (Commercial Building Telecommunications Wiring Standard) subsequently replaced and updated all addendums, including addendum 5. This release established Category 5E as the minimum performance level for all new installations. Starting in 2002, the TIA, the International Organization for Standardization, and the International Electrotechnical Commission (ISO/IEC) no longer recognized Categories 4 and 5 for new installations. The recommendation is now Category 5E UTP cable.

UTP Category Designations

UTP cabling is graded according to the quality of construction and the immunity to EMI. The source document for UTP category ratings is ANSI/TIA/EIA 568. The category ratings you will need to remember are as follows:

➤ *Categories 1 and 2*—These are obsolete and are not used for structured-wiring applications.

➤ *Category 3*—This was originally used as the media for 10BASE-T Ethernet LANs as well as the early 4Mbps Token-Ring LANs. This cable type has a rated bandwidth to 16MHz and a twist specification of three or four twists per foot.

➤ *Category 4*—This has a rated bandwidth to 20MHz and has seldom been used outside of IBM Token-Ring LANs. Category 4 UTP cable has virtually disappeared from the market.

➤ *Category 5*—This has a rated bandwidth of 100MHz. It is manufactured to provide three or four twists per inch. It was originally used for Fast Ethernet (100BASE-T) LAN designs.

➤ *Category 5E*—This specification (ANSI/TIA/EIA-568-A-addendum 5) was officially ratified as a standard on December 13, 1999, and updated with the release of ANSI/TIA/EIA-568-B.1-2000. This specification provides "enhanced" parameters for Category 5 wiring, including an extra three dB (decibels) of headroom—the amount of theoretical bandwidth the installation can support across the entire frequency domain (up to 100MHz)—in several critical crosstalk parameters. However, it has improved specifications for near end crosstalk (NEXT), power sum equal level far end crosstalk (PSELFEXT), and attenuation. The specification also enforces several attributes that were optional in the original Category 5 specification. Category 5E cable is usually tested to a bandwidth of 350MHz, despite its 100MHz specified bandwidth.

NOTE: Category 5E stands for Category 5 enhanced; it's recommended for all new installations and was designed for transmission speeds of up to 1 gigabit per second (Gigabit Ethernet).

EXAM ALERT: The category ratings for UTP are partly dependent on the number of twists per unit of length in the cable pairs. Category 3 is normally manufactured with 3 or 4 twists per foot, whereas Category 4 has 10 twists per foot. Category 5 cable has 3 or 4 twists per inch. The higher category ratings require a tighter twist in the cable pairs. Also, UTP cable is more susceptible to EMI than ScTP cable, coaxial cable, or fiber-optic cable. Category 5 cable pairs are twisted a minimum of 36 times per foot.

Advanced International UTP Cable Standards

A new Category 6/class E standard describes a performance range for unshielded and screened twisted-pair cabling in the maximum bandwidth region of 250MHz. Category 6/class E is intended to specify the best performance that UTP and ScTP cabling solutions can be designed to deliver. Category 6 performance has been published as ANSI/TIA/EIA 568-B.2-1. An example of Category 6 cabling and connectors is illustrated in Figure 10.2.

Figure 10.2 Category 6 cable.

Another new category of communications cabling called Category 7 class F is being developed. It is reported to contain a new performance range for fully shielded (overall shield plus individually shielded pairs) twisted-pair cabling. It is anticipated that Category 7/class F will be specified in the frequency range of 600MHz and that its requirements will be supported by an entirely new modular plug and socket interface design, as illustrated in Figure 10.3.

Figure 10.3 Proposed Category 7 connectors.

NOTE Class E and class F are performance specifications for UTP cabling specified in ISO and IEC specifications. These standards are explained later in this chapter, in the section "International Standards Organizations." These design standards closely parallel the U.S. ANSI/TIA/EIA 568-B cabling design standard for cable performance categories. The ANSI/TIA/EIA 568-B.2-1 standard published in 2002 contains the U.S. standard specifications for Category 6 cable. The TIA has not published any performance standard for Category 7 cable.

Coaxial Cable

Coaxial cable is an important element in the design of residential low-voltage structured wiring. Although no longer used for new LAN installations, it is used for video signal distribution in residential structured wiring as well as CATV and satellite terminal applications. Structured cabling standards defined in ANSI/TIA/EIA 570 recommend coaxial cable installation between each room and a central location in accordance with grade 1 and grade 2 requirements.

Several varieties of coaxial cable are available for transporting video and high data rate digital information. Cable supply organizations offer various sizes and impedance options to match a particular application, such as cable TV signal distribution, home video systems, and radio frequency applications. The preferred type of coaxial cable for the design of residential structured cabling is RG-6.

Coaxial Cable Manufacturer Specifications

Coaxial cable is available in a number of inner conductor sizes and outer layer sizes. Outer diameters range from .195" to .405", and inner diameters for the copper core range from 14-gauge to 24-gauge AWG solid or stranded copper wire. Some examples of common coaxial cable types are listed in Table 10.1.

Table 10.1	Coaxial Cable Design Specifications		
RG #	Impedance	Center (AWG)	O.D. (inches)
6/U	75	18ga	.275
8A/U	50	10ga	.405
8X	50	19ga	.242
11/U	75	14ga	.405
58	50	24ga	.195
59	75	22ga	.245

The *RG* designation stands for *radio grade*, a term used in military specifications. The *U* suffix indicates that the cable type is general utility, and the *A* suffix designation indicates that the copper core wire is stranded instead of solid.

Coaxial Cable Shield Construction

Shielding for coaxial cable is available in single, double-triad, and quad-shield layers. Quad-shield contains four layers (see Figure 10.4). Note that in Figure 10.4, the braid and aluminum foil layers are both doubled, which provides the best protection from EMI among all types of coaxial cable. Quad-shield RG-6 coaxial cable is also recommended for video distribution in homes by the TIA/EIA-570-A residential cabling standard.

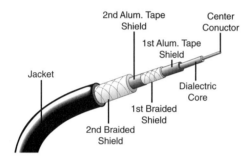

Figure 10.4 Quad coaxial shield.

Bundled Cabling

Addendum 3 to the ANSI/TIA/EIA-568-A standard has introduced a new term—*bundled cables*. As mentioned earlier, wiring new homes with grade 2 cabling includes four cables. The addendum describes four-pair cable assemblies that are not covered by an overall sheath (as specified for hybrid cables), but by any binding method such as speed-wrap or cable ties. A standard bundled cable for grade 2 structured cabling design consists of two Category 5E cables and two RG-6 quad-shield cables, as shown in Figure 10.5.

Figure 10.5 Bundled cabling.

The construction of a quad-shield coaxial cable contains two layers of aluminum foil and two layers of braid. You might also encounter the term *bundled cabling*, which is a convenient packaging configuration for grade 2 structured cabling that includes two RG-6 coaxial cables and two Category 5E cables.

Fiber-optic Cable

Fiber-optic cable is used for connecting audio/video equipment and is available under several proprietary brand names, including Toshiba's TOSLINK patch cable. Single-mode and multimode fiber-optic cable is used primarily for commercial applications. Fiber-optic cable provides the greatest immunity to EMI and is used for high-bandwidth digital transmission applications.

Multimode Fiber-optic Cable Transmission

Multimode fiber is designed so that light will travel in many paths from the transmitter to the receiver. Light rays that enter the cable reflect off the cladding at different angles as they move along the length of the cable. These rays disperse in the cladding and are useless as signals. Only those light rays introduced to the core of the cable within a range of critical angles travel down the cable. Even with the excessive signal attenuation created by this method, it is still the most commonly used cable type because it is cheaper and transmits light over sufficient distances for use in LANs.

Single-mode Fiber-optic Cable Transmission

In a single-mode fiber-optic cable, the diameter of the core is reduced so that just one wavelength of the light source travels down the wire. The light source for this type of cable is a laser. Laser diodes (LDs) produce the in-phase, single-frequency, unidirectional light rays required to travel down such a cable. These cables are normally reserved for use in high-speed, long-distance cable runs.

Design Options for Fiber-optic Interconnecting Cables

High-definition television (HDTV) and digital audio components have started to appear on the market that use digital input/output jacks. Cables for interconnecting these devices require a high-quality media. Products such as TOSLINK have appeared as a solution. TOSLINK uses fiber-optic cable with an integrated LED transmitter and photoelectric detector. These cables provide high performance and immunity from noise. Other manufacturers are also offering similar designs for low-loss optical cable connectors. TOSLINK cable is discussed later in the section "Fiber-optic Cable."

When comparing the various types of low-voltage structured-wiring components and cabling, fiber-optic cable provides the greatest immunity to EMI. UTP cable is the least resistant to EMI.

Conduits

Design considerations for new construction should include a plan for using conduits in accordance with local and national building codes. Wiring and electrical equipment must be enclosed in conduits in areas that are considered to be hazardous areas. Theses areas are described in the National Electrical Code in articles 500–517. Hazardous areas described in the NEC are more applicable to industrial settings and are not usually encountered in residential environments. Hazardous areas and applicable definitions are treated in more detail later in this chapter, in the section "The National Electrical Code."

Conduits are not required by any standard for indoor, residential, low-voltage circuits except for plenum areas. However, many contractors recommend installing PVC conduits as a "future proof" feature in new construction.

Future proof is a term used in the home automation industry to describe any home design or installation feature that goes beyond the minimum required standard. As an example, placing structured wire or cable in PVC conduit enables future upgrades for new or modified wiring to be installed by pulling the wire through PVC pipe between floors instead of drilling new holes for additional or upgraded wiring in the future.

Conduits should also be used for low-voltage wiring in outdoor locations. This protects landscape lighting or outdoor speaker wires from rodents or being cut accidentally while digging in flower beds or outside areas around the home.

Daisy-chain Wiring Topology

Daisy chain is a term used to describe one of three topologies found in residential wiring schemes. The other two are called *home-run* and *mixed*. Daisy-chain wiring, or *serial* wiring, is a type of wire routing using a single chain of low-voltage cable throughout a home. It is most often found in telephone line routing in existing homes.

Daisy-chain wiring designs are traditional in older homes for connecting phone jacks and intercom systems. It is also the topology used for the serial bus Ethernet LAN standard called 10BASE-2 using a single coaxial cable. It is simple and economical. However, it has several disadvantages as a general design plan for low-voltage wiring in new or upgraded home construction. Daisy-chain routing is not recommended as a wiring design for new or remodeled homes.

Home-run Wiring Topology

Home-run is the popular term used to describe the newest concept for modern home low-voltage wiring. This design is associated with structured wiring and uses a star topology.

In a *star* topology, each outlet in the home is connected by its own individual home run of cabling extending back to a central distribution point where all wiring in the home is accessible on a terminal block. Figure 10.6 shows an example of a modern structured-wiring design in which all wiring for both coaxial cable and Category 5E cable are run from a central location.

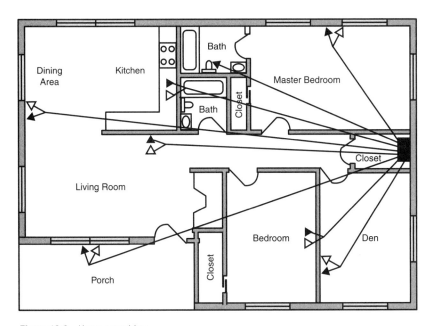

Figure 10.6 Home-run wiring.

Category 5E cable serves as the data communications outlet, and the coaxial cable handles the video distribution throughout the home. Some clear advantages to using structured home-run wiring are as follows:

➤ *Each outlet in the home can be dedicated to a single, independent source*—One phone line can be dedicated to a home office, while other phone lines selected at a future time can be dedicated to the other members of the family. Any future changes in distribution of services can be quickly and easily made at the common connection point.

➤ *Troubleshooting is simplified because each home-run circuit is independent from the others.*

▶ *Installing an outlet for all services in every room allows maximum flexibility*—
This also avoids having to pull new wires into the home as floor plans change. When new audio or video equipment or telephone systems are relocated, the dedicated home-run circuit is prewired and available in every room. The changes can therefore be easily made at the central distribution point.

Structured-wiring Location Considerations

This section addresses the location considerations for structured wiring that include both residential remodeling/existing structures and new construction project settings.

Remodeling/Existing Wiring Upgrades

Wiring choices available for remodeling an existing home depend on the number and type of home automation features planned, the electrical load requirements for added features, the floor plan for additional space, and the budget for replacing old low-voltage wiring. In most existing homes, no wiring exists for room-to-room signaling or control functions other than security wiring to sensor locations. Video cabling for television viewing is often routed to a few rooms using splitters. Network cabling such as Category 5 might exist for owners who have a home office.

The ideal solution when feasible is to install Category 5E cable and RG-6 coaxial cable to each room with new outlets for either single or double outlets for each. This includes the installation of 2" PVC pipe from the attic to the basement for ease of pulling new wires while leaving space for future wiring. Wiring can be routed between rooms through the attic for upper stories and through the basement for the ground floor. Home-run installations in which each outlet's wiring bundle runs to a central hub instead of being wired in series are essential.

Existing In-home Wiring Location Options

Existing homes do not always offer an economical solution for a complete upgrade to structured wiring. However, some technologies are available that use existing power-line wiring or telephone wiring for signaling to compatible plug-in modules and home networking interface equipment. Wireless devices also offer an excellent economical solution for home networking as well as control of home appliances and lighting systems.

Existing power-line wiring can be used as an alternative communications media to installing new low-voltage cable. Three protocols are available for sending digital commands over existing power-line wiring. Special interface modules that plug in to standard AC power outlets provide isolation from the voltages present on the power line. Then the modules use the copper paths in the power wiring to communicate with other modules plugged in to wall outlets. The protocols and compatible hardware for power-line systems are summarized as follows:

➤ *X10*—A popular protocol for home power-line networking. Receiving devices are plugged in to the residence's traditional electrical outlet. When commands are transmitted, the receiver responds to its address and ignores all other commands. The X10 specification sets an address structure supporting a total of 256 different addresses.

➤ *CEBus*—Another standard adopted as ANSI/TIA/EIA-600 that uses existing wiring to communicate with other home compatible appliances. CEBus can be used on a number of media, including power lines, video cable, telephone cables, and wireless transmissions.

➤ *LonWorks*—A multipurpose networking protocol developed by Echelon Corporation that can be implemented over basically any medium, including power lines, twisted-pair cable, wireless (RF), infrared (IR), coaxial cable, and fiber-optics. It has no central controller but uses a system of intelligent nodes that communicate with each other. The LonWorks protocol was published as ANSI/TIA/EIA-709 in 1998.

At the time of this publication, no standards exist for ensuring interoperability between different power-line communications products.

New Construction

Structured-wiring locations for new residential construction are described in ANSI/TIA/EIA-570-A. Minimum grades of residential structured cabling for each location in a new home are outlined in the standard. The essential elements for compliance for structured-wiring locations are summarized as follows:

➤ *Grade 1*—This grade is designed to support basic residential telephone and video services. The standard recommends using one four-pair Category 3 UTP cable and one 75-ohm coaxial cable.

- *Grade 2*—This grade is designed to support enhanced residential voice, video, and data services. The standard recommends using two Category 5 cables and connectors that meet or exceed Category 5 performance and two RG-75 ohm coaxial cables plus, as an option, optical fiber cabling. One Category 5 cable is used for voice and the other for data. One 75-ohm cable is normally used for input video signals to each room, and the second coaxial cable is used for output of signals with both input and output video routed to the central distribution panel location. Also, the 570A standard recommends—but does not require—the use of Category 5E rather than Category 5 cable in residential systems.

- *The structured-wiring central distribution panel*—The ANSI/TIA/EIA 570-A standard indicates that a central location within a home or multi-tenant building be chosen at which to install a central cabinet or wall-mounted rack to support the centralized wiring. This location should be close to the telephone company demarcation point and near the entry point of cable TV connections.

- *Home-run wiring location*—A home-run wiring location plan should be implemented for all wiring. This includes low-voltage wiring referenced as grade 1 or grade 2 recommendations. Home-run wiring requires that all the wiring runs to the grade 1 or 2 outlets be routed from a central distribution panel in the house, usually in a closet or basement. At the distribution panel, the wires terminate patch panels or cross-connect panels.

- *Conduit*—For new, multilevel residential construction, PVC (plastic) conduit should be run vertically for low-voltage wiring between floors. This enables wiring to be clustered in a centrally located conduit, which allows for future expansion and accessibility. New wiring can be pulled through the conduit when modifications are necessary.

- *Security system wiring locations*—When planning for the location of security system wiring, most contractors recommend keeping the sensor and control wiring location separate from video, network, and telephone structured wiring. Although it is considered to be low-voltage wiring and can be run in close proximity to the other wires, security wiring is usually installed, along with the associated alarm control panel, motion detectors, and magnetic switches, by an alarm system professional. The security system usually does not need any interconnection with these other systems, so it doesn't need to be installed in the same location. Security systems are therefore excluded from the bundling associated with structured wiring for new construction.

The ANSI/TIA/EIA 570A is the source for all residential structured wiring requirements. Memorize the minimum requirements for both grade 1 and grade 2 structured cabling. Although the 570A standard does not require the use of Category 5E cable, it is the preferred cable for new installations. RG-6 is also used in structured-wiring installations for compliance with the 75-ohm coaxial cable requirement.

Equipment Component Placement

The components of a structured wiring system include all the residential low-voltage cabling runs, a central distribution panel, a gateway device, multipurpose outlet boxes, and computer networking hardware and software. A further breakdown of the functions and locations of components is included in the following sections.

Structured Cabling Location Considerations

As you are aware from previous discussions regarding grade 1 and grade 2 cabling requirements, the recommended cabling components of a structured-wiring system include two Category 5 UTP cables and two 75-ohm coaxial cables routed from a central distribution panel to each room in a home. This enables maximum flexibility for locating telephones; video and audio equipment; computers; and cable TV, satellite TV, distributed audio, doorbell/intercom, and video surveillance equipment. When the grade 2 wiring components are installed in a star topology, virtually any of the components can be located in any area of the home. Locating the wiring from the distribution panel to a PVC pipe installed vertically from the basement to the attic is the usual practice for new construction. This enables cabling to be routed to each room via the attic for upper floors or from the basement area for single-story homes.

Distribution Panel Location

The distribution panel is the location at which all outside services enter the home; this includes cable TV, satellite TV, or off-the-air TV antenna, telephone, broadband Internet access (cable or DSL), Internet, and so on. It contains distribution devices for all these services and should be located in the home in a place that is readily accessible to cabling maintenance. The usual location is the basement or a wiring closet on the ground floor. All the coaxial (RG-6) cables and Category 5 cables are terminated in the distribution panel. The panel also contains the punch-down blocks for all the home-run wiring.

Multipurpose Outlets

The multipurpose outlets in each room determine which services are available in that room. With structured cabling, each room can be wired with grade 2 outlets, as illustrated in Figure 10.7, which provides maximum flexibility for locating audio, video, and telephone systems.

Figure 10.7   A grade 2 outlet.

Each outlet can be customized to a consumer's specific needs based on which services are desired in each room (cable, Internet access, telephone, audio, video, LAN, and so on).

Gateways

Gateways are special interface devices similar to routers that allow a single Internet access line, such as DSL, cable, or telephone dial-up access, to be connected to the home computer network. Most gateway products also include a firewall. The gateway is typically located near the distribution panel to simplify the connections from the gateway to the home network patch panel and the incoming Internet access port.

Home Computer Network

A home network can contain computers, fax machines, printers, scanners, video surveillance cameras, routers, gateways, or any number of associated peripheral devices. Computers are usually located in a home office where they can be easily interfaced with peripheral equipment through the Category 5 wall outlets.

Physical Structured-wiring Connection Components

A low-voltage structured-wiring system requires components designed to interface with the components and cabling in the home. In this section, you will become acquainted with the cables, connectors, and other components required for a structured cabling residential design.

Distribution Panels

Structured-wiring distribution panels make for a safe, convenient, and professional-looking installation. They provide the central location for all Category 5E data and phone network cables and audio and video distribution coaxial cable, and are a good place for incoming feeds from the local telephone company, satellite TV receiving antenna, or CATV service provider. The home-run topology is easily implemented with a central control panel designed to accommodate low-voltage structured-wire and cable designs.

RJ-45 Connectors

The RJ-45 jack is one of the types in the series of RJ-*XX* telephone jacks. *RJ* stands for registered jacks. These jacks were originally a type of interface derived from AT&T's Universal Service Order Codes (USOC) and were adopted as part of FCC regulations.

Two versions of a standard describe the color codes and pin assignments for RJ-45 cable terminations—they are the ANSI/TIA/EIA-568-A and 568-B standard. Both are permitted under the wiring standards. The only difference between the two color codes is that the orange and green pairs (pairs 2 and 3) are interchanged. The 568-A standard is the preferred method of terminating UTP cable, shown in Figure 10.8.

> As described in this section, the ANSI/TIA/EIA 568-A and 568-B standards are the two color codes used for wiring eight-position RJ-45 modular plugs. Both are allowed under the ANSI/TIA/EIA-568 wiring standards; the only difference between the two color codes is that the orange and green pairs are interchanged. ANSI/TIA/EIA-568-A jack wiring is recognized as the preferred scheme for this standard because it provides backward compatibility to both one-pair and two-pair USOC wiring schemes. Some confusion results from the fact that the ANSI/TIA/EIA-568-B standard is older than the 570-A counterpart. The ANSI/TIA/EIA-568-B standard matches the older ATA&T 258A color code and was previously the most widely used wiring plan, but it provides only single-pair backward-compatibility to the USOC wiring scheme. The U.S. government requires the use of the preferred ANSI/TIA/EIA-568-A standard for wiring done under federal contracts.

The RJ-45 interface handles four twisted pairs (eight wires total) of wires and is used for terminating UTP cable. When terminating RJ-45 connectors with Category 5 UTP cable, it is important to untwist no more than 1/2" of the cable to ensure that it will maintain the required performance at fast Ethernet (100Mbps) speeds.

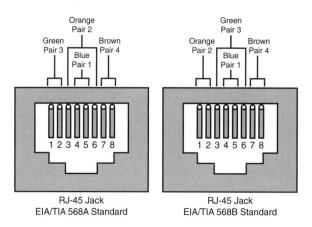

Figure 10.8 RJ-45 connector color codes.

 The RJ-45 jack and connector are used for both data and telephony in a structured-wiring design. Pins 4 and 5 (blue, pair 1) and pins 7 and 8 (brown, pair 4) are reserved for voice. The remaining two pairs are used for data. When terminating Category 5 cable with RJ-45 connectors, no more than 1/2" of the cable should be untwisted for the termination.

RJ-11 Connectors

The most common RJ series jack is the RJ-11. It is the connector used for terminating standard telephone cable. RJ-11 plugs are shown in Figure 10.9; they are keyed similarly to an RJ-45 connector bit but are smaller.

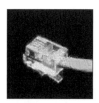

Figure 10.9 An RJ-11 phone plug.

RJ-11 connectors have six pins but are normally wired with four-conductor cable. As indicated in Figure 10.10, the four center pins (2–5) are used. The red and green wires (pins 3 and 4) are used for line 1, and the black and yellow wires (pins 2 and 5) are used for a second line.

Telephone terminology labels the green wire as the *tip* and the red wire as the *ring* connection. A similar notation is used for the second pair where the black wire is the tip and the yellow wire is the ring connection.

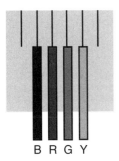

Figure 10.10 RJ-11 color codes.

RG-6 Coaxial Cable

RG-6 coaxial cable is the preferred type of coaxial cable for residential grade 1 and grade 2 structured wiring, discussed earlier. Figure 10.11 shows an RG-6 coaxial cable with a type F connector. It is used for video distribution, connecting satellite receiving antenna systems to standard TV, and digital and HDTV receivers. It is available in dual-shield, tri-shield, and quad-shield configurations. Quad-shield RG-6 has two foil wrapped and two braid type shields, which provide maximum protection from external EMI.

Figure 10.11 RG-6 coaxial cable.

BNC Connectors

Bayonet Neill Concelman (BNC) is a type of connector used with smaller types of coaxial cable. As shown in Figure 10.12, it has a center pin connected to the center cable conductor and a metal tube connected to the outer cable shield. A rotating ring outside the tube locks the cable to any female connector. It features two bayonet lugs on the female connector; mating is achieved with only a quarter turn of the coupling nut.

BNC connectors are ideally suited for cable termination for miniature to subminiature coaxial cable, such as RG-58, RG-59, RG-62, RG-179, RG-316, and so on.

Figure 10.12 A BNC connector.

 NOTE The BNC connectors were named after their creators: Paul Neill of Bell Labs and Amphenol engineer Carl Concelman. Neill designed the N-type connector, and Concelman designed the C-type connector. The BNC is a hybrid N/C-type with the bayonet as an added feature.

Amplifiers

Amplifiers are used in home audio/video systems to boost the signal levels of distributed audio and modulated video information between the source equipment (VCRs, camcorders, cameras, and DVD and CD players) and the destination components (television receivers, intercoms, security system monitors, and speaker systems).

Filters

Digital subscriber line (DSL) service uses a modem to send high-speed digital information over a standard telephone line. To prevent the data signals from interfering with the analog voice signals, a low-pass filter must be installed on the phone line. A *low-pass filter* is a device that allows the low-frequency signals (voice) to pass through the telephone lines in the home while blocking the high-frequency signals (digital data) from traveling through the phone lines to your telephone. Without this filter, the high-frequency data signals would be heard on your phone as static and noise and would diminish the quality of your voice and data services.

Binding Posts

All speakers and amplifiers have a binding post or spring clip terminal, usually found on the rear of the enclosure or amplifier chassis, for terminating the speaker cable.

Spring Clip Terminals

Spring clip terminals, shown in Figure 10.13, are usually found on lower-priced speakers and low- to medium-priced receivers. They secure the speaker wire after it has been stripped of its installation. They are secured by inserting the bare wire into the hole where they are secured by a spring clip.

Spring clip terminals

Figure 10.13 Spring clip terminals.

Binding Post Terminals

Binding post terminals, shown in Figure 10.14, are a sturdier, more reliable type of connector for speaker wire. They are found on higher-quality speakers and receivers and on most amplifiers. They use a treaded post that allows the wire to be bound by wrapping it under the cap. Binding-post connectors are designed to accept not only bare wire, but also pin connectors, spade lugs, single plugs, and double banana plugs.

Binding post terminals

Figure 10.14 Binding post terminals.

Connectors

Connectors for home low-voltage wiring systems serve a number of roles. They are used to connect video and audio components, computer systems, and telecommunications equipment. This section describes the features of the most frequently used low-voltage connectors.

Coaxial Video Cable Connectors

Coaxial cable is used primarily for transporting video information between components such as television receivers, DVD players, personal video recorders, VCRs, cable box converters, satellite antenna systems, and video amplifiers. RG-6 coaxial cable is used for one of the elements of a whole home structured-wiring system and serves as the basic cable for connecting and distributing video signals. The most common connectors and cable for home video are listed in Table 10.2.

Table 10.2 Video Connectors and Cables

Connector Type	Application
F Type	Standard connector for RG-6 and RG-59 coaxial cable.
S Video(Y/C)	Connector for video devices equipped with S video inputs and outputs.
Component	Three-conductor coax connector used for component video (Y, Pb, and Pr).
RCA	Used for audio input and output jacks for TV, CD, VCR, and DVD components.
UHF	Found in radio communications applications up to 300MHz using larger 50-ohm cable. It's known more commonly as PL-259 (Male) and SO 239 (Female).
Barrel	Adapter used for connecting two coaxial cables (male).
RCA Y	A cable and an adapter using RCA jacks to convert a single audio source to a double source for routing audio to multiple locations.
BNC	Rugged 75-ohm impedance matched coaxial connector used for video cable.
BNC to RCA	Adapter for converting a BNC cable termination to an RCA connector.

Configuration and Settings for Structured-wiring Designs

The configuration and settings of a low-voltage structured-wiring system include the distribution panel, structured cabling (including Category 5E UTP and RG-6 coaxial cable), and grade 1 or grade 2 outlets installed in each room in a home. In the next two sections, you will learn about the general topology of a structured-wiring residential system.

Distribution Panels

To meet the requirements of the ANSI/TIA/EIA 570A standard, every structured cabling system must have a central distribution device or cabinet. It should be centrally located and be as close as possible to the entry and demarcation points of the telephone and cable television service providers. In no case can the furthest outlet be located more than 492 feet from the demarcation point. The cabinet should be properly grounded and be within 5 feet of a dedicated duplex power outlet capable of supplying a load of 15 amperes. Distribution panels are the core element of a low-voltage structured wiring system and provide the focal point for cable distribution to every room in the home for video, telephone, audio, and broadband telecommunications circuits. Structured cabling is routed from the panel to room wall outlet boxes for coaxial cable, twisted-pair cabling, and other home network cabling. Distribution panels are usually located in a wiring closet or basement area.

Termination Points

Termination points are the locations in a structured-wiring system at which all the cabling is terminated in room outlets at one end and at a central location at the other end. In home-run wiring, all cables are run from the room outlets to a central location such as a control panel or a punch-down block. A *punch-down block* is a convenient type of termination block to which telephone cables and networking cable such as UTP can be connected. This serves as the termination point for incoming telephone lines from outside the home.

In Figure 10.15, a punch-down block using type 66 connectors has four rows of termination points available. In the example shown, four pins of an RJ-11 telephone jack are wired to the four pins of the punch-down block. The residential phone lines are terminated as shown.

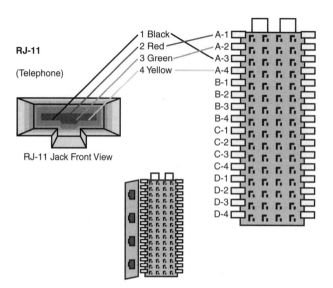

Figure 10.15 A punch-down block.

Column 4 is commonly used as an outgoing column, with each row connected to a pin on a cable leading to a phone in the home. Column 1 is used as an incoming column. To wire straight through, connecting pin 1 to pin 1, you must use a bridge clip to connect column 2 to 3 on the row for pin 1. A *bridge clip* is a small, metal clip just the right size to connect columns 2 and 3 together. *Crossover* is the function performed to connect a wire from column 2 in one row to column 3 in a different row. This feature has made 66 blocks very popular in residential telephone and network structured cabling management.

The color-coding for the two lines is illustrated in Table 10.3.

Table 10.3	Line 1 and Line 2 Color Codes	
Pair	Colors	Use
1	Red and green	Telephone line 1
2	Black and yellow	Telephone line 2

Device Connectivity for Structured-wiring Design

A structured-wiring topology also needs cables to connect the individual rooms to the distribution panel and punch-down blocks. Some types of short, high-performance cables are also needed to connect audio/video components. This section describes the characteristics of the media types required to complete the residential structured-wiring design.

Coaxial Cable

Coaxial cable is recommended for video signal distribution for all residential low-voltage cabling. It is used for connecting satellite receiving antenna systems to standard TV, digital, and HDTV receivers. Type RG-6 coaxial cable is recommended for all video signal connectivity in new as well as remodeled residential construction because RG-6 has lower attenuation characteristics than RG-59.

RG-6 cable is also used for connecting DVDs, VCRs, and personal video recorders to the A/V distribution system.

According to the ANSI/TIA/EIA-570A standard, RG-59 coaxial cable can be used only for short jumpers and RG-11 can be used for backbone video distribution cables in a very large home or a multitenant building such as an apartment or condominium complex.

Category 5 UTP Cable

Category 5E UTP cable has become the standard for data and telephone connectivity in low-voltage structured-wiring systems. Previously, telephone wiring in older homes was run in serial fashion from one outlet to the next in each room.

Structured-wiring low-voltage systems for new construction use home-run cabling. Separate Category 5E cabling is run from the central control panel to each room to accommodate both telephone wiring and home computer networking applications on the same cable. As explained earlier, for the pin connections for RJ-45 connectors, pairs 2 and 3 are dedicated to computer networks and pairs 1 and 4 provide the capacity for two separate telephone lines to any room in the home. The connectivity and wire pair color codes as prescribed in ANSI/TIA/EIA 568-A for RJ-45 connectors and Category 5E UTP cable (see Figure 10.16).

As mentioned in Chapter 4, "4.0—Telecommunications Standards," you will need to memorize the RJ-45/Category 5E plug and jack wire termination pinouts for both LAN and telephone connections in a structured-wiring home design configuration. Pins 4 and 5 (blue, pair 1) and pins 7 and 8 (brown, pair 4) are reserved for voice (telephony). Pins 1 and 2 (green, pair 3) and pins 3 and 6 (orange, pair 2) are used for inter-room connection to an Ethernet LAN home office network. The wiring scheme for ANSI/TIA/EIA 569-A and ANSI/TIA/EIA 568-B are exactly the same except the color codes for pairs 2 and 3 are reversed.

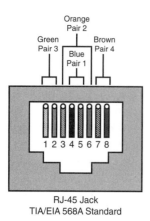

Figure 10.16 Category 5E color codes.

Fiber-optic Cable

Fiber-optic cable is beginning to have a major role in home structured wiring, primarily for audio/video systems. HDTV, digital TV, and digital audio components require a newer broadband type of cables and connectors than the type used for analog video and audio connections. Cable types such as TOSLINK are being used for connecting digital outputs of audio/video components. Fiber-optic cables are easy to use and are available in patch cord lengths of 2–10 feet. An example of a TOSLINK optical connector is shown in Figure 10.17.

Figure 10.17 A TOSLINK fiber-optic connector.

Plenum Cable

According to the definition from the National Electrical Code 1999, a *plenum* is "a compartment or chamber to which one or more air ducts are connected and that forms part of the air distribution system."

Building Industry Consulting Services International (BICSI) defines a *plenum* as "a designated closed or open area used for transport of environmental air."

According to the NEC, when installing cables in plenums, you must use plenum cables that are listed as Type CMP (communications plenum cable). Article 800-53 of the NEC states, "Cables installed in ducts, plenums, and other spaces used for environmental air shall be Type CMP." A permitted substitution for Type CMP cables is Type MPP cables (multipurpose plenum cable).

The NEC requires that Type CMP plenum cable be listed as being suitable for use in ducts, plenums, and other spaces used for environmental air and should also be listed as having adequate fire-resistant and low smoke-producing characteristics. When purchasing UTP cable for residential construction, in addition to the category, you need to know whether the cable is PVC or plenum rated and whether the cabling scheme is compliant to EIA/TIA-568 A or B. For example, plenum-rated UTP Category 5E cable has a higher heat-resistance rating so that it can be run in HVAC plenums. PVC-rated UTP without the plenum rating should be used when the cabling will be routed through walls, subflooring, or overhead—but not through a plenum.

NOTE: The plenum space is rarely used in residential construction to route low-voltage communication cables; however, the use of plenum areas for cable storage in any commercial or residential setting poses a serious hazard in the event of a fire. When fire reaches the plenum space, few barriers exist to contain the smoke and flames. Plenum cable is coated with a fire-retardant coating (usually Teflon) so that, in case of a fire, it does not give off toxic gasses and smoke as it burns.

Audio Wire

Audio wire is used to connect sound equipment in home theater systems. Two types of cable are required for audio connectivity. Coaxial shielded cable is used for audio component connectivity, and speaker wire is used to connect the amplifier output terminals to the speaker systems. Typical types of audio cable used for low-voltage structured cable connectivity are described as follows:

- ▶ *Shielded twisted pair*—This audio cable is commonly used to carry balanced microphone signals and balanced line level signals. Balanced connections and good STP cables are needed to transfer audio signals over relatively long distances.

- ▶ *Shielded single conductor*—This cable is used to carry unbalanced audio signals. It has one central coaxial signal conductor and a ground shield. This cable type is sometimes called *high impedance cable* because it typically is used with high-output impedance equipments such as tape and CD players and musical instruments (such as an electric guitar).

- ▶ *Unshielded twisted-pair cable*—This cable is used in telephone (voice) and home network wiring (structured cabling systems) such as Category 5E UTP. It is not very suitable for professional audio use. With suitable adapters, Category 5E cable can be used to carry digital audio signals and analog audio signals, but with limited performance.

- ▶ *75-ohm coaxial cable*—This cable is used to carry digital audio signals between audio equipment that requires a 75-ohm unbalanced connection.

- ▶ *TOSLINK cable*—This fiber-optic, digital audio cable was discussed earlier in the section "Fiber-optic Cable." It transfers audio information in a digital format rather than an analog format. TOSLINK cable can be used with all audio components equipped with a fiber-optic connector.

Security Wire

Security wiring connectivity is normally a dedicated system and is not part of the structured-wiring connectivity associated with Category 5E UTP and RG-6 coaxial cable.

The reason for this is that most security systems are installed by a professional, licensed security systems contractor. Security system signals are not transmitted over UTP or coaxial cable. Again, the ANSI/TIA/EIA 570A standard has recently added requirements for home security wiring. The new addendum 1 to the standard has been recently published as "Residential Telecommunications Cabling Standard—Addendum 1—Security Cabling for Residences (ANSI/TIA/EIA-570-A-1-2002)."

Termination Points

Termination points in a low-voltage structured-wiring system involve the outlets in each room, punch-down blocks, and the central distribution panel.

The outlets serve as the termination points for a structured-wiring system. As indicated earlier, grade 2 structure wiring recommends a four-outlet connector in each room. Each outlet provides a connection point for two Category 5E UTP cables and two RG-6 coaxial cables. The punch-down blocks provide the demarcation point for all telecommunications cables entering the home, and the central distribution panel provides the other termination location for all structured UTP and coaxial cable in a home.

Industry Standards for Structured-wiring Design

The following sections describe the various standards and organizations that have published documents relating to residential low-voltage structured wiring. In this chapter, standards sponsored by the American National Standards Institute (ANSI) and developed or published by the TIA and the Electronic Industries Alliance (EIA) organizations are identified by the long title "ANSI/TIA/EIA" because they reflect the correct composite title for this group of standards. You might encounter questions on the exam related to international standards groups discussed in this section. These standards are applicable to countries that do not develop their own domestic wiring standards.

Standards and Organizations

The most important standard you will need to know about when preparing for the infrastructure and integration examination is ANSI/TIA/EIA 570A and its three addendums. It sets guidelines for residential single-family and multifamily telecommunications equipment and wiring. Among other aspects, the standard covers specifications for the network interface and demarcation point, auxiliary disconnects, a distribution device and cross-connects, cables and cable installation, outlets and connectors, and equipment cords.

Although codes imply a legal requirement, standards are recommendations designed to ensure a minimum level of quality and good practices. Although not legally required, they ensure that the installed product will work properly and be compatible with related products and systems. The most applicable standard-producing organizations in the United States are

➤ *The Electronics Industries Alliance (EIA)*—Writes consumer-related electronics standards

> *The Telecommunications Industry Association (TIA)*—Writes telecommunications and cabling standards

> *The Institute of Electrical and Electronic Engineers (IEEE)*—Writes performance and technology standards that apply to low-voltage systems, including computer and data networks

International Standards Organizations

The International Organization for Standardization and the International Electrical Commission together form a system for worldwide standardization as a whole. National bodies that are members of ISO or IEC participate in the development of international standards through technical committees established by the respective organization to deal with particular fields of technical activity.

The ISO 11801 standard referenced earlier in this chapter is the principal design standard for structured cabling systems for all countries of the world that do not have more specific standards, such as ISO 11801 (European Union), TIA 568-A (United States), AS 3300 (Australia), and CSA T529 (Canada). The ISO/IEC standards for class E and class F circuits are listed in Table 10.4, illustrating a comparison with the ANSI/TIA/EIA categories.

Table 10.4 ISO Standards				
ISO/IEC11801	ANSI/TIA/ EIA-568	Frequency (MHz)	Applications	Comments
Class C	Category 3	Characterized up to 16MHz	802.5: 4Mbps Token-Ring 802.3: 10BASE-T	Typically used to support voice.
—	Category 4	For TIA/EIA only; characterized up to 20 MHz	802.5: 16Mbps Token-Ring	No longer recognized by TIA/EIA.
Class D	Category 5	Characterized up to 100MHz	155Mbps ATM 1000BASE-T	No longer supported as the minimum standard by the TIA/EIA.

(continued)

Table 10.4 ISO Standards *(continued)*				
ISO/IEC11801	ANSI/TIA/ EIA-568	Frequency (MHz)	Applications	Comments
Class D	Category 5E	Characterized up to 100MHz	155Mbps ATM 1000BASE-T	Recommended as the minimum for all future installations by ANSI/TIA/ EIA-569B 1 2000.
Class E	Category 6	Characterized up to 250MHz	All applications previously listed and future emerging technologies	Applications are currently being developed within various standards organizations.
Class F	Category 7	Characterized up to 600MHz	All applications previously listed	Fully shielded, nonstandard.

The National Electrical Code

The most well-known electrical standard for safety is the National Electrical Code. Published by the National Fire Protection Association (NFPA), the code provides "practical safeguarding of persons and property from hazards arising from the use of electricity." According to the NFPA Web site, the NEC protects "public safety by establishing requirements for electrical wiring and equipment in virtually all buildings." The NEC is dynamic, long, and complex; however, it is important for you, as an HTI+ certified technician, to understand its purpose and key sections, which are summarized in the following paragraphs.

Classes of Power-limited Circuits and Wiring

The NEC describes classes of circuits, how they are to be installed, and the type of wiring required for specific locations and circuit types.

Article 725-2 of the NEC provides definitions of class 1, class 2, and class 3 circuits. They are differentiated from each other by power limitations and maximum current permitted. Class 1 circuits might or might not be supplied by power-limited sources, whereas class 2 circuits are limited to voltage and current values that will not usually present a shock or fire hazard. Class 3 circuits are allowed to have higher voltage and current values than class 2 circuits. The voltage and current levels of class 3 circuits might present a shock hazard but generally do not present a fire hazard.

Class 1 circuits can be either class 1 power-limited circuits or class 1 remote-control and signaling circuits. They operate at less than 30v and not over 1000v amperes (VA). Power-limited class 1 circuits are necessary when the energy demands of the system exceed the energy limitations of class 2 or class 3 circuits of 100 VA.

Class 2 circuits generally operate at 30v or less at 100 VA. They include wiring for thermostats, programmable controllers, and security systems, as well as limited-energy audio systems. In addition, cables (twisted-pair or coaxial) that interconnect LANs are considered a class 2 circuit. Class 2 power limitations with maximum ampere ratings are as follows:

- 0v–20v at 100 VA (5 amperes maximum)
- 21v–30v at 100 VA (3.3 amperes maximum)
- 31v–150v at 0.5 VA (5 milliamperes maximum)

Class 3 circuits generally operate at voltages over 30v and are used for circuits when energy demands exceed 5 milliamperes but not over 100 VA total. Examples of class 3 circuits include signaling circuits, security systems, telephone systems, intercom, and sound and public address systems.

Section 725-61 of the NEC lists the cable types and their permitted uses along with acceptable substitutions. For example, type CL3 (class 3) cable can be used as CL2 (class 2) cable, but the reverse is not permitted.

Power-limited Circuit Wiring

Article 800 of the NEC covers rating requirements for low-voltage power-limited cables for class 2 (CL2) and class 3 (CL3) power-limited circuits. The NEC requires that cables used in premises, both commercial and residential, be "listed for the purpose" by a Nationally Recognized Testing Laboratory (NRTL).

The NEC book also describes many types of cable ratings. The eight most widely used types of power-limited wiring are listed in Table 10.5.

Table 10.5 NEC Cable Types

Type	Use
MPP, MPR, MPG, MP	Multipurpose cables
CMP, CMR, CMG, CM, CMX	Communications cables
CL3P, CL3R, CL3, CL3X, CL2P, CL2R, CL2, CL2X	Class 2 and class 3 remote-control, signaling, and power-limited cables
FPLP, FPLR, FPL	Power-limited fire alarm cables
CATVP, CATVR, CATV, CATVX	Community antenna television and radio distribution cables
OFNP, OFNR, OFNG, OFN	Nonconductive optical fiber cables
OFCP, OFCR, OFCG, OFC	Conductive optical fiber cables
PLTC	Power-limited tray cables

Each level from the applicable sections of the NEC is listed in Table 10.6 in the order of most to least burn resistance. The Underwriters Laboratories (UL) testing criteria for each of the four levels are also listed. The NEC ratings are hierarchical—in other words, from the top down. A cable can be substituted for any cable lower in Table 10.6.

Table 10.6 NEC Cable Fire Resistance Level

Fire Resistance Level	Test Requirement	NEC Article 800	NEC Article 725	NEC Article 760
Plenum cables (highest)	UL910 (Steiner Tunnel) CSA-FT6 (Steiner Tunnel)	MPP CMP	CL3P CL2P	FPLP
Riser cables multiple floors	UL-1666 (Vertical Shaft) CSA-FT4 (Vertical Tray)	MPR CMR	CL3R CL2R	FPLR
General-purpose cables	UL-1581 (Vertical Tray)	MP CM	CL3	FPL
Residential cables (lowest) restricted use	CSA-FT4 (Vertical Tray) UL-1581 VW-1	CMX	CL2 CL3X	

Table 10.7 lists the naming conventions and cable markings for communications cable contained in NEC Article 800-50, "Listings, Markings, and Installation of Communications Wires and Cables."

Table 10.7 NEC Markings

Marking	Meaning
MPP	Multipurpose plenum [cable]
CMP	Communications plenum
MPR	Multipurpose riser
CMR	Communications riser
MPG	Multipurpose general-purpose
CMG	Communications general-purpose
MP	Multipurpose
CM	Communications general-purpose
CMX	Communications, limited-use
CMUC	Communications, under-carpet wire and cable

It is important to keep in mind that the NEC ratings have nothing to do with cable performance. A plenum-rated speaker cable and a plenum-rated high-definition video cable only have their fire ratings in common.

Hazardous Locations

The NEC articles 500–517 define hazardous locations as those areas "where fire or explosion hazards may exist due to flammable gases or vapors, flammable liquids, combustible dust, or ignitable fibers and flyings." Although hazardous areas aren't often found in residential settings, when present, they require special consideration during installation of any type of residential wiring, including low-voltage wiring. If ignition of flammable substances or vapors from a damaged wire is possible, conduit is recommended to protect any electrical wiring. Local fire and building codes must be consulted to meet the requirements for enclosing wires in conduit.

Federal Communications Commission Rulings

On July 8, 2000, the Federal Communications Commission (FCC) established a policy for all copper telephone wire installed, new or retrofit, residential or nonresidential. The FCC ruling states (which is also the law) that, for new installations and modifications to existing installations, copper conductors should be, at a minimum, solid, #24-gauge or larger, twisted pairs that comply with the electrical specifications for Category 3.

The provision was suggested to the FCC by a BICSI committee in 1995. Since then, Category 5E UTP cable has largely replaced Category 3 cable for residential and commercial structured wiring installations at almost the same cost. As of this publication, the FCC ruling is not yet well-known by builders or installers of communications cabling.

Installation Plans and Procedures for Structured Wiring

Cable runs and outlet placement for structured wiring components are explained in this section.

Installing Structured-wiring Outlets

One of the best contributions of the ANSI/TIA/EIA 570A standard is a guideline on the number and locations of low-voltage structured-wiring outlets. The standard states that a minimum of one dual or quad outlet for UTP and coaxial cable should be placed in the kitchen, each bedroom, the family or great room, and the den or study. It also recommends that additional outlet locations be provided within unbroken wall spaces of 12 feet or more.

Installation Tips for UTP and Coaxial Cable

When installing the four-pair twisted-pair cables, you should not exceed a pulling tension of 25 lbs. Also, the minimum bend radius of the cable is four times the cable diameter. Because the TIA recommends only a maximum 1/4" cable diameter, the minimum bend radius is typically 1". RG-6 coaxial cable has a maximum pulling tension of 35 lbs. and an installed minimum bend radius of 10 times the cable diameter, which typically equates to 3". Normally, while pulling cable, you should observe twice the installed minimum bend radius to avoid degrading cable performance. Leave a minimum of 8" of slack cable at both ends to allow for termination.

Remember the parameters involved in installing structured cable. The maximum pulling tension for UTP four-pair cable is 25 lbs. and the cable bending radius is four times the diameter. Always leave at least 8" at each end for termination.

Maintenance Plans and Procedures for Structured-wiring Design

A professionally installed low-voltage structured-wiring system can be expected to last approximately 15 years. It has the longest life cycle of components in a low-voltage system. However, several steps are listed here that should be taken to ensure trouble-free operation over the expected life cycle of the cabling system:

➤ *Keep records of any changes to the control panel terminations or punch-down block wiring*—One of the main advantages for structured cabling is fault isolation. When problems occur, an updated log of any changes can be critical when looking for the source of a problem.

➤ *Examine wall outlets periodically for loose connections or damaged connectors*—Loose connectors can be a source of degraded performance, particularly for video systems. RJ-45 and RJ-11 plugs have locking tabs to hold them in place. When moving telephones or computers, you should ensure that the connectors are securely attached to the wall outlets. Make sure the tabs are locked and the cable is secure.

➤ *Conduct more in-depth cabling tests on the network cables in a home office*—Wire mapping is the most basic test and requires a tester with a remote unit. The tester's main unit then transmits a signal over each wire and detects which pin at the remote receives the signal. Only split pairs might not be immediately detected. These can cause an excess of near-end crosstalk that degrades the performance of the cable at high speeds. Therefore, you should check the capability to detect split pairs when evaluating a tester.

➤ *Measure all structured low-voltage network cable length*—You do this to ensure that specified maximums are not exceeded during the initial construction phase before the drywall is installed. A time domain reflectometer (TDR) can also help locate shorts, breaks, and defective terminators.

Exam Prep Questions

Question 1

> Which type of UTP cable is recommended for new residential as well as commercial installations?
> - A. Category 5 UTP
> - B. Category 3
> - C. Category 5E
> - D. Category 7/class F

The correct answer is C. Category 5E UTP cable is recommended for new low-voltage structured wiring construction. Answer A is incorrect because Category 5 cable is no longer recommended by the ANSI/TIA/EIA as the minimum required cable for new installations. Answer B is also incorrect because Category 3 cable is no longer recommended by the ANSI/TIA/EIA standards group for new wire installations. Answer D is incorrect because, although Category 7/class F is an advanced type of cable, it is currently not approved or adopted in any U.S. standard and is not the recommended cable for new installations.

Question 2

> What is the maximum pulling tension permitted when installing four-pair UTP cable?
> - A. 25 lbs.
> - B. 35 lbs.
> - C. 15 lbs.
> - D. 10 lbs.

The correct answer is A. The ANSI/TIA/EIA 570A standard indicates 25 lbs. is the maximum permitted pull tension for four-pair UTP cable. Answer B is incorrect because 35 lbs. is above the maximum pulling tension permitted for four-pair UTP cable. Answers C and D are both incorrect because 15 lbs. and 10 lbs. are each below the maximum pulling tension permitted for UTP cable.

Question 3

When terminating a structured-wiring Category 5E cable outlet with an RJ-45 receptacle, which ANSI/TIA/EIA-568-A color codes and pair numbers are reserved for voice circuits?

- A. Pair 1 (blue) and pair 2 (orange)
- B. Pair 2 (orange) and pair 3 (green)
- C. Pair 4 (brown) and pair 3 (green)
- D. Pair 1 (blue) and pair 4 (brown)

The correct answer is D. Pairs 1 (blue) and 4 (brown) are reserved for voice. Answer A is incorrect because pair 1 (blue) is used for voice but pair 2 (orange) is used for data. Answer B is incorrect because pairs 2 (orange) and 3 (green) are used for data and are not reserved for voice circuits. Answer C is incorrect because pair 4 (brown) is used for voice but pair 3 (green) is used for data.

Question 4

Which of the following NEC cable types is required for plenum installation when using a UTP Category 5E cable?

- A. CMP
- B. CMR
- C. CMX
- D. CMG

The correct answer is A. CMP cable is rated as communications cable for plenum installation. Answer B is incorrect because CMR is rated as riser cable and not plenum cable. Answer C is incorrect because CMX is the lowest-rated cable for general use but is not approved for use in plenums. Answer D is also incorrect because CMG is the marking for general purpose and is not rated for plenum installation.

Question 5

Which U.S. standards organization is responsible for writing telecommunications and cabling standards?

- A. EIA
- B. NEC
- C. TIA
- D. ANSI

The correct answer is C. The Telecommunications Industry Association (TIA) writes telecommunications and cabling standards. Answer A is incorrect because the EIA prepares consumer-related electronics standards and not telecommunications and cabling standards. Answer B is incorrect because the NEC publishes standards related to fire safety. ANSI is a standards-sponsoring organization for a wide range of U.S. standards organizations but is not the organization directly responsible for writing standards, so answer D is also incorrect.

Question 6

What is the rated bandwidth of the new ISO/IEC standard for Category 6/class E cable?

- A. 100MHz
- B. 600MHz
- C. 800MHz
- D. 250MHz

The correct answer is D. Answer A is incorrect because 100MHz is the rated bandwidth for Category 5 cable. Answer B is incorrect because 600MHz is the rated bandwidth for Category 7/class F cable, not Category 6. Answer C is incorrect because 800MHz exceeds the rated bandwidth for any existing ISO/IEC standard cable.

Question 7

Which type of cable offers the least immunity to EMI?

- ○ A. STP cable
- ○ B. Fiber-optic cable
- ○ C. Coaxial cable
- ○ D. UTP cable

The correct answer is D. UTP cable offers the least amount of immunity to EMI when compared to the other types of cable. Answer A is incorrect because STP cable has a higher immunity to EMI than UTP cable. Answer B is incorrect because fiber-optic cable has the highest immunity to EMI when compared to all types of communications cable. Answer C is incorrect because coaxial cable has a higher immunity to EMI than UTP cable.

Question 8

Which standard sets guidelines for residential single-family and multifamily telecommunications equipment and wiring?

- ○ A. ANSI/TIA/EIA 568-B
- ○ B. IEEE 802.3
- ○ C. ANSI/TIA/EIA 570-A
- ○ D. IEEE 802.11b

The correct answer is C. The most important standard you will need to know when preparing for the HTI+ infrastructure and integration exam is ANSI/TIA/EIA 570-A and its three addendums. It sets guidelines for residential single-family and multifamily telecommunications equipment and wiring. Answer A is incorrect because ANSI/TIA/EIA 568-B is the standard titled "Commercial Building Telecommunications Cabling Standard." Although many recommendations contained in the 568-B commercial standard apply to modern residential structured wring principles, it is not specifically focused on residential telecommunications standards. Answer B is incorrect because the IEEE 802.3 standard applies to Ethernet LANs and does not address residential wiring. Answer D is also incorrect because the IEEE 802.11b standard is applicable to wireless LANs and is not a standard specifically oriented to residential wiring.

Question 9

> What are the minimum wire size and category rating specified by the FCC for home wiring?
> - A. Category 5, solid copper #16-gauge or larger
> - B. Category 3, solid copper #24-gauge or larger
> - C. Category 4, solid copper #12-gauge or larger
> - D. Category 6, solid copper #18-gauge or larger

Answer B is correct. The FCC ruling states that, for new installations and modifications to existing installations, copper conductors shall be, at a minimum, solid, #24-gauge or larger, twisted pairs that comply with the electrical specifications for Category 3. Answer A is incorrect. Although Category 5 exceeds the performance rating of Category 3 cable, it is higher than the minimum rating called for in the FCC order. Answer C is incorrect because Category 4 exceeds the minimum performance rating of Category 3 cable. Answer D is incorrect because Category 6 is a new type of communications cable that exceeds the minimum rating specified by the FCC.

Question 10

> What is a grade 1 structured-wiring connection as defined in the ANSI/TIAEIA 570A standard?
> - A. A Category 3 cable and 75-ohm coaxial cable at each location
> - B. A Category 5 cable and a 75-ohm coaxial cable at each location
> - C. Two Category 5 cables and two 75-ohm coaxial cables at each location
> - D. Two Category 3 and two 75-ohm coaxial cables at each location

The correct answer is A. Grade 1 is designed to support basic residential telephone and video services. The standard recommends using one four-pair Category 3 UTP cable and one 75-ohm coaxial cable. Answer B is incorrect because the grade 1 rating for UTP cable is Category 3 and not Category 5. Answer C is incorrect because two Category 5 cables and two 75-ohm coaxial cables are the requirement for grade 2 structured wiring. Answer D is incorrect because the rating called for in grade 1 structured wiring is for one Category 3 cable and one 75-ohm coaxial cable and not two of each type.

Need to Know More?

 Gerhart, James. *Home Automation and Wiring*. Berkeley, CA: McGraw-Hill, 1999. (This book will help you install or plan a structured-wiring system for a home.)

 See http://www.smarthome.com/9372.html for "Wiring Your Home for the Future" (VHS tape Stock Number 9372).

 See http://www.smarthome.com/9372.html for "Structured-wiring Design," Manual Number 9356.

 See http://www.tiaonline.org/standards/ for a complete catalog and price list for ANSI/TIA/EIA standards.

 See http://www.tiaonline.org/standards/overview.cfm for a historical overview of how the ANSI/TIA/EIA standards are developed and published.

 See http://www.nfpa.org/catalog/home/index.asp for an online catalog, prices, and descriptions of the NEC/NFPA standards.

 See http://www.ansi.org/news_publications/latest_headlines.aspx?menuid=7 for the latest information on publications and news releases from the ANSI organization.

 See http://www.eia.org/ for news and a list of standards available from the Electronic Industries Alliance.

11.0—Structured Wiring—High-voltage

Terms you'll need to understand:
- ✓ Ground
- ✓ Kilowatt-hour
- ✓ Three-way switch
- ✓ UPS
- ✓ Romex
- ✓ BX
- ✓ MC cable
- ✓ VA
- ✓ Power factor
- ✓ Ballast
- ✓ Entrance panel
- ✓ GFCI
- ✓ Load

Techniques you'll need to master:
- ✓ Calculating true power and apparent power when the power factor is known
- ✓ Selecting the correct wire color code for wiring an AC outlet
- ✓ Drawing a diagram of a three-way switch lighting circuit
- ✓ Knowing how to reset a circuit breaker

Structured-wiring and high-voltage cabling are topics included in the second (HT0-102) HTI+ examination, "Systems Infrastructure and Integration." This chapter covers the essential points you will need to know for this portion of the HTI+ exam.

The term *high-voltage* might appear to those of you familiar with home wiring as the generic term used by electric utility companies to describe transmission systems that distribute electric power over large grids throughout the country. HTI+ objectives use the term *high-voltage* to include all home AC power wiring. In the HTI+ exam objectives, *high-voltage* and *low-voltage* structured wiring are terms used to differentiate the two fundamental structured wiring concepts included in the second HTI+ exam. In Chapter 10, "10.0—Structured Wiring—Low-voltage," you were introduced to the concepts and designs for structured wiring low-voltage, domain 1A. This chapter introduces the second portion of the structured wiring objectives, called structured wiring high-voltage, domain 1B.

High-voltage structured wiring follows the same model you learned about in Chapter 10. All wiring is distributed throughout a home from a central location called a *circuit breaker panel*. High-voltage wiring includes all two-phase 240v alternating current (AC) service provided to a home by an electric utility company. Two-phase high-voltage service includes two hot wires of opposite phase and a neutral wire. The two main wires provide 240v service and two 120v circuits are available between each hot wire and a neutral wire. Three-phase electric service is widely used for commercial facilities but typically is not used for residential construction.

This chapter introduces the types of circuits, switches, wiring, load determination, and distribution methods used in residential high-voltage wiring. Although you will need to know the basic terms and standards included in this chapter, many of the authoritative codes covering residential high-voltage wiring are dictated by local residential electric inspection criteria and local building codes.

High-voltage Wiring Design Considerations

Planning the design for a residential high-voltage wiring system involves adding the total amount of current (*amperes*) used by all the electrical devices in a home. This includes making a table of the load requirements, installing the ground system, planning for surge protection and uninterrupted power supplies, and keeping safety issues in mind. In this section, you will learn the basics of design considerations for home wiring.

NOTE: *High-voltage* usually refers to voltages used for power-line transmission systems measured in the kilovolt range. The term *high-voltage* is used in the HTI+ exam objectives to describe the requirements for residential AC 240v/120v service. This is in contrast to the objectives described in Chapter 10, for low-voltage structured wiring. This chapter covers the exam objectives for HTI+ high-voltage wiring.

Load Requirements

The *load* of a residential electrical system is the total amount of electrical power that is used by all appliances and HVAC and lighting systems. The load is determined during the initial design of a home and is based on the size of the home, number of outlets, HVAC system, appliances, and lighting systems. The load for a residential system varies with the time of the year as well as the time of day. Electrical contractors size the load based on the peak demand calculations covered earlier in Chapter 5, "5.0—Home Lighting Control and Management," for residential lighting systems.

Electrical contractors use a table in which each appliance, light fixture, HVAC, entertainment system, and so on is listed with the power in watts/hour. Some items might be used for less than 1 hour at a time, such as a garage light. In this example, the garage light would be calculated for fractions of a watt-hour. A 120-watt light multiplied by .2 hours (120 × .2 hours = 24 watt/hours). A watt-hour is a relatively small amount of power; therefore, it is easier to express power consumed in kilowatt-hours (kWh), which equal 1000 watt-hours. Electrical power companies charge consumers for the amount of kilowatt-hours consumed during a month.

Grounding

As mentioned in Chapter 5, an electrical ground in an AC power system is a wire that is connected to the earth, hence the name *ground*.

Figure 11.1 illustrates how the ground connection is used to establish the common point location for the earth ground in an AC power distribution system.

A step-down transformer provides a conversion from the power utility 7200-volt (v) main power line to 240v. The 240v service is extended from a pole-mounted or in-ground transformer to the residential service entrance. The transformer windings shown in Figure 11.1 show primary and secondary windings. The secondary, or service, side includes two wires plus the center tap. The voltage between the two outside wires provides a two-phase 240v service. The neutral is taken from the center tap of the transformer, which serves as the return for the 120v service that is used for the standard 120v circuits in the home.

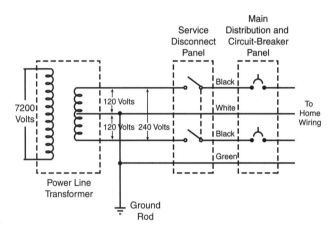

AC Power Distribution

Figure 11.1 AC power distribution.

Earth Ground Rods

The center tap connection of the power pole transformer is connected to the earth using a copper bar driven into the ground. From this ground connection, two wires are taken to the service entrance. One of these wires is called the *safety ground* or *green* wire, and the other is called the *neutral* wire. The neutral wire is designed to serve as the return path for the home wiring to the center of the transformer. The safety ground path is not designed to carry any current unless a fault in the wiring occurs due to an appliance failure. The safety ground wire appears to be redundant because the neutral wire is derived from the same point. In electrical equipment that has a safety ground connection (as evidenced by a three-prong plug), the safety ground is always connected to any exposed metal parts of the equipment in case of a wiring fault inside the appliance or lamp fixture. If a fault were to occur, such as an accidental connection between the hot wire and the case, the safety ground connection would cause the hot connection to be directly connected to earth, which would result in the activation of protective devices such as a ground fault circuit interrupter (GFCI), fuse, or circuit breaker shutting down the power circuit.

In the past, earth grounds for the home service entrance used the water pipe as the earth ground reference point. Because an increased use of plastic water pipe and nonconducting fittings has made the effectiveness of grounding to plumbing systems questionable, the method does not meet current safety standards. Although water lines can be used in some circumstances, the National Electrical Code (NEC) now states that the home might also require one or more "supplemental grounding electrodes" buried in the house foundation or in the earth outside the home. An example of a ground rod installation with the grounding wire installed is shown in Figure 11.2. One or more copper-clad grounding rods several feet in length are often recommended. Local code-enforcement also might require that grounding rods be added to existing homes when new electrical work is done or when the home is sold.

Figure 11.2 Ground rod.

Hot and *neutral* are terms frequently used to describe the black and white wires respectively in residential high-voltage wiring. The green wire referred to as the *equipment grounding conductor* is returned to the ground rod and normally does not carry any current unless a fault occurs in the wiring. It is important to remember that the NEC indicates the equipment grounding conductor used in residential power wiring must be green or bare.

Surge Protection

Surge suppressors provide protection to home appliances and computers from potentially destructive voltage surges and spikes on power lines that occur primarily as a result of lightning strikes. A *spike* is a short burst of energy superimposed on the AC line, usually of very short duration. Magnitudes can be from 0.5v to 2000v and last for 8–10 milliseconds. A *surge* is also an unexpected change in the nominal line voltage, usually no more than two or three times the normal voltage. Surges are characterized as being longer than 10 milliseconds. Surges and spikes can be hazardous to equipment that contains semiconductor components, such as computers.

Power Backup (UPS)

A residential backup power system is also referred to as an *uninterruptible power supply (UPS)*, which is used to provide emergency AC power for the home when commercial power service is unavailable. An AC generator powered by a gasoline engine is required for larger loads for an extended power outage. Commercial power outages are often unpredictable and can occur due to storms, accidents, or emergency maintenance activity by the power utility company.

Backup power supplies are available for powering the entire home or for the home computer.

Residential Backup Power Options

Designed to be permanently installed outside the home, an emergency backup power generator is fueled by gasoline, diesel, natural gas, or propane. It can be activated automatically or manually to supply power after a utility power interruption or outage.

Residential backup power systems can be sized and wired for a home in one of two power distribution designs described a follows:

➤ *Whole house power distribution*—This type of power generator is sized for powering all devices within the home.

➤ *Essential power distribution*—This type of power system is designed for isolated essential devices such as a security system, a freezer, a refrigerator, HVAC, and limited emergency lighting.

Residential power backup systems use a transfer switching device to transfer the electrical load to the generator and disconnect the commercial power source. Automatic switches provide an automatic signal to start the generator and switch over the load to the generator when commercial power has failed. When commercial power is restored, the switch automatically returns the residential load to the commercial power source and shuts down the generator. Most residential backup systems have voltage regulators to maintain a constant voltage under varying load conditions.

Computer Backup Power Systems

The most widely used backup power system is an uninterruptible power supply. A UPS is designed to provide power to client computers and servers in a business location or home office for a brief period after a power failure to allow a gradual planned shutdown of computer systems.

UPS specifications indicate the length of time as well as the amount of electrical power the unit can deliver when commercial power is lost. The most important rating to be aware of is the volt-ampere (VA) rating. The VA rating is the product of the volts and amperes delivered to the load. The VA rating of a UPS unit is considered to be the full load rating. Half load is simply a VA load that is half of that figure. For example, a typical UPS can support an 800 VA load for 7 minutes or a 400 VA load for 23 minutes.

Switching mode power supplies (SMPSs) are widely used to power all modern computers and present a power factor to the source in the range or 0.6–0.7, depending on their design and rating. This is the basis for UPS specifications listing both VA ratings and watts ratings.

> VA ratings for UPS units are the product of volts and amperes. The product of the voltage and current measured gives the apparent power of the load on a UPS. This apparent power is made up of a part that does the useful work (measured in watts) and a part that does no work but does generate heat. The part that does useful work is called *real power*, and the part that does no useful work is called *reactive power*.
>
> To convert from VA to watts in an AC system, *power factor (pf)* is used. The power factor is a number between 0.0 and 1.0 representing the fraction of the apparent power delivered to the load that does useful work. Power factor is defined as follows:
>
> $$\frac{Watts}{VA}$$
>
> To determine the real power of the load, simply multiply the VA by the pf, like so:
>
> Watts = VA * Power Factor

Converting Between VA and Watts

The VA rating of a computer is used to select the proper VA rating required for an UPS. The watt and VA ratings can differ significantly with computers, with the VA rating always being equal to or larger than the watt rating. Most computers equipped with an SMPS use capacitor input switching power supplies, which (due to the input characteristics of the switching type converter) exhibit a power factor of approximately 0.6. Thus, the watts drawn by a typical computer are 60% of the VA rating. VA and watts can be converted using the equations presented in the previous section. An example of a computer and monitor VA and wattage calculation follows:

➤ A typical monitor has a VA rating of 252 (120v × 2.1 amperes = 252 VA). Converting to watts using the monitor power factor of 0.6 calculates as 252 × 0.6 = 151.2 watts.

➤ A typical computer has a VA rating of 378 VA (120v at 3.15 amperes = 378 VA). Converting to watts using the computer power factor of 0.6 calculates as 378 × 0.6 = 226.8 watts.

➤ The combined VA rating for this computer system is 252 + 378 = 630 VA.

➤ When selecting a UPS for this computer system, the UPS should have a VA rating equal to 630 or higher.

A UPS system is designed to provide temporary power to a computer system after a commercial power failure. It can provide uninterrupted power because it is always providing AC power to the protected system during normal operation as well as during emergency operation; therefore, no switching time is involved when commercial power is lost. It also isolates the computer from transient voltage disturbances and spikes during normal operation.

Safety Considerations

Working with electrical wiring requires you to have an understanding of basic electrical safety procedures. The installation of new wiring in a home should be performed by a qualified licensed electrician. This ensures that the remodeled or newly constructed home wiring installation will be done according to local building codes, and it is required to pass electrical wiring inspections in most areas.

Any house that has been properly wired by a qualified electrician will have a circuit breaker panel that is used to shut off circuits in the event that they draw too much current. It is the current capacity of the circuit breaker (in amperes) that determines how much current a circuit can supply. The breaker size is chosen relative to the type of cabling and connector used for the circuit because each has a different capacity.

When making repairs to the electronic wiring system or equipment, that part of the wiring of equipment needs to be disconnected from the main voltage so no dangerous voltages are present when the work is done. This is the case in all electronics and electrical repairs.

Ground Fault Circuit Interrupters

A special type of electrical outlet designed to guard against accidental electric shock is called a *ground fault circuit interrupter (GFCI)*. A GFCI works by comparing the amount of electrical current coming into a circuit (on the black wire) with the amount leaving the circuit (on the neutral or white wire). If more current enters the circuit than leaves, a current leak or ground fault occurs. A GFCI is designed to detect as little as a 0.005 amp ground fault current and can shut down a circuit within 0.025 seconds to prevent serious electrical shocks. A GFCI might be required by some local electrical building codes to protect an entire branch of circuits such as an outdoor pool area.

A typical GFCI outlet is shown in Figure 11.3.

You should be prepared to describe how a GFCI outlet works and how it protects the user against electrical shock. Also, know the meaning of a ground fault.

Figure 11.3 A ground fault circuit interrupter.

Audio and Video Equipment Locations

This section describes the location considerations for audio and video systems and the associated AC power wiring needed for handling the loads. The topics you will need to study include new and remodeled wiring and equipment component placement for home entertainment systems.

Project Settings

Project settings include the planning for high-voltage power distribution as well as whole home audio and video distribution discussed in Chapter 2, "2.0—Audio and Video Fundamentals." Residential remodeling and new construction involve planning for the number of AC power circuits and the locations of electronic components.

Remodeling/Existing Structure

When remodeling an existing structure that includes new electrical service upgraded to 200 ampere service, you might need to check with the local power company for placement of the service entrance. You might have to move it to a new location if the breaker panel is installed in a different area.

Older homes might use a fuse panel containing replaceable fuses. Homes built many years ago also might have no ground wires, with only the hot and neutral conductors present. Upgraded wiring is recommended in this example to bring an older home up to code with ground wires and grounded outlets in every room.

The service entrance includes the masthead, conduit, and wiring from the top of the masthead into the home. The service panel or breaker panel is normally located as close as possible to the service entrance; however, in some cases you might need to locate it farther away depending on the extent of the remodeling activity. The NEC provides for minimum construction standards, including conductors and electrical load requirements. The size of electrical wiring is directly related to its capacity to conduct electricity. The larger the wire, the more electricity it can safely conduct. Wire gauge size is measured inversely—that is, 14-gauge wire is smaller than 12-gauge wire. The amount of electrical current a conductor can carry is measured in *ampacity*, a shorter term for ampere-capacity. A 14-gauge wire is rated at 15 amps, whereas a 12-gauge wire is rated at 20 amps. If electrical wiring is subject to electrical current in excess of its capacity, it generates heat and eventually causes a fire. Because of this situation, all electrical circuits are required to be provided with circuit breaker or fuse protection rated at not more than the wire's maximum ampacity. New lighting and outlet circuits in a remodeled home require #12 American Wire Gauge (AWG) wire to safely handle 20 amperes. Special outlets and circuits for appliances require larger wiring and breakers. For example, electric furnaces vary in size, and the breaker and wiring size depend on the furnace requirements. Central air conditioning requires a breaker and wiring of the size recommended by the manufacturer.

New Construction

New home construction typically requires a minimum of 200-amp service to meet the load requirements of a modern residential electrical system. If the home has electric heat, a larger service might be required. Larger homes can require 400-amp service or greater.

The number of lighting circuits required for new construction depends on the size of the home. You should have at least two circuits, staggering the connections so that if a breaker trips, adjoining rooms will still have lights.

Power outlets for audio and video equipment should be located in the area planned for the home theater TV, video distribution, and sound components. Typical design specifications for powering a home theater equipment rack include two dedicated 20-amp circuits. One circuit is dedicated to the equipment rack, and the other is dedicated to powering the whole-house audio amplifiers and the media room wall outlets. Many electrical contractors recommend running separate power circuits from the circuit breaker panel to sensitive audio/video equipment. This helps keep large current-drawing items such as furnaces, air conditioners, refrigerators, and so on from causing voltage fluctuations on the audio/video outlets that power the electronic equipment.

Equipment Component Placement

Home audio/video component placement should be planned in advance with the electrical outlets necessary to meet all the power requirements for VHS/DVD players, TV, audio and video amplifiers, tuners, and settop satellite TV decoders. Most components of a typical home theater system sit in cabinetry against one wall in the room with the main three speakers to the side, under, or on top of the cabinets. The surround speakers often mount on walls, mount on stands, or rest on furniture. Figure 11.4 illustrates the normal component location of home theater components.

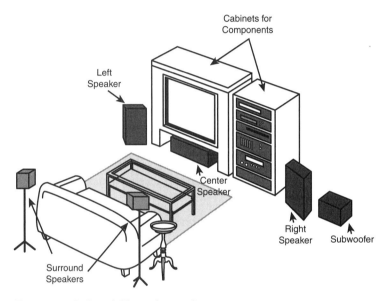

Figure 11.4 Audio and video equipment placement.

Physical High-voltage Structured-wiring Components

High-voltage structured-wiring components used in the typical automated home are covered in this section. Components you will need to know more about include standard AC wall outlets with the identifying nomenclature, GFCI outlets, incandescent and fluorescent light dimming modules, light switches (including standard, three-way, and four-way switches), light fixtures, and AC power source equipment.

Outlets

The National Electrical Manufacturers Association (NEMA) has developed standards for the physical appearance of AC power wall outlets. Figure 11.5 shows the configuration for standard wall outlets. The large slot identifies the neutral (white) wire, and the narrow slot is connected to the hot side. The round hole is connected to the ground wire.

Figure 11.5 Wall outlet configuration.

Outlets have two terminal screws on each side. One pair is black or brass in color and the other screws are silver. The black or red (hot) wire always connects to the brass terminal, whereas the white wire always connects to the silver terminal. The ground wire connects to the green screw. When the break-out link or fin between the brass terminals is removed, the power is connected to only one of the outlets. This is often used where it is desired to use one of the receptacles for a switched outlet controlled from a light switch for turning on or off an appliance or light from another location.

A back-wired option is available on most outlets. They have openings in the rear that provide openings for inserting the wires. The same color code should be observed for inserting the black and white wires to the proper side of the receptacle. Be aware that some local electrical codes might not allow the use of the back-wired option for outlets. The identification of terminal connections for both standard wall outlets and ground fault circuit interrupter (GFCI) outlets are shown in Figure 11.6. Note the location of the break-out fins discussed earlier.

Ground-fault circuit interrupter protection should be used with all bathroom, kitchen, and outdoor circuits such as pools, spas, and adjacent areas where moisture is present. This type of device, discussed earlier in the section "Safety Considerations," immediately disconnects the power in case of a ground fault.

The purpose of the break-out fins is to provide isolation between duplex outlets. Be sure you remember the correct connections on a wall outlet for the black wire (brass screw) and the white wire (silver) and that the green screw is used for connecting the green (or bare) ground wire.

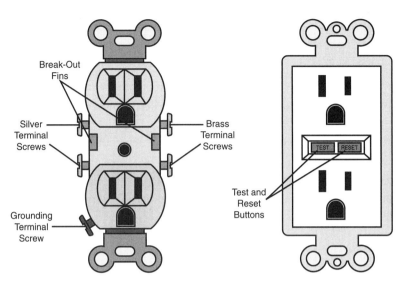

Figure 11.6 Wall outlet and GFCI nomenclature.

Dimming Modules

In many residential lighting systems, various lighting areas and zones might not need to be operated at full brightness levels. They might be required to fade in and out and be used at different light levels at different times. A device called a *dimming module* or *dimmer* is required to regulate the amount of electrical voltage sent to each light fixture, thereby allowing the intensity of the light to be varied.

A dimmer contains a current limiting feature that can vary the voltage distribution between the dimmer and the lamp. Dimmers are available from numerous vendors in many styles to control different types of loads. Older dimmers use a variable resistor to limit the current and resulting voltage available to the lamp.

More modern electronic light dimmers use a variable phase control principle where the voltage is switched on and off with special components inside the dimmer module. The components are called silicon controlled rectifiers (SCRs) and thyrister for AC applications (TRIACS). A SCR can be thought of as a device that is either on or off. After it is turned on, it only turns off when no current is flowing through it. It has three terminals and schematically looks just like a diode with an extra wire coming off the anode. This extra wire is the gate, and it requires very little current to switch on the SCR. An SCR conducts current only one way, so if the current changes polarity, it turns off because it doesn't allow any current to flow.

A TRIAC schematically resembles two SCRs, back to back with only one gate coming from one anode. This enables AC operation, but every time the current changes polarity, the TRIAC turns off unless the gate is held high.

Variable phase control, mentioned earlier, refers to allowing only portions of the AC cycle to go through the load. It is usually performed with a TRIAC or two back-to-back SCRs. AC current sources have a zero-crossing two times every cycle. This happens 120 times per second for 60 hertz (Hz) AC service.

The TRIAC turns off 120 times per second. By varying the turn-on period, the amount of power applied to a lighting circuit can be varied. This produces this waveform show in Figure 11.7.

NOTE — TRIAC dimmers are designed to work exclusively with incandescent lighting. Fluorescent lighting requires special ballast designed for dimming and a special dimmer (one that indicates it is for a dimming fluorescent fixture). Ordinary fluorescent fixtures are not designed for dimming. Additional information on fluorescent lighting is covered later in this chapter in the section "Fluorescent Lighting Fixtures."

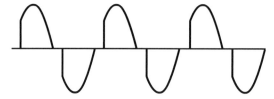

Figure 11.7 TRIAC waveform.

To deliver only half the power, the TRIAC is fired between each zero-crossing, producing the AC waveform shown in Figure 11.8.

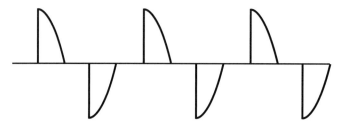

Figure 11.8 TRIAC half-power waveform.

This type of module is normally used for light dimming because lights are resistive loads and do not flicker even if they only get voltage spikes, as illustrated in Figure 11.8's waveform.

Light Switches

Light switches are used to control lighting circuits from locations that are convenient to home residents. The types of switches available for residential installations vary according to where they are installed, the type of circuit they are controlling, and the current rating of the load connected to them.

Wall switches are the most popular type of switch for residential light and appliance control. They are designed for easy installation in a wall switch box. The switch contacts are available in various configurations depending on the type of control required. The *pole* is the number of circuits the switch makes or breaks, and the *throw* is the number of positions to which each pole is switched. The most common types of poles and contact configurations for switches use the following abbreviations and nomenclature.

Standard Switch Configurations

The standard types of switch configurations are shown schematically in Figure 11.9. Each configuration is explained in the following list:

▶ *Single pole single throw (SPST)*—This type of switch is used for operating a single light fixture. The incoming hot wire is hooked to one terminal screw, and the outgoing hot wire is connected to the other screw.

▶ *Single pole double throw (SPDT)*—This type of switch is referred to as a *three-way* switch and is used for controlling a light from two locations. It has a double-throw capability and can change the current path between two different wires. A description of a three-way installation for controlling a light fixture from two locations is discussed in the following paragraphs.

▶ *Double pole single throw (DPST)*—This switch is another type of on-off control and is used to interrupt a two-wire circuit.

▶ *Double pole double throw (DPDT)*—This type of switch is another changeover type that is normally used to control 240v, two phase circuits located at the service entrance panel.

 Remember the nomenclature and feature for each switch configuration, including DPDT, DPST, SPDT, and SPST.

430 Chapter 11

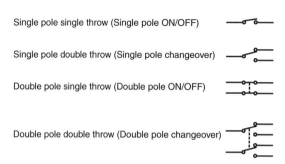

Single pole single throw (Single pole ON/OFF)

Single pole double throw (Single pole changeover)

Double pole single throw (Double pole ON/OFF)

Double pole double throw (Double pole changeover)

Figure 11.9 Switch configurations.

Three-way Light Switches

The three-way switch circuit is used frequently in residential installations to operate a light from two locations with one three-way switch located at each position.

Three-way switches look the same as the common single pole switch, but instead of having only two screws on which to make connections, they have two connections on one side and two on the other side (see Figure 11.10).

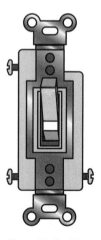

Figure 11.10 Three-way switch.

One terminal is referred to as the *common* terminal, one terminal is the ground terminal, and the other two are known as *travelers*. This is because the electrical connection goes from the common screw to one of the travelers, depending on the switch position. This is illustrated in the wiring diagram in Figure 11.11 for controlling a light fixture from two locations. Notice how moving either switch to the opposite position has full on/off control. As you can observe from the three-way switch diagram, there is no "on and off" marking for a three-way switch. Moving either switch to the opposite position changes the status of the light from on to off or off to on.

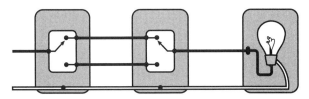

Figure ... Three-way switched circuit.

A light switch used to operate a light from two locations is called a three-way switch. Remember that a three-way type switch is required to be located at both positions.

As indicated in this section, a three-way switch is used when two switches are required to control one light from two locations. The term can be misleading because the switch controls a single light from two locations. Memorize the wiring diagram for a three-way switch and remember how it is used.

Fixtures

120v circuits are used to provide power to light fixtures as well as outlet boxes. Light fixtures are always mounted on round or octagon junction boxes and should be mounted in the center of a room, unless for a specific use such as a dining room table. Fixtures such as chandeliers and ceiling fans that weigh more than 25 lbs. require a special mounting box with a bracket that is supported by ceiling joists. Fixtures can include both incandescent and florescent illumination.

Incandescent Light Fixtures

As mentioned previously, incandescent lamps are the least energy-efficient electric light source and have a relatively short life (750–2500 hours). Incandescent lamps contain a tungsten filament enclosed in a glass bulb with a small amount of low-pressure, inert gases to preserve filament life. Light is produced by passing a current through a tungsten filament, causing it to reach incandescence and glow to produce light. With extended use, the tungsten material in the filament slowly evaporates, eventually causing the filament to fail.

Three-way bulbs provide three settings for brightness and contain two filaments. They must be installed in lamp fixtures that have a switch that can select one or both of the three-way bulb filaments. When the switch is operated, it selects the lowest wattage filament (30 watts as an example). The next position selects the second filament (70 watts), and the third position of the switch connects both filaments to the line voltage to produce a 100-watt light output.

Halogen lamp fixtures are a type of incandescent lamp. They have a longer life than conventional light bulbs, but they are only slightly more efficient. Halogen lamps are best suited for lighting areas in which a direct focus of light is required.

Fluorescent Lighting Fixtures

A fluorescent lamp is a low-pressure or low-intensity discharge lamp. The lamp consists of a closed tube that contains two cathodes, an inert gas such as argon, and a small amount of mercury. A florescent lamp produces light when AC line voltage is supplied to the lamp in the correct amount. An electrical arc strikes between the two cathodes and emits energy that the phosphor coating on the lamp tube converts into usable light.

A fluorescent lamp tube has argon combined with a minuscule amount of mercury. At the low pressure within the lamp, it becomes mercury vapor, even at temperatures only slightly above room ambient. An electrical discharge ionizes the mercury vapor, which emits UV radiation that stimulates phosphors that coat the interior of the lamp's glass envelope. The phosphors then convert essentially all the UV radiation to visible light. The conversion of electrical energy to light is much more efficient than in an incandescent lamp, and a considerably smaller fraction of the input energy is converted to heat. Generally, fluorescent fixings produce approximately three times as much light per watt as halogen lights. The color of the light a fluorescent lamp produces depends on the composition of the lamp's phosphors.

Prior to the 1990s, the magnetic ballast was the most commonly used ballast for all fluorescent lighting systems. A magnetic ballast consists of aluminum or copper wire wound around a laminated steel coil, a power factor correction capacitor in high-power factor units, and an appropriate thermal potting compound. Other ballast technologies that have gained acceptance in the past decade include hybrid magnetic (low-frequency electronic) ballasts, and high-frequency electronic ballasts.

Hybrid ballasts are similar in construction to magnetic ballasts and include circuitry that removes power from the lamp electrodes after the lamp has ignited. This technology provides an additional 10% energy savings and cooler operating temperatures, while at the same time maintaining or improving lamp life.

Hybrid ballasts are available for F32T8, F40T12, and F96T12/HO lighting systems.

Electronic ballasts consist of electronic components and small magnetic devices that change the 60Hz line input frequency to 20,000Hz or greater output frequency to the lamp. Electronic ballasts are generally 10%–25% more efficient than magnetic systems, produce less waste heat, and have a sound output that is less than half of existing magnetic ballast designs. Communications cabling should always be located as far away from fluorescent fixtures as possible due to the potential for electromagnetic interference from ballasts.

Fluorescent Tube Specifications

Fluorescent tube specifications include letters and numbers used to identify physical and electrical features. The marking nomenclature describes the tube type, wattage, color, and diameter using the form FSWWCCC-TDD. Each of the letters identifies one or more of the tube specifications according to the following codes:

- ➤ *F*—Fluorescent lamp.
- ➤ *S*—Style—No letter indicates a straight linear tube. C indicates a circular tube.
- ➤ *WW*—lists the power in watts such as 4,5,8,12,20,39, and 40.
- ➤ *CCC*—Color includes W=White, CW=Cool White, WH=Warm white, BL/BLB=Black light.
- ➤ *T*—Tubular bulb.
- ➤ *DD*—Diameter of the fluorescent tube in eighths of an inch. T8 is 1 inch, T12 is 1.5 inches, etc.

For the most widely available tube such as a T12, the wattage rating is normally 5/6 of the length of the tube. Therefore, an F40 tube is 48 inches long.

Source Equipment

The electrical power for AC circuits is distributed around the home from a central source point, which is often called a *circuit breaker box*. Circuit breakers are a special type of switch installed to detect over-current conditions on the various circuits distributed throughout the home. A circuit breaker automatically opens an electrical circuit and disconnects power when a problem occurs that results in excessive current being drawn from fixtures or devices plugged in to outlets. The breaker box is a central distribution point for all high-voltage (120v and 240v AC) wiring.

Configuration and Settings for High-voltage Structured Wiring

This section addresses the types of termination points, entrance panels, and power consumption for residential high-voltage wiring systems.

Distribution Panels

The drop (or service line) from the outside power transformer is connected to the household system at a service head, usually on the outside wall of the house. Distribution panels for AC power wiring include the main disconnect box located outside the home under the electric watt-hour meter. This panel contains the main circuit breaker providing a point where power to the home can be disconnected. This distribution panel is equipped with either a 100-amp or 200-amp circuit breaker(s), depending on the type of service installed in the home.

Circuit Breaker Panels

Circuit breaker panels (also referred to as *entrance panels*) have two sets of breakers. One set, for the main power supply, is usually at the top of the panel, and these breakers control power to all the circuits. The other breaker set is usually arranged in two columns running down the panel and control individual circuits throughout the house. Circuit breakers with a single toggle serve 120v circuits. Two-toggle breakers, on the other hand, serve 240v breakers have two hot wires that run through the circuit—if a problem occurs on either line, both breakers trip.

Some older homes are equipped with fuses instead of circuit breakers. They perform the same service, except a fuse must be replaced if it detects an overcurrent condition. A circuit breaker can be reset after being tripped and therefore is a reusable type of protection device. Both types of entrance panels are shown in Figure 11.12.

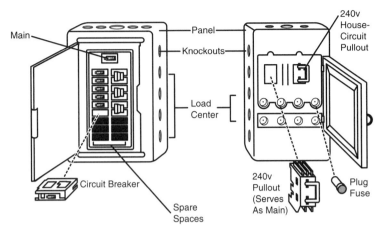

Figure 11.12 Circuit breaker and fuse panels.

Termination Points

Termination points for AC power wiring include wall outlet boxes, wall switches, and junction boxes. Each of these electrical devices provides termination points for connecting AC power to outlet receptacles, light fixtures, ceiling fans, GFCI outlets, and wall switches. All switches and outlets require a properly sized junction box. For example, a 2" × 3" box with three wires (14-gauge) should be 2 1/2" deep. The same box with five wires must be 3 1/2" deep. A junction box is installed next to a wall stud, as shown in Figure 11.13.

Figure 11.13 Junction box installation.

Power Consumption

Utilities measure energy by the kilowatt-hour—that is, how many thousands of watts of power are consumed per hour. Monthly electric utility bills are based on the amount charged per kilowatt-hour.

Power consumption is the amount of electrical power in watts used by the various appliances. To determine the cost of power consumption in a home, estimates are made of the average time in watts per hour used for various appliances. Times can vary between households; however, most consumer products publish information on the cost of operation based on the average number of watt-hours/minutes used per day. Some appliances are used for short intervals, whereas others are used throughout a 24-hour period.

As an example, a ceiling fan that consumes 100 watts per hour would use .1 kilowatt-hours of energy each hour. If the utility company charged 10 cents per kilowatt-hour, it would cost the consumer 1 cent per hour to operate the fan. A refrigerator is used continuously and would require the kilowatt per hour consumption rate to be multiplied by 24 to determine the daily cost of operation.

Typical wattage ratings for electrical appliances and their corresponding costs for an assumed rate of 8 cents per kilowatt-hour are listed in Table 11.1.

Table 11.1 Residential Electrical Power Consumption Estimates @ 8 cents/kWh

Appliance	Usage	Rate
Air cleaner	30 watts per hour	.24 cent per hour (.03 kWh × 8 = .24)
Can opener	50 watts per minute	.4 cent per minute (.05 kWh × 8 = .4)
Clock	2 watts per hour	.38 cent per day (.002 kWh × 8 × 24 = .384)
Electric heating pad	60 watts per hour	.48 cent per hour (.06 kWh × 8 = .48)
Fans (ceiling)	100 watts per hour	.8 cent per hour (.1 kWh × 8 = .8)
Fax machine	90 watts per hour	17.28 cents per day (.09 kWh × 8 × 24 = 17.28)
Home computer	200 watts per hour	1.6 cents per hour (.2 kWh × 8 = 1.6)
Iron	1100 watts per hour	8.8 cents per hour (1.1 kWh × 8 = 8.8)
Microwave	1500 watts per hour	12 cents per hour (1.5 kWh × 8 = 12)
Stereo	100 watts per hour	.8 cent per hour (.1 kWh × 8 = .8)
Television	150 watts per hour	1.2 cents per hour (.15 kWh × 8 = 1.2)
Toaster	1100 watts per hour	.146 cent per minute (1.1 kWh × 8/60 = .1467)
Lights	100-watt bulb	.8 cent per hour (.1 kWh × 8 = .8)
Oven (electric)	1500 watts per hour	12 cents per hour (1.5 kWh × 8 = 12)
Range top stove	300 watts per hour	2.4 cents per hour (.3 kWh × 8 = 2.4)
Refrigerator	220 watts per hour	42.2 cents per day (.22 kWh × 8 × 24 = 42.24)

Device Connectivity for High-voltage Structured Wiring

The selection of the proper wire type is necessary for various environments for home wiring. You need to know the abbreviations and code names for the interior nonmetallic cable as well as metallic-covered cable used in residential construction.

NM Cable

Nonmetallic (NM) cable is the most widely used cable for indoor AC power wiring. Romex is a brand name for a type of plastic insulated wire. It is often called nonmetallic sheath, although the formal code name is NM cable. This type of wire is suitable for stud walls and on the sides of joists that are not subject to mechanical damage or excessive heat. Newer homes are wired almost exclusively with NM wire.

NM cable comes in two types: NM and NMC. NM cable has a flame-retardant and moisture-resistant cover, whereas NMC cable is corrosion-resistant. Its covering is flame-retardant, moisture-resistant, fungus-resistant, and corrosion-resistant.

NM cable is an example of cable in which several wires are wrapped together. When a cable such as NM has two number 14 wires, it is known as 14-2 (fourteen-two) cable; in the case of three number 12 wires, it would be identified as 12-3 (twelve-three). An example of NM 14-2 with a ground wire cable construction is shown in Figure 11.14.

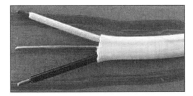

Figure 11.14 NM cable.

MC Cable

MC (metal-clad) cable is a type cable that includes its own flexible metal covering, as shown in Figure 11.15.

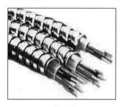

Figure 11.15 MC cable.

MC cable is composed of THHN soft-drawn, copper wire conductors, and an insulated grounding conductor. It is suitable for branch, feeder, and service power distribution in commercial and industrial applications as well as in multifamily buildings, theaters, and other populated structures. The letters *THHN* refer to heat-resistant thermoplastic (90° C) for dry locations. This is a reference to the type of insulation on wiring, coded per the NEC. All the NEC designations for wire insulation letter codes are referenced later in this chapter.

MC cable can be installed in cable trays and approved raceways and as aerial cable. It is suitable for use in wet locations (when it has a PVC jacket on it) or in dry locations, at temperatures not exceeding 90° C.

Many installers prefer the use of MC cable instead of conduit because it installs quickly and does not require the level of experience necessary for installing rigid conduit.

AC Cable

Armored cable is identified by the code name AC cable and is also often referred to as BX cable. Basic armored cable was developed in the early 1900s by Harry Greenfield and Gus Johnson, who called their product BX cable. It has become a generic term for all armored cable. Armor clad and metal clad might sound like the same cable, but some important differences exist between them. AC cable uses the interior bond wire in combination with the exterior interlocked metal armor as the equipment grounding means of the cable. MC cable is manufactured with a green insulated grounding conductor, and this conductor, in combination with the metallic armor, comprises the equipment ground. Armored cable is suitable for wet locations and is also used where extra protection from damage to wiring is desired.

AC cable can have up to four insulated conductors only; a fifth insulated conductor is allowed by Underwriters Laboratories, Inc. (UL) specifications if it is a grounding conductor. Each conductor in AC cable is paper wrapped.

Termination Points

Termination points for home electrical wiring are established in the circuit breaker panel, wall outlets, and light fixtures. Each circuit is identified by the circuit breaker that protects the outlets, appliances, and lighting circuits connected to it. A standard stud-mounted outlet box and wall outlet termination point is illustrated in Figure 11.16.

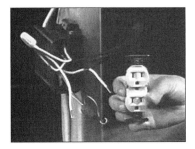

Figure 11.16 Stud-mounted outlet box and receptacle.

Standards and Organizations

This section summarizes the roles of several standards organizations referenced in this chapter. When preparing for the HTI+ examination, you should be able to describe the name and technology area served by each organization.

National Electrical Code

The National Electrical Code is a document that basically describes recommended safe practice for the installation of all types of electrical equipment. The NEC is not a legal document unless it is so designated by a municipality as its own statute for safe electrical installations. It is revised and published every three years. The NEC is *national* only in the fact that it is the only document of which all or part is accepted by all states as an electrical guide. It is the only document of its kind written with national input supplied by 20 panels of advisors containing several hundred experts in the electrical field from all parts of the country. The sponsoring agency of the NEC is the National Fire Protection Association (NFPA). In this chapter, the following documents were used in the development of material on cable types:

- *NEC Article 333*—Covers armored cables. Armored cables are manufactured in accordance with UL 4.

- *NEC Article 334*—Covers metal clad cables. MC cables are manufactured in accordance with UL 1569.

Wiring Types and the National Electrical Code

The NEC sets standards for home wiring types. A set of standard terms used to identify different wire types is listed here. Electrical contractors use these designations when selecting the correct wire type for each environment:

- *T*—Thermoplastic (60° C) for dry locations

- *TW*—Moisture-resistant thermoplastic for wet locations

- *THHN*—Heat-resistant thermoplastic (90° C) for dry locations

- *THWN*—Heat-resistant (75° C) for dry and wet locations

- *W*—Moisture-resistant

- *H*—Heat-resistant

- *HH*—More heat-resistant

The *N* means the wire has a nylon outer jacket (which helps reduce abrasion damage when pulling through conduit).

EIA/TIA Standards

The Electronic Industries Alliance/Telecommunications Industry Association (EIA/TIA) referenced in previous chapters publishes standards for telecommunications systems installation, cabling, manufacturing, and performance. They are developed under the joint sponsorship of the ANSI/EIA/TIA organizations. The EIA/TIA standards focus primarily on telecommunications technology and are not applicable to high-voltage residential wiring.

IEEE Standards

The Institute for Electrical and Electronic Engineers (IEEE) is a nonprofit, technical professional association of more than 380,000 individual members in 150 countries. Through its members, the IEEE is a leading authority in technical areas ranging from computer engineering, biomedical technology, and telecommunications to electric power, aerospace, and consumer electronics, among others.

Underwriters Laboratories

Underwriters Laboratories, Inc. (UL) is an independent, nonprofit product safety testing and certification organization. Each year, more than 17 billion UL marks are applied to products worldwide. In this chapter, the following UL standards were referenced:

➤ *Armored cable*—AC cable is manufactured in accordance with UL 4.

➤ *Metal clad cable*—MC cable is manufactured in accordance with UL 1569.

Installation Plans and Procedures

Installation planning should start with an analysis of the capacity needed for electrical service for the home. For example, anyone expanding the electrical capacity in her home should consider installing a higher-capacity panel such as 150- or 200-ampere service. Most electrical contractors use 12-AWG copper wiring as a minimum size for all new electrical installations, even for those circuits protected by 15-amp breakers. Although thinner 14-AWG wire is allowed for most residential applications, the heavier 12-AWG wire runs cooler and is more energy-efficient. This also brings the added benefit of an extra margin of safety.

The Residential Electrical Code presently allows electricians to calculate a home's power needs (in total watts) by multiplying the living space, or square-foot area, by a factor of three. Planning wall outlets is the key to a good basic design for each room.

Installing Wall Outlets

Duplex wall power outlets are usually installed 6–8 feet apart in newer homes, depending on local building codes. The design is intended to provide enough circuits so that extension cords are unnecessary. Ideally, each room should be on a separate circuit.

Maintenance Plans and Procedures

Maintenance is typically performed by a licensed electrician. Some home electrical maintenance tasks, however, can be performed by the homeowner when required, such as resetting and inspecting circuit breakers.

Reactivating Breakers

Three shut-off modes exist for circuit breakers depending on their brand. These modes are as follows:

- *The toggle is in the center position*—It can be reset by flipping the switch all the way off, waiting a couple of seconds, and then turning the switch to the on position.

- *The toggle is in the center with a red flag indicating the breaker has been tripped*—It can be reset by flipping the toggle all the way off, waiting a couple of seconds, and then switching it to the on position. A couple of clicks might be heard as the breaker resets.

- *The toggle switch is all the way in the off position*—It is reset by flipping the toggle switch all the way to the on position.

- *GFCI outlets have a spring-loaded reset button that pops out when a ground fault is detected*—If the ground fault condition is removed, the reset button can be reset by pushing it in to restore power to the outlet.

A circuit breaker might occasionally need to be checked to determine whether it is seated securely in the breaker box bus bar. When resetting a breaker, you might find that it's loose in the breaker box, wobbling when you switch it. Usually it's just that the individual breaker is loose of its mounts. The procedure for securing it is as follows:

1. Turn off power to the circuit breaker box using the main breaker switch (or fuse).

2. Remove the breaker box's metal front-cover panel by unscrewing its mounting screws.

3. Press the loose breaker back into position in the box. You should hear a click or snap when the breaker unit properly seats itself. If it won't reseat properly, it might be broken and need to be replaced.

4. Replace the front-cover panel and turn on the main power.

GFCI Breaker Maintenance

When power shuts off on a GFCI circuit, it might be due to a ground fault. If no water is present to cause a ground fault, reset the circuit. If power is restored, the fault is corrected. But if the breaker trips after a few seconds, it might be due to a ground fault. Test GFCI outlets once a month by pushing the test button. The power should shut off in the circuit, indicating that the GFCI is working properly. Press the reset button to restore power.

Exam Prep Questions

Question 1

What is the real power delivered in watts by a UPS that is connected to a computer rated at 185 VA and a power factor of .7?

- A. 185
- B. 26.4
- C. 129.5
- D. 70

Answer C is correct. The apparent power of the computer receiving power from the UPS is the product of the volts × amperes. The real power is found by multiplying the apparent power by .7 (the power factor) to arrive at the correct value of 129.5 watts. Answer A is incorrect because 185 is the apparent power. True power is calculated by multiplying 185 by the power factor (.7). Answers B and D are incorrect values that are not the result of multiplying the apparent power by .7.

Question 2

What is the per hour cost of operation for a lighting fixture that is rated at 120 watts? The utility company rate is 8.6 cents per kilowatt-hour. (Assume a pf of 1.0.)

- A. 10.3 cents per hour
- B. 1.03 cents per hour
- C. 36 cents per hour
- D. .36 cent per hour

The correct answer is B. The rate is 8.6 cents per kilowatt-hour; therefore, the cost is calculated as .12 kWh used by the lighting fixture × 8.6 cents = 1.03 cents per hour. Answers A, C, and D are incorrect values for the proper solution of .12 × 8.6 = 1.03 cents.

Question 3

> Which type of cable is also referred to as BX cable?
> ○ A. NM cable
> ○ B. Coaxial cable
> ○ C. MC
> ○ D. AC

The correct answer is D. Armored cable (AC) was developed in the early 1900s by Harry Greenfield and Gus Johnson, who called their product BX cable. It has become a generic term for all armored cable. Answer A is incorrect because NM cable is nonmetallic cable and is not referred to as BX cable. Answer B is incorrect because coaxial cable is not referred to as BX cable. Answer C is incorrect because MC cable is the abbreviation used to identify metal-clad cable and is not referred to as BX cable.

Question 4

> Which electrical cable is the most widely used type for interior home wiring?
> ○ A. NM
> ○ B. AC
> ○ C. MC
> ○ D. Armored cable

The correct answer is A. Nonmetallic (NM) cable is the most widely used cable for indoor AC power wiring. Romex is a brand name for a type of plastic insulated wire. It is often called nonmetallic sheath; however, the formal code name is NM cable. This type of wire is suitable for stud walls and on the sides of joists that are not subject to mechanical damage or excessive heat. Newer homes are wired almost exclusively with NM wire. Answer B is incorrect because AC cable is used in residential construction where special protection for electrical wiring is needed. It is a special-purpose cable and not as widely used as NM cable. Answer C is incorrect because MC cable is suitable for branch, feeder, and service power distribution in commercial and industrial applications as well as in multifamily buildings, theaters, and other populated structures. It is not the most widely used cable for interior home wiring. Answer D is incorrect because armored cable is used primarily in special applications where additional protection from damage is required.

Question 5

> Which type of switch is normally used to control 240v two-phase service at a disconnect panel?
> - A. SPST
> - B. SPDT
> - C. DPST
> - D. DPDT

The correct answer is D. Double pole double throw (DPDT) switches are used to control 240v, two-phase circuits at the service entrance. Answer A is incorrect because an SPST switch is not designed for closing and opening a two-phase service entrance point because it has only one set of contacts. Answer B is incorrect because an SPDT switch will not work with a two-phase disconnect point. Answer C is incorrect because, although a DPST has paired contacts, it is not usually utilized for two-phase main disconnect service.

Question 6

> What is the purpose of the break-out fins (or break-out link) on an AC wall outlet?
> - A. They are used to limit the current for appliances plugged in to the outlet.
> - B. They are used to protect against ground faults.
> - C. They're used when it is desired to use one of the AC sockets for a switched outlet.
> - D. They're used for a surge protector circuit.

The correct answer is C. When the break-out link or fin between the brass terminals is removed, the power is connected to only one of the outlets. This is often used when it is desired to use one of the receptacles for a switched outlet controlled from a light switch for turning on or off an appliance or light from another location. Answer A is incorrect because break-out fins are not current-limiting devices. Answer B is incorrect because GFCI outlets are used to prevent electrical shock from ground faults. This is not the function performed by break-out fins. Answer D is incorrect because break-out fins are not used for surge protection.

Question 7

> Which electrical component allows the AC voltage to be switched on and off at variable levels inside the dimmer module?
>
> ○ A. A thermostat
> ○ B. A TRIAC
> ○ C. A UPS
> ○ D. A ballast

The correct answer is B. A TRIAC is used to switch the AC voltage at various phase settings controlled by a dimmer switch. TRIACs are found in most light dimmer control modules. Answer A is incorrect because a thermostat is used as a temperature sensor and controller for HVAC systems. Answer C is incorrect because a UPS is used to supply power to a computer system when a power interruption occurs. Answer D is incorrect because a ballast is a component used in a fluorescent light fixture and is not used with a dimmer.

Question 8

> Which gauge wire should be used for wall outlets to safely handle 20 amperes of AC current?
>
> ○ A. #18-gauge
> ○ B. #14-gauge
> ○ C. #22-gauge
> ○ D. #12-gauge

The correct answer is D. #12-gauge wire is recommended for 20-amp home circuits. Answer A is incorrect because #18-gauge wire is smaller than the recommended gauge wire size for 20-amp circuits. Answer B is incorrect because #14-gauge wire is below the wire gauge size recommended for 20-amp circuits. Answer C is incorrect for the same reason; #22-gauge wire is too small for a 20-amp circuit.

Question 9

According to the NEC, what is the correct color code used for the grounding conductor in residential wiring?

○ A. Red or bare
○ B. Green or bare
○ C. White or black
○ D. Red

The correct answer is B. The NEC indicates the equipment grounding conductor used in residential power wiring must be green or bare. Answer A is incorrect because the red wire is used to indicate the traveler connection on a three-way switch. Answer C is also incorrect because white is used for the color of the neutral wire and black is used as the color for the hot wire. Answer D is incorrect because red is used to identify the traveler terminal connection on a three way switch.

Question 10

What is the minimum amp capacity of electrical service recommended for modern residential construction?

○ A. 200-amp service
○ B. 335-amp service
○ C. 100-amp service
○ D. 50-amp service

Answer A is correct. The current practice is to install 200-amp service as the minimum for new residential construction. Answer B is incorrect because it is above the recommended 200-amp service. Answers C and D are incorrect because they are both less than the recommended 200-amp service.

Need to Know More?

 http://www.epanorama.net/links/wire_mains.html

 http://www.jetcitymaven.com/0002feb/handgwiring.html

 http://www.geocities.com/dsaproject/electricals/power.html

 http://innovations.copper.org/2002/April02/pq_home.html

 Gerhart, James. *Home Automation and Wiring*. Berkeley, CA: McGraw-Hill, 1999. (This book will help you install or plan a high-voltage wiring system for a home.)

12.0—Systems Integration

Terms you'll need to understand:
- ✓ Web pad
- ✓ Touchscreen
- ✓ Infrared receivers
- ✓ Patch panel
- ✓ Gateway
- ✓ Network address translation
- ✓ Control system programming
- ✓ Sniffer
- ✓ Time domain reflectometer
- ✓ Spectrum analyzer

Techniques you'll need to master:
- ✓ Making a list of rules for installing a central controller
- ✓ Listing the most common types of test equipment and their functional capabilities
- ✓ Compiling a list of required final testing tasks to be performed after the home installation is complete
- ✓ Listing the options available for programming most gateways

This chapter covers the topics you will need to know for the systems integration domain included in the HTI+ exam 2. The chapter includes the tools and techniques for completing the integration of an automated home design. The exam elements discussed in this chapter include the functional performance features, and installation location considerations for gateways; control processors; and distribution panels, which serve as the main control interfaces for input devices such as touchscreens, keypads, and wireless RF/IR devices.

System Integration Design Considerations

The integration of system components is the main task associated with ensuring the correct operation of an automated home. In the previous chapters, you learned the characteristics of various whole home automated systems. In this section, you will become familiar with the overall integration and design features of home automation.

Web Pads

A *Web pad* is a portable handheld computer system that provides wireless access to the Internet from any location in the home. The design features include a touchscreen interface, a microprocessor, flash memory, and a wireless network interface card allowing an Internet connection using the home PC. The walk-around Web access is made possible by a wireless networking connection technology known as HomeRF. Several types of wireless Web pads have been developed by prominent PC manufacturers. They typically use the Windows Me operating system with the Internet Explorer browser. An example of a Web pad design is shown in Figure 12.1.

Tablet PCs

Tablet PCs are also a popular form of portable computer system used in home networking. They resemble laptops and notebooks in size and allow digital pen and ink input like a personal digital assistant (PDA) does. The pen is an obvious choice for an input device on a tablet and is the basis for the name *tablet PC*. Tablet PCs provide a method for using all the standard computer applications, such as word processing and spreadsheets with the convenience of a pen as the method of inputting information. Because a tablet is often used in a horizontal position, and you can't be sure of a firm place to rest something like a mouse, a pen is appropriate for use as a pointing and input device.

Figure 12.1 A Web pad.

Keypads

Keypads provide the operational interface for most automated home technology systems. They are available in portable or wall-mounted designs with a small LCD display and backlit keys for operation in low-light conditions.

Keypads are also used to enter programming commands to a central processor/controller integrated with such systems as home security, light management, and audio/video distribution systems. An example of a programmable keypad design is shown in Figure 12.2.

Figure 12.2 A key pad.

Touchscreens

A *touchscreen* is an input device that enables users to operate home automated systems by simply touching icons on a display screen. Touchscreens are available as standalone large-screen panels similar to a computer monitor, wall-mounted units, or handheld remote controllers using wireless infrared technology. A typical touchscreen remote controller is shown in Figure 12.3.

The touchscreen interface has been widely available for several years and has been recently adopted by the home networking industry. It enables users to navigate a home network management system by touching icons or links on the screen. Touchscreen controllers can be programmed to meet the requirements of most custom integrated home systems such as audio and video systems, security systems, and home light management systems.

Figure 12.3 Touchscreen remote controller.

Some of the newer home controller systems include a color touchscreen interface. This provides the homeowner with a consumer-friendly solution that's simple to operate. The size of a standard screen varies from 8" to 10" in size. They are available in panel format and can be flush-mounted in a wall or installed in attractive real-wood tabletop valets.

The more advanced touchscreen controllers include a high-quality color monitor that enables you to view video sources such as cable TV, satellite TV, and closed circuit TV cameras. Although they are expensive, used in the right locations, they avoid the installation of several multibutton switches and keypads.

In a distributed management configuration, the touchscreen interfaces can be located wherever the homeowner needs system control capabilities.

Wireless RF/IR Receivers

In addition to user interfaces and a centralized control unit, a fully integrated home networking system includes devices for receiving and processing particular events. They offer low power consumption and a high level of flexibility that enables HTI+ technicians to optimize home networking designs. They normally capture infrared (IR) or radio frequency (RF) signals and pass this information to the controller. They are generally easy to install, secure, reliable, and affordable. Sensors are the most common types of components used to provide the controller with event information and are available in various types, ranging from motion and temperature to fire and light.

Equipment Location Considerations

After the selection of a home control system has been completed, the physical locations for its main components must be planned. The planning for the location of various system modules such as the centralized controller, keypads, Web pads, touchscreens, and wireless RF/IR receivers is essential for the proper operation of the completed system. All the installation location choices should be done patiently, correctly, and with good planning.

The location of devices depends on a number of factors, including the special requirements for new home construction or existing building needs. The following sections outline the location considerations for key components.

Web Pads

As described earlier, Web pads are portable units that provide the capability to access the Internet from any location in the home. They can be used in any location that is convenient for the user, which normally includes the kitchen, den, home office, or outside patio locations during the daytime and the master bedroom during evening hours.

Keypads

Keypads need to be located for maximum access by the home residents, such as in main entrance doorways, patio access areas, near the garage door entrance to the home, and in the master bedrooms. Security keypads need to be located inside the home entrance and exit areas to avoid tampering by a potential intruder. For very large installations, keypad user interfaces are typically located on each level of the construction.

Various keypads for audio/video remote operation should be located in the auxiliary areas where the homeowner needs to operate sound and video components when away from the home theater area.

Touchscreens

Touchscreens are usually mounted in permanent locations in the home. The location criteria are similar to keypads. Each touchscreen should be located in the area chosen by the homeowner during the planning phase because the preferences among families can vary due to personal lifestyles and home interior designs.

RF/IR Receivers

RF and IR receivers must be located in areas where the wireless RF and IR handheld transmitter units are most frequently used by the residents. IR transmitters must be capable of accessing the wall-mounted receiver in a clear unobstructed path from the transmitter to the receiver. They operate similarly to the more common IR remote handheld units used for TV control. The IR receiver needs to be permanently located in an area free from blockage or other areas that can restrict the use of the transmitter unit. The IR receiver must also be capable of being connected to the wiring inside the wall to transmit signals to the end item being controlled, such as an X10 module or central control unit.

Core Components Found in System Integration Designs

The core components are those items you will need to become familiar with for this portion of the HTI+ exam. They range from the input devices, such as Web pads, keypads, touchscreens, and RF/IR receivers, to other core components, such as distribution panels, control processors, and patch panels.

Keypads

The dominant user interface currently in use for home control systems is the alphanumeric keypad. Keypads are mounted on walls throughout a residence, and their main purpose is to provide the central controller with input data. This input data is then used by the controller to decide which actions need to be taken.

Numerous low-voltage keypads are available for home automation. The main features of these devices include the following:

➤ An audible feedback sound whenever a keypad button is pressed.

➤ Removable covers, which enable you to label each button with its appropriate function. (Some models do, however, support custom engraving.)

➤ Backlit alphanumeric keypad for low-light conditions.

➤ Single and multiple gang configurations.

➤ LCD display that shows status information.

RF/IR Receivers

RF receivers and IR receivers use wireless technology to send commands over home wiring to control various home automation components. The receivers are mounted on the wall in areas of the home where they receive commands from remote handheld units similar to TV remote control units. Whereas RF units use radio signals as the wireless medium, IR systems send a beam of light with coded signals for operating automated system components. IR signals cannot go through solid objects, but RF signals are more tolerant and can be used over a wider range than IR transmit/receiver units where line of sight is required.

Distribution Panel

The terms *patch* and *distribution panel* are used interchangeably to describe a piece of equipment that is used by HTI+ technicians to control cable runs between the central controller and the various system interfaces. The functional description of a distribution panel is discussed later in this chapter in the section "Patch Panel."

Interface Location

When all components of a home automated control system have been purchased, the next step involves the selection of the physical location of all the devices. First, you must locate the centralized equipment and wiring cabinet in an area that is in close proximity to external cable feeds (cable, satellite, and telephone services). Ideally, this will be in a central part of the house that is easily accessible.

 Know that the centralized cabinet needs to be located close to the center of the house and have enough space for any future equipment needed.

The next phase of integrating a home automation system is to decide in conjunction with the homeowner where to locate sensors, audio equipment, user interfaces, and controller modules. After the physical locations are agreed on, you can physically install the home network and associated components. This involves the following tasks:

1. Mount the equipment cabinet in the selected location.
2. Run wires between sensors, keypads, and the central controller.
3. Mount and wire the central controllers.
4. Mount and wire the various automation control systems.

Control Processor

Control processors have been in existence for many years. Most of you using this guide to prepare for the HTI+ exam will be familiar with controllers that use the AC wiring for the home networking media described earlier in Chapter 1, "1.0—Computer Networking Fundamentals." Commonly called *power-line controllers*, these systems are easy to install and operate but have limited capabilities. Newer control processor systems are more complex and are capable of executing a wide range of control functions.

Modern systems are designed to have multiple subsystems integrated into one controller. For example, when connected to the security system, a controller enables the homeowner to remotely unlock doors when remote access is needed at the residence. The door phone is linked via the dialer to the homeowner's mobile phone or place of work, allowing two-way conversation with the person at the door. From a technical perspective, a home networking controller is defined as a software-enabled device used to manage devices connected to an in-home network.

Inside a Typical Home Controller

Because home networking technology is relatively new and continuously evolving, the functionality of a home controller must also evolve and its design must remain relevant for the next 5–10 years. This means building it in a modular, future-proof fashion where software and hardware are concerned. To better understand the home controller, you need to understand its key components:

- A processor
- Memory, including volatile (RAM) and nonvolatile (flash memory)
- Physical interfaces
- Indicators
- A backup battery unit
- Application software

The processor (a shortened term for microprocessor; it's also often called the central processing unit [CPU]) is the central component of the home controller responsible for almost every task the controller performs. The processor is a major determinant of overall cost of the in-home controller.

Much of the configuration flexibility is made possible by the central controller or gateway's support for a range of interfaces. Most modern controllers support the following:

12.0—Systems Integration   457

- F-connectors for video distribution
- RJ-45 ports for interconnection with the home office Ethernet network
- A serial data port (RS232 or RS485) for remote-location communications
- An interface to the residential power-line system (typically X-10)
- A port for a backup processor unit
- An expansion port that supports off-the-shelf Type II compact flash memory

Gateway and *central controller* are terms often used interchangeably by the home automation industry. The *gateway* serves as the primary interface between the external communications facilities and the home network. The *controllers* are more elaborate and are designed to manage multiple whole home technologies such as irrigation, lighting, HVAC, and security systems.

The RJ-45 port on a gateway controller supports both static and dynamic IP addressing.

A number of indicator lights might be included on the front panel of a home control processor. The most common ones are as follows:

- *Power*—This indicator illuminates when the unit receives power (from any source).
- *Network*—This indicator illuminates and blinks when data traffic is traversing the home network.
- *Errors*—This LED illuminates when an error condition is detected. This can be the result of a particular fault that occurred in the controller due to a hardware or software failure. Incorrect programming of a controller can also result in the illumination of this indicator.

All controllers include a battery backup to maintain programming configurations in the event of a power failure.

Software running on the controller enables the smooth inter-operation of devices and services within the home. Software hides the complexities of the system elements from the homeowner. The software included in the controller enables you to program the unit to complete particular home control tasks. Modern processor control units also come with software-based firewall systems to control remote access. The firewall systems used by control processors (and routers) are based on a security system called network address translation (NAT). NAT is a system that enables computers using multiple IP addresses on the home network to connect to the Internet using one publicly visible IP address.

Several manufacturer-specific controllers are available on the market. Some homeowners, however, use a PC for home control purposes. A variety of hardware interfaces are available that allow a home control system to interface directly with a PC. Using a PC to manage the in-home automation provides the homeowner with extra levels of control. For example, the PC can also be used to interface the home network with the Internet.

The firewall systems used by control processors or routers are based on a security system called network address translation (NAT). Remember the capability for hiding IP addresses offered by the NAT protocol for home network users.

Patch Panel

Patch panels are located in the center of a home where the cable runs can be limited to the shortest possible routing. Patch panels are often called *distribution panels*. Both terms refer to a central location where changes can be easily implemented with patch cords and jumper cables. The main benefits of patch panels for homeowners include the following:

➤ They allow people to make changes to a home network easily and quickly.

➤ They support the interconnection of a wide array of cable types including 2-, 4-, 6-, or 8-conductor cable with RJ-11 and RJ-45 jacks.

➤ They help minimize wiring mistakes.

➤ They increase system flexibility by fast and simple offline equipment bypassing.

Patch panels are available in a wide range of sizes. Most panels include port numbering and write-on labeling areas. They are usually mounted in standard 19" racks, which can be wall-mounted on a plywood backboard or a freestanding rack system.

Configuration and Settings

The integration of the components of an automated home involves programming the gateways and controllers and each of the user interfaces. This section covers the tasks involved in each of the final steps in configuring controllers and user interface input devices.

User Interface Programming

The configuration and settings for home user interface programming includes the following key tasks:

➤ Configure and program a residential gateway.

➤ Configure and program controller systems.

➤ Configure and program keypad-based controllers.

➤ Program touchscreen user interfaces.

➤ Program intelligent remote controls.

Configuring and Programming a Residential Gateway

The residential gateway serves as the communication hub for the entire suite of home networking devices deployed throughout the home. Gateways provide integrators with two functions—one is to connect the home network to the Internet, and the other is to act as a central communications interface resource for all in-home digital devices, ranging from lighting, security, and HVAC to phones, home entertainment, and personal digital assistants. Various types of residential gateways are available based on the gradients of functionality required. Most provide capability for advanced data, voice, and video routing within the home. Advanced digital set-top boxes, broadband modems, and PCs are the main types of hardware devices used to provide homeowners with residential gateway functionality. Most basic gateway products allow you to configure the following operational features:

➤ Internet content filtering

➤ Port mapping

➤ Establishment of a demilitarized zone (DMZ)

➤ Log files

➤ Backup configuration settings

➤ Remote administration

➤ Length of passwords

Internet Content Filtering

Specific IP ports on the gateway can be set to block particular types of traffic emanating from the Internet. With this capability, you can block FTP, instant messaging, and email traffic entering the home network. In addition to port blocking, some gateways support the definition of groups with specific access rights. As an example, you might be requested to establish a group for the homeowner's children. After the group is defined, lists of full or partial Web addresses that need to be blocked are then assigned to this group.

Port Mapping

This gateway feature enables you to configure the gateway to map the IP address of the gateway to computers on the home network that access the Internet. In other words, people on the Internet will see only the address of the gateway. Keep in mind that all devices and PCs connected to the gateway must be configured to use DHCP.

Establishment of a Demilitarized Zone

This is another useful feature that enables you to allocate a home computer unprotected access to the Internet. A DMZ is usually configured for homeowners who want to run applications such as online gaming. Keep in mind that, after you configure a computer as a host within a DMZ, you effectively turn off all security features for that particular machine. Such an action makes the DMZ-based machine prone to hacker attacks. The solution is to install a gateway that allows you to control traffic to and from the DMZ port.

Log Files

Most gateways create log files that list Web site activity. You might be required to configure a gateway to send a copy of the file to a specific email address.

Backup Configuration Settings

Providing a backup system for gateway settings is an important part of the configuration process. Some homeowners will adjust the configuration settings and will, in some instances, make changes that affect the performance of the device. In such instances, you might be called to restore the original working configuration.

Remote Administration

A typical gateway enables you to specify the address of a remote PC for administration purposes. In most cases, the original factory IP address settings are 192.168.1.1. This is a useful feature for homeowners who turn to their HTI technician for external support. Configuration of the remote administration settings involves changing the IP address from the original factory setting to the new host address and changing the default administrator name and password to prevent any unauthorized access to the system. The remote administration feature is not enabled by default.

Length of Passwords

Modern gateways enable you to define the maximum and minimum lengths of passwords that can be used by home occupants. You might need to restrict the use of certain types of characters that can be used in passwords. You can also set the frequency at which home occupants must change their passwords.

Configuring and Programming Controller Systems

Depending on the level of sophistication involved with a particular home networking installation, you need to have a fundamental understanding of configuring and programming home networking controllers. You will be expected to configure a controller to allow the homeowner to adjust settings and monitor her home network locally and over the Internet. Technical information to configure the control system can come from many sources, including the owner of the house, a specification document that is included with a device, or possibly the service provider.

After the required system integration information is gathered and analyzed, you can configure the system. This can be done onsite or remotely. Most systems enable you to connect remotely to the controller via a small router or hub installed in the home. Configuration is typically done through a software application that lets you perform the desired modifications. Home control software applications are available in various types; some software providers accept configuration changes through a basic command-line interface. Other applications offer an icon-based environment that enables you to program the system via predefined windows and dialog boxes. These types of systems are relatively easy to use and do not require programming expertise. Common configurable elements include the following features:

➤ *Scheduling particular events*—This option is used to schedule events to occur within the home at a specific time. For instance, the homeowner might want the system configured to open the drapes at 8 a.m. every day.

- *Creating groups of home networking devices*—This option lets you group together similar networking devices. For example, all the motion sensors within the house can be defined as a group and all be activated or deactivated at once.

- *Logging events*—This option allows you to define a particular file for recording events that occur on the home network over a predefined time period. This feature is widely used by HTI+ technicians as a troubleshooting tool.

- *Defining house modes*—Home automation control systems simplify the process of defining response events to occur under particular conditions. Most systems allow you to define different modes of operation for a home network. For instance, the way in which devices on the security subsystem operate when in home mode can differ from their operation when configured to operate in away mode.

Most controllers receive their inputs from several types of interface devices. Similar to the main controller, these input devices can also be customized to reflect the homeowner's needs.

Configuring and Programming Keypad-based Controllers

Keypads are typically preprogrammed by the HTI+ technician to provide the homeowner with specific automation functions. They usually support an IR detector, which accepts commands from a remote control and passes these commands to the central PC-based controller for processing. The main drawbacks of these types are as follows:

- They are normally proprietary.
- Homeowners might find the labels difficult to read, particularly in poorly lit areas of the room.
- They are limited to one function per key.

Programming Touchscreen User Interfaces

Modern systems include some type of communications port that lets you use an external terminal such as a notepad or a standard PC to configure the interface for the homeowner. The desired interface is loaded into the touchscreen via the communications port.

Touchscreen files can also be uploaded via the home controller unit.

You must spend time ensuring that the navigation pages are functionally correct and the look and feel of these pages blends in with the surrounding interior design. Every icon is configurable for single or multiple functions.

Most of your time configuring touchscreen interfaces will be devoted to programming functionality, but it is important that the touchscreen pages are professionally designed and easy to use.

Programming Intelligent Remote Controls

A standard remote control that comes with a piece of equipment is preprogrammed with all the necessary codes already recorded.

Intelligent and portable controls can now be purchased that consolidate multiple handheld remotes into a single module. They have intelligence built in that enables the unit to learn commands from existing remote controls. All the buttons can be programmed for individual commands. In addition, many of the models enable the remote control to be programmed to send multiple commands simultaneously. For example, an intelligent remote control can be programmed to turn on all the components within the home theater system at the touch of a button. Programming is achieved manually or through the use of a PC software package. Different command configurations can be designed and stored for downloading at a future date. A typical IR programmable remote controller is shown in Figure 12.4.

Figure 12.4 Programmable remote controller.

Device Connectivity

Many interconnection technologies are used with automated home designs, including various types of communications cable, low-voltage wiring, and wireless systems. The following sections include the characteristics of each type of media used to connect the components of a whole home design.

Communications Cable

Table 12.1 includes the specification naming conventions and markings you will need to know for communications cabling used in the installation of automated home systems.

Table 12.1	NEC Markings
Marking	Naming Convention
MPP	Multipurpose plenum [cable]
CMP	Communications plenum
MPR	Multipurpose riser
CMR	Communications riser
MPG	Multipurpose General-purpose
CMG	Communications General-purpose
MP	Multipurpose
CM	Communications general-purpose
CMX	Communications, limited-use
CMUC	Communications, under-carpet wire and cable

The naming conventions and cable markings for communications cable in Table 12.1 are contained in NEC Article 800-50, "Listings, Markings, and Installation of Communications Wires and Cables."

Low-voltage Wiring

As discussed in Chapter 10, "10.0—Structured Wiring—Low-voltage," a low-voltage structured-wiring design uses cables to connect the individual rooms to the distribution panel and punch-down blocks. Low-voltage media types include coaxial cable, unshielded twisted-pair cable, and fiber-optic cable.

Coaxial Cable

Coaxial cable is recommended for video signal distribution for all residential low-voltage cabling. RG-6 coaxial cable is recommended for all video signal connectivity in new as well as remodeled residential construction. RG-6 cable is also used for connecting DVDs, VCRs, and personal video recorders to the A/V distribution system.

According to the ANSI/TIA/EIA 570-A standard, RG-59 coaxial cable can be used only for short jumpers and RG-11 can be used for backbone video distribution cables in very large homes or multiple-tenant buildings such as apartment or condominium complexes.

Category 5E UTP Cable
Category 5E UTP cable has become the standard for data and telephone connectivity in low-voltage structured-wiring systems. Separate Category 5E cabling is run from the central control panel to each room to accommodate both telephone wiring and home computer networking applications on the same cable.

Fiber-optic Cable
As discussed in Chapter 10, fiber-optic cable is beginning to play a major role in home structured wiring, primarily for home audio/video systems. High-definition television (HDTV), digital TV, and digital audio components require a newer broadband type of cables and connectors than the type used for analog video and audio connections.

Wireless
Wireless home networking solutions originate from four simple consumer demands:

- ➤ No new wiring infrastructures should be required.
- ➤ The solutions must be simple to install and easy to use.
- ➤ Interoperability with other existing home wiring, such as phone line-based systems, is essential.
- ➤ Solutions need to be economical, and home security cannot be compromised.

Wireless networking offers solutions to most of these requirements. The two main core technologies used in the design of wireless home networking products are infrared and radio frequency.

Infrared
In a typical home networking configuration, many of the home automation devices are controlled via one of the power-line standards. The choices are X10, CEBus, or LonWorks.

System integrators often use IR control systems to improve the flexibility levels of the final network configuration.

As discussed earlier in this chapter, IR transmission is categorized as a line-of-sight wireless technology. This means the transmitter units and the receivers must be in a direct line of sight to operate. An IR-based network suits environments where all the digital appliances that require network connectivity are in one room. IR receivers pick up infrared signals from IR remote controls and then encode these signals and convert them into electrical impulses. These electrical impulses can be carried around the home on standard cabling.

Radio Frequency

The other main category of wireless technology is radio frequency. RF technology is a flexible technology, allowing consumers to link appliances distributed throughout the house. Several remote handheld units are available that use radio technology instead of IR to remotely operate devices. The main benefit of using RF technology to control in-home devices is its capability to penetrate solid objects, such as interior walls and ceilings. If deciding whether to use RF-based devices to control a home network, be aware of the following disadvantages:

- Distances are of concern for consumers who own large households. Some of the RF technologies have limited coverage—for example, the range of Bluetooth is only 10 meters.
- Most of the RF systems operate in the unlicensed spectrum in the 2.4GHz band where there is no protection from interference from other devices operating in the same frequency band.
- RF mobile control devices can be expensive.
- RF devices also bring an added concern about security. Eavesdropping on a wireless network is much easier than on a wired system.
- Careful installation of the equipment is required to maintain the integrity of the radio signal.

Termination Points

Termination points are the locations in a structured wiring system where all the cabling is terminated in room outlets at one end and a central location at the other end. In home-run wiring, all the cables are run from the room outlets to a central location such as a control panel and a punch-down block. A *punch-down block* is a termination block to which telephone cables and networking cable, such as UTP, can be connected. This serves as the termination point for incoming telephone lines and TV coaxial cable from outside the home.

Industry Standards

Making home networking and automation a reality requires the cooperation of a variety of industries and companies, along with the development of many new standards. Most standards organizations create formal standards by using specific processes: organizing ideas, discussing the approach, developing draft standards, voting on all or certain aspects of the standards, and then formally releasing the completed standard to the general public. Some of the best-known international organizations that contribute to the standardizing of home automation and management include

- National Electrical Code (NEC)
- EIA/TIA
- IEEE
- The National Electrical Contractors Association
- Underwriters Laboratories, Inc. (UL)

Their contributions to the standardization process are explained in the following paragraphs.

National Electrical Code

The most well-known electrical standard for safety is the National Electrical Code (NEC). Published by the National Fire Protection Association (NFPA), the code provides "practical safeguarding of persons and property from hazards arising from the use of electricity." According to the NFPA Web site, the NEC protects "public safety by establishing requirements for electrical wiring and equipment in virtually all buildings."

ANSI/TIA/EIA Standards

As discussed in Chapter 10, the most important standard you will need to know about when preparing for the infrastructure and integration examination is ANSI/TIA/EIA 570-A and its three addendums. It sets guidelines for residential single-family and multifamily telecommunications equipment and wiring. Among other aspects, the standard covers specifications for the network interface and demarcation point, auxiliary disconnects, a distribution device, cross-connects, cables and cable installation, outlets and connectors, and equipment cords.

Two specific EIA standards deal with connecting user interface devices with a centralized home controller processor unit: the CEBus and RS-485 interface specification. CEBus (Consumer Electronics Bus) is an open standard for home automation released by the EIA as EIA-600. The CEBus standard gives HTI installers a uniform way to link household devices. It defines a set of rules for transporting messages between CEBus-compliant products throughout the home.

Know the titles and purposes of the EIA-600 CEBus standard and the ANSI/TIA/EIA 570-A standard.

IEEE Standards

As previously discussed in this book, the Institute of Electrical and Electronics Engineers (IEEE) is a nonprofit, technical professional association of more than 377,000 individual members in 150 countries. As discussed previously, the association is responsible for publishing a wide range of industry standards.

HTI and other residential systems cabling installers should be familiar with these standards that relate directly to the transfer of data between user interface components and the home control processor. The IEEE 802 series and IEEE 1394 are examples of standards that are widely used to interconnect such devices.

Underwriters Laboratories, Inc.

As previously discussed, the Underwriters Laboratories, Inc. (UL) is an independent, nonprofit product safety testing and certification organization. All components of the wiring system used to connect the user interfaces with the central home control processor should be UL listed.

National Electrical Contractors Association

The National Electrical Contractors Association (NECA), founded in 1901, is the leading representative of a segment of the construction market comprised of more than 70,000 electrical contracting firms.

Because electrical contracting firms are moving into the provision of home networking services, HTI technicians must be aware of the various services offered by this organization.

Installation Plans and Procedures

The physical installation of an integrated home system normally involves running wires between the centralized controller unit and the various system interface modules. Historically, the daisy-chaining of wiring between user interfaces was a common occurrence. This is now considered an obsolete method of wiring a home for low-voltage circuits and has largely been superseded by cabling systems that comply with the ANSI/TIA/EIA 570-A standard. This standard specifies that each low-voltage interface connection point needs to have a cable run back to a centralized distribution panel.

Installing Central Home Controllers

The control processor and gateway are the heart of the automation control system, linking touch panels, keypads, wireless remotes, and the Internet together.

Home control processors need to be air cooled; therefore, you should mount the unit in a well-ventilated area. When installing, remember to verify that the ambient temperature range does not fall below 32° F or exceed 110° F.

 When installing control processors and gateways in a rack-mounted system, the ambient temperature of the rack environment can be greater than the room ambient temperature.

You also need to ensure that the relative humidity of the installation area remains low at all times. Installing a control processor in an area that is prone to high humidity can result in problems.

Not only should you verify temperature and humidity conditions before installing components, but you also need to allow adequate clearance in front of the control unit for servicing purposes.

Installing User Interfaces

When installing user interface devices, the following procedures are important:

- When calculating the wire gauge for a network run between the controller unit and a particular user interface, the length of the run and the power consumption of the user interface need to be considered.
- Category 5E UTP cable must be installed in the interior of walls between the controller location and each user interface.

Maintenance Plans and Procedures

Maintenance planning involves establishing testing procedures and becoming familiar with test equipment used during maintenance and troubleshooting.

Testing

The final step of an installation is testing the system interfaces and the main controller to ensure that everything works properly. Testing verifies that no faults occurred during the installation of the system. When testing, remember to adhere to the following rules:

1. Before using test equipment, perform a full and thorough visual inspection of the integrated system.
2. Identify any faults in the cable runs.
3. Document the description and location of the device you tested.
4. Record the tools you used to test the device.
5. Record the results from the test and the date on which the test was performed.
6. If the system is reconfigured, make sure that a retest is performed on the affected components.

The availability of detailed test results will help you troubleshoot the system if the need arises in the future. In addition to test result documentation, the following items should also be updated and stored in a safe place:

- *Cabling diagrams*—These show the hardware locations and descriptions of cabling used in the network.
- *Controller system details*—These will include configuration details, operating system descriptions, and hardware specifications.
- *Licensing*—Licensing and support agreements might be required in the future to resolve network problems.

In addition to testing and documenting the final network configuration, you must train the homeowner in the use of the system prior to finishing the project.

Troubleshooting Tools

Despite the fact that the installation of the home networking hardware and software components were installed according to plan, problems can still occur. Problems can occur across any one of the following subsystems or components:

- *The PC and telecommunications network*—PC-centric subsystems are comprised of the modems, scanners, digital cameras, modems, PBXs, servers, residential gateways, and printers that are connected to the local home network.
- *The lighting control system*—Lighting control subsystems are comprised of outlets, dimming modules, light switches, and various types of fixtures.
- *The HVAC and water control systems*—HVAC and water control subsystems include a wide range of electromechanical devices, such as thermostats, air handlers, control valves, keypads, relays, heaters, pumps, and solenoids.
- *The home security and surveillance system*—Security and surveillance subsystems contain keypads, sensors, security panels, cameras, monitors, and the infrastructure that interconnects all these devices.
- *The multimedia network*—Multimedia subsystems include set-top boxes, digital televisions, digital video recorders, speakers, stereos, and DVD players.
- *The wireless network*—Wireless home systems can include personal digital assistants (PDAs), mobile phone sets, Web pads, and RF/IR remote controls.

A number of specialty tools are available to help in the troubleshooting process. The most common types are listed in the following paragraphs.

Voltmeter/Ohmmeters

The volt/ohmmeter is a multipurpose tester familiar to most electronics technicians. It is the most basic test measuring tool you will use to troubleshoot a home network. It is primarily used to measure the amount of resistance that exists on an electrical circuit as well as measuring both AC and DV voltages.

Time Domain Reflectometers

A time domain reflectometer (TDR) is defined as an electronic device that can be used to detect and locate breaks, shorts, or open circuits on the cable. The capability to locate breaks on a network is made possible through the use of sonar-like pulses, which are sent along the cable. Modern TDRs can precisely pinpoint the location of the problem. This tool is invaluable when troubleshooting a home networking cabling system.

An example of a time domain reflectometer is shown in Figure 12.5.

Figure 12.5 A time domain reflectometer.

Integrated Cable Testers

This type of troubleshooting device is more advanced and provides a complete solution for testing, certifying, and documenting cabling systems. Integrated cable testers have the following capabilities:

➤ Provide a graphical analysis of the problem.

➤ Manage test results with a single software application.

➤ Ensure cable standards compliance.

➤ Export the data collected into a PC for further analysis.

A typical integrated cable tester is shown in Figure 12.6.

Network Sniffers

A network sniffer enables you to examine data packets on a home network. A typical protocol analyzer/sniffer includes a PC, a network card, and an application that analyzes problems on the network. Real-time network analysis helps detect and resolve network faults and performance problems quickly.

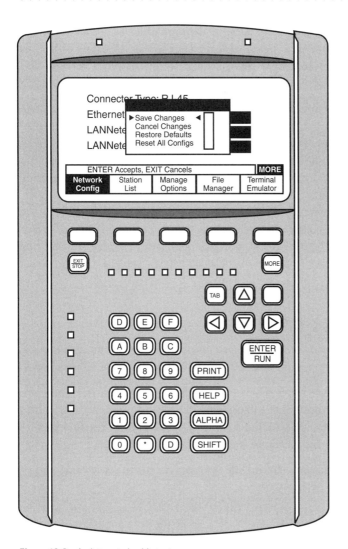

Figure 12.6 An integrated cable tester.

RF Spectrum Analyzers

One of the problems associated with implementing a home network that includes wireless technologies is the testing and analyzing propagation of RF signals. The random dispersal and propagation of signals can be problematic and can often impact the performance of a home network. HTI technicians often use a test tool called a spectrum analyzer to observe the magnitude and frequency of RF signals to locate the source of interfering emissions. Traditionally, spectrum analyzers were an expensive item and were primarily used in engineering labs and by large telecommunications service organizations. Today, the price of a spectrum analyzer is still in the range of $5,000, but analyzers provide excellent service when troubleshooting wireless network problems.

In simple terms, a spectrum analyzer is an electronic device whose main purpose is to measure the frequency spectrum and power of a signal. It automatically scans or sweeps across a particular range of frequencies. The display shows any interfering signals that are occupying the scanned frequency range. They are particularly useful in analyzing problems associated with 802.11b wireless home network elements.

A spectrum analyzer can be used to pinpoint wireless LAN problems in a home office installation.

Maintenance Issues

After the physical installation of all systems and cabling is complete, the following tasks will help ensure that the project has been completed to the customer's satisfaction:

- ➤ Conduct a walk-through to identify any discrepancies and areas that the customer is unhappy with. Additionally, it presents you with an opportunity to introduce the customer to the new equipment.
- ➤ It is also a good practice to provide the customer with some basic training on how to use the system.
- ➤ Provide the customer with relevant warranties, test results, and network documentation.

Exam Prep Questions

Question 1

> When troubleshooting a home wiring system, which type of test equipment is used to detect and locate breaks on a cable?
> - A. A volt/ohmmeter
> - B. A sniffer
> - C. A spectrum analyzer
> - D. A TDR

The correct answer is D. A TDR is an electronic tester that can be used to detect and locate breaks, shorts, or open circuits on a cable. Answer A is incorrect because a volt/ohmmeter is used to measure volts and ohms. Although an ohmmeter can detect an opening in a cable, it is not designed to provide the location of a cable anomaly. Answer B is incorrect because a sniffer is a test instrument used to analyze the content of a packet or frame on a network. A typical protocol analyzer/sniffer includes a PC, a network card, and an application that analyzes problems on the network. Answer C is incorrect because a spectrum analyzer is a test instrument whose main purpose is to measure the frequency spectrum and power of a signal. It automatically scans or sweeps across a particular range of frequencies. The display shows any interfering signals that are occupying the scanned frequency range.

Question 2

> Which standard is used by the home technology industry and establishes guidelines for residential single-family and multifamily telecommunications equipment and wiring?
> - A. EIA 600
> - B. IEEE 1394
> - C. ANSI/TIA/EIA 570-A and its three addendums
> - D. ANSI/TIA/EIA 568-B

The correct answer is C. This important standard sets guidelines for residential single-family and multifamily telecommunications equipment and wiring as well as security wiring and home-run wiring design guidelines. Answer A is incorrect because the EIA 600 standard is the CEBus protocol standard. Answer B is incorrect because the IEEE 1394 standard describes the FireWire high-speed serial bus. Answer D is incorrect because ANSI/TIA/EIA 568-B is the commercial building telecommunications standard.

Question 3

> Which type of portable handheld computer system is designed to provide wireless access to the Internet from any location in the home?
> - A. A Web pad
> - B. A gateway
> - C. A keypad
> - D. An IR receiver

The correct answer is A. The design features include a touchscreen interface, a microprocessor, flash memory, and a wireless network interface card allowing an Internet connection with the home PC. The walk-around, portable Web access is made possible by a wireless networking protocol known as HomeRF. Answer B is incorrect because a gateway is a device for connecting external telecommunications services with home networks and is not designed for portable use. Answer C is incorrect because a keypad is a wall-mounted or handheld alphanumeric data entry device for interfacing with a receiver. It is not designed for Internet access or portable computing. Answer D is incorrect because an IR receiver is used with handheld IR controllers for sending commands to audio/video equipment.

Question 4

> Which type of jacks are usually found on home controllers or gateways for interfacing with the home network?
> - A. RCA jack
> - B. RJ-11 and RJ-31X
> - C. RJ-45
> - D. F-type connector

The correct answer is C. This is used to connect the home office LAN to the Ethernet port on a controller. Answer A is incorrect because an RCA jack is used with audio/video equipment and is not the type of jack used with controllers. Answer B is incorrect because RJ-11 jacks are used for telephone extensions and RJ-31X jacks are special telephone jacks used with security systems. Answer D is incorrect because F connectors are used with audio/video equipment as input/output connectors.

Question 5

Which programming feature of a home controller can be useful when troubleshooting a home controller or gateway problem?

- ○ A. Event logging
- ○ B. Creating groups of home networking devices
- ○ C. Port mapping
- ○ D. Internet content filtering

The correct answer is A. Analysis of events that have occurred over a specific period as recorded by the event logger can often assist you in analyzing when and how a problem developed. Answer B is incorrect because a home controller does not create groups of home networking devices as a tool for troubleshooting. Answer C is incorrect because port mapping is not a useful troubleshooting aid. Answer D is incorrect because Internet content filtering is not a feature used for troubleshooting a home controller or gateway.

Question 6

What is the primary purpose of a keypad in the design of an automated home?

- ○ A. It is used to control remote video surveillance cameras.
- ○ B. It is used only for controlling audio/video distribution systems.
- ○ C. It is primarily used to control irrigation systems.
- ○ D. Its main purpose in an automated home is to provide input data to the central controller.

The correct answer is D. Even though keypads are used in a wide variety of home technology designs, their primary function in all situations is to provide input data from the user to some type of central controller. This input data is then used by the controller to decide which actions need to be taken. Answer A is incorrect because keypads are not used to control remote video surveillance cameras. Answer B is incorrect because a keypad does not typically have a role in controlling audio/video distribution systems. Answer C is incorrect because, although keypads are used in some types of irrigation system controllers, their main purpose is to function as standalone devices in home automation systems.

Question 7

> Which type of cable used for low-voltage integrated home designs has become the standard for telephone wiring and data communications in the home?
>
> ○ A. RG-6 coaxial cable
> ○ B. Category 3 UTP
> ○ C. Category 5E UTP cable
> ○ D. Fiber-optic cable

The correct answer is C. Category 5E UTP cable has become the standard for data and telephone connectivity in low-voltage structured-wiring systems. Separate Category 5E cabling is run from the central control panel to outlets in each room, which accommodates both telephone wiring and home computer networking applications on the same cable. Answer A is incorrect because RG-6 coaxial cable is used for video signal connectivity and is not used for voice service. Answer B is also incorrect. Although the FCC has mandated Category 3 wiring as a minimum grade for all future residential structured wiring, Category 5E has replaced it as the preferred choice at essentially the same price with a higher quality of performance. Answer D is incorrect because fiber-optic cable is not used in residential network designs for voice and data applications.

Question 8

> A customer is experiencing some type of RF interference that is causing the home wireless network computers to operate erratically. Which type of test equipment would you recommend to find the source of the interfering signals?
>
> ○ A. A sniffer
> ○ B. A TDR
> ○ C. A volt/ohmmeter
> ○ D. A spectrum analyzer

The correct answer is D. A spectrum analyzer can observe the waveform, magnitude, and frequency of any radio emissions in a selected area. It is very useful in locating sources of radio frequency interference as well as ensuring that components on a home wireless LAN are operating properly. Answer A is incorrect because a sniffer is a test instrument used to analyze the content of a packet or frame on a network. It is not used to observe wireless transmissions or radio interference. Answer B is incorrect because a TDR is used as a cable tester to locate cable breaks or improper connectors and is not suitable for analyzing wireless problems. Answer C is incorrect because a volt/ohmmeter is used to measure volts and resistance and is not designed to measure RF interference.

Question 9

Which feature found in most home controllers enables computers using multiple IP addresses on the home network to connect to the Internet using one publicly visible IP address?

- ○ A. NIC
- ○ B. Remote access
- ○ C. NAT
- ○ D. Log files

The correct answer is C. The firewall systems used by control processors, controllers, gateways, and routers are based on a security system called network address translation (NAT). NAT enables computers on a home network to connect to the Internet using one visible IP address while hiding their own IP addresses. Answer A is incorrect because a NIC is installed in a home computer to provide the physical interface to the network media. Answer B is incorrect because remote access is the feature provided in home controllers for accessing or administering a home network from a location outside the home. Answer D is incorrect because log files are a feature used by most gateways that list Web site activity.

Question 10

Which type of coaxial cable is commonly used only for short jumpers?

- ○ A. RG-58
- ○ B. RG-59
- ○ C. RG-11
- ○ D. RG-8

Answer B is correct. According to the ANSI/TIA/EIA 570-A standard, RG-59 75-ohm coaxial cable can be used only for short jumpers. Answer B is incorrect because RG-58 is 50-ohm cable and is not used for short jumpers. The basic purpose of coaxial jumper cables is to interconnect audio/video equipment that uses 75-ohm input/output connector ports. Answer C is incorrect because RG-11 coaxial cable is used primarily for backbone video distribution cables in very large homes or multiple-tenant buildings, such as apartment or condominium complexes. Answer D is incorrect because RG-8 cable is 50-ohm cable and is not applicable to most home audio/video distribution applications.

Need to Know More?

 Gerhart, James. *Home Automation and Wiring.* Berkeley, CA: McGraw-Hill, 1999. (This book will assist you in reviewing the main issues associated with systems infrastructure and automation.)

 http://www.hometoys.com/htilinks.htm. (This Web site is a source for additional information on HTI infrastructure and integration issues.)

PART III
Practice Exams

13 Practice Exam 1: Residential Systems

14 Answers to Practice Exam 1: Residential Systems

15 Practice Exam 2: Systems Infrastructure and Integration

16 Answers to Practice Exam 2

Practice Exam 1: Residential Systems

Now it's time to put to the test the knowledge you've learned from reading this book! Write down your answers to the following questions on a separate sheet of paper. You will be able to take this sample test multiple times this way. After you answer all the questions, compare your answers with the correct answers in Chapter 14, "Answers to Practice Exam 1: Residential Systems." The answer keys for both exams immediately follow each Practice Exam chapter. When you can correctly answer at least 90% of the 70 practice questions (63) in each Practice Exam, you are ready to start using the PrepLogic Practice Exams CD-ROM at the back of this *Exam Cram 2*. Using the Practice Exams for this *Exam Cram 2*, along with the PrepLogic Practice Exams, you can prepare yourself well for the actual HTI+ certification exam. Good luck!

Question 1

Which of the following is a standard targeted at the residential user that uses the SWAP protocol for home networking?

○ A. Bluetooth
○ B. Home RF
○ C. IEEE.802.11
○ D. IEEE.802.11b

Question 2

Which items are needed to control automated systems in a home using a Web page? (Select all that apply.)

❏ A. Modem
❏ B. Web browser
❏ C. Password
❏ D. IrDA

Question 3

How is the HomePNA standard capable of transmitting voice and data communications over a telephone line at the same time with no interference?

○ A. Frequency division multiplexing
○ B. Twisted-pair wiring
○ C. RJ-11 jacks
○ D. HPNA USB adapters

Question 4

The EIA-600 is a(n) _____ standard.

○ A. X10
○ B. CEBus
○ C. Router
○ D. PLC

Question 5

Which of the following uses Category 3 unshielded twisted-pair cable and permits up to 10Mbps over an Ethernet?

○ A. 100BASE-TX
○ B. 100BASE-T4
○ C. 10BASE-T
○ D. 100BASE-FX

Question 6

Which cable is specified as having data transfer rates of 100Mbps, 200Mbps, and 400Mbps over a maximum length of 4.5 meters?

○ A. RS-232
○ B. USB 1.1
○ C. IEEE 1284
○ D. IEEE 1394

Question 7

Which type of jack is used with home security alarm systems?

○ A. RJ-31x
○ B. RJ-11
○ C. RJ-45
○ D. RJ-14

Question 8

Class _____ addresses are reserved for large networks and use the last 24 bits.

○ A. B
○ B. A
○ C. C
○ D. D

Question 9

A _____ audio/video system places all the audio and video equipment in one location.

- ○ A. Distributed
- ○ B. Whole home
- ○ C. Dedicated
- ○ D. Remote access

Question 10

The Dolby Digital surround sound 5.1 format uses five surround speakers and a sixth _____ channel as the subwoofer speaker.

- ○ A. LFE
- ○ B. Center
- ○ C. Right
- ○ D. Left

Question 11

The _____ part of a stereo or A/V receiver is used for selecting AM/FM radio stations.

- ○ A. Receiver
- ○ B. Amplifier
- ○ C. Tuner
- ○ D. Volume control

Question 12

Which of the following causes clipping of the output audio signal?

- ○ A. The amplifier power rating is too large for the speakers.
- ○ B. An overdriven low-powered amplifier is used.
- ○ C. The speakers are the same power rating as the amplifier.
- ○ D. The monitor is too large.

Question 13

Which of the following is not a video display used with a portable laptop computer system?

- ○ A. Cathode-ray tube display
- ○ B. Liquid crystal display
- ○ C. Plasma display
- ○ D. Flat-panel display

Question 14

Which type of jack uses Category 5 UTP cabling?

- ○ A. RCA
- ○ B. F-type
- ○ C. RJ-45
- ○ D. Optical

Question 15

_____ is a popular technology for capturing audio and video files and playing them on a computer while they are being downloaded from an Internet Web server.

- ○ A. Streaming
- ○ B. PVR
- ○ C. PPV
- ○ D. VOD

Question 16

Which type of service technology does the DISH Network provide?

- ○ A. Cable service
- ○ B. Digital service
- ○ C. Satellite service
- ○ D. Antenna service

Question 17

Which protocol has been adopted by the home entertainment industry?

○ A. FTP
○ B. HAVi
○ C. HTTP
○ D. TCP/IP

Question 18

Which of the following video recording formats provides 400 lines of resolution?

○ A. VHS
○ B. 8mm
○ C. Mini DV
○ D. HI-8

Question 19

Which of the following converts a whole band, or block, of frequencies to a lower band?

○ A. NEC
○ B. LNB
○ C. DBS
○ D. HDTV

Question 20

Where is the best location to put the security system keypad?

○ A. Hidden in a closet
○ B. In front of the house
○ C. On the inside wall near the door
○ D. In the back of the house

Question 21

Temperature sensors are normally used to monitor _____ and _____ temperature values in homes, water pipes, furnaces, and heating vents. (Select two.)

- ❏ A. High
- ❏ B. Right
- ❏ C. Low
- ❏ D. Left

Question 22

Computer systems can be integrated with control protocols, such as _____, to operate and manage security systems.

- ○ A. CEBus
- ○ B. TCP/IP
- ○ C. HTTP
- ○ D. FTP

Question 23

The ANSI/TIA/EIA 570A standard requires at least _____ inches of separation between parallel runs of security wire and AC power wiring.

- ○ A. 6
- ○ B. 32
- ○ C. 12
- ○ D. 24

Question 24

What is the degree angle of the crossover when security wires and cabling is crossed with AC power wires?

- ○ A. 360°
- ○ B. 180°
- ○ C. 60°
- ○ D. 90°

Question 25

How many pins does an RJ-31x telephone jack have?

- ○ A. 8
- ○ B. 6
- ○ C. 9
- ○ D. 12

Question 26

All alarm systems that use the telephone lines in a home to automatically dial a central monitoring station must have an _____ jack installed.

- ○ A. RJ-44
- ○ B. RJ-12
- ○ C. RJ-31x
- ○ D. RJ-31s

Question 27

How many wires are needed for a motion detector alarm circuit?

- ○ A. 1
- ○ B. 2
- ○ C. 3
- ○ D. 4

Question 28

In a security system, passive sensors need two wires and active sensors require _____ wires.

- ○ A. 4
- ○ B. 6
- ○ C. 3
- ○ D. 2

Question 29

What is the generic term used in the telecommunications industry to describe any system that uses a mix of media and transmission standards?

- ○ A. Mixer
- ○ B. Scrambler
- ○ C. Hybrid
- ○ D. Multiplex

Question 30

The _____ communications systems are designed for systems transporting voice information.

- ○ A. Digital
- ○ B. Satellite
- ○ C. Cable
- ○ D. Analog

Question 31

_____ is a product line of equipment used primarily by businesses that need their own internal telephone-switching interface with the external telephone network.

- ○ A. CTI
- ○ B. PBX
- ○ C. ACD
- ○ D. KSU

Question 32

Which protocol is used to encapsulate a message when it is transmitted over the Internet using a virtual private network?

- ○ A. FTP
- ○ B. TCP
- ○ C. UDP
- ○ D. PPTP

Question 33

Which of the following is used to terminate wire using a 66-type block? (Select all that apply.)
- ❑ A. A punch-down tool
- ❑ B. A hole punch tool
- ❑ C. An impact tool
- ❑ D. Wire cutters

Question 34

Which type of block is used to block the phone from dialing a 1 or a 0?
- ○ A. 800 number block
- ○ B. 011 number block
- ○ C. Collect call block
- ○ D. Long distance block

Question 35

Which grade requires two Category 5 UTP cables and two 75-ohm coaxial cables to each location?
- ○ A. Grade 1
- ○ B. Grade 2
- ○ C. Grade 3
- ○ D. Grade 4

Question 36

What is another name for a star wiring topology in which all cabling is routed from each room to a common termination location?
- ○ A. Home run
- ○ B. Hit and run
- ○ C. Grade A
- ○ D. Grade 1

Question 37

Which protocols are used to send signals through handheld wireless keypad devices and to remotely control the lighting scenes and illumination levels in any area of the home? (Select all that apply.)

- ❏ A. X12
- ❏ B. X10
- ❏ C. GEBus
- ❏ D. CEBus

Question 38

Which of the following is a thin wall conduit that can be used for exposed or concealed electrical installations?

- ○ A. EMT
- ○ B. CCR
- ○ C. MET
- ○ D. RCC

Question 39

A(n) _____ is one group of lights controlled by one switch or a dimmer module.

- ○ A. Cluster
- ○ B. PPP
- ○ C. Zone
- ○ D. ASB

Question 40

What is the correct terminal connection for the red or black (hot) wire when connecting to a standard wall outlet?

- ○ A. Brass terminal
- ○ B. Silver terminal
- ○ C. Black terminal
- ○ D. Green terminal

Question 41

Which type of dimmer switch can be used with an incandescent light?
- ○ A. SPST
- ○ B. DPST
- ○ C. X-10-compatible two-wire
- ○ D. SPDT

Question 42

Which type of switch is used to control 220v, two-phase circuits?
- ○ A. SPST
- ○ B. SPDT
- ○ C. DPST
- ○ D. DPDT

Question 43

Which type of switch is used for controlling lights depending on the light intensity?
- ○ A. An automated dimmer switch
- ○ B. A photoelectric switch
- ○ C. A three-way switch
- ○ D. A clock switch

Question 44

Which of the following is not designed to work with a furnace?
- ○ A. Plastic
- ○ B. Wood
- ○ C. Coal
- ○ D. Fuel oil

Question 45

Which type of switch must be mounted horizontally and depends on gravity to achieve proper operation?

- ○ A. A magnetic switch
- ○ B. A mercury switch
- ○ C. A solar switch
- ○ D. A legacy switch

Question 46

A(n) _____ coil is the portion of a heat pump or central air conditioning system that is located inside the home.

- ○ A. Condenser
- ○ B. Evaporator
- ○ C. Heat
- ○ D. Cooling

Question 47

The _____ is a measure of the pool water's overall capability to eliminate waste.

- ○ A. ORP
- ○ B. PH
- ○ C. CCP
- ○ D. POR

Question 48

Which type of pump is used to remove water from basements or flooded areas?

- ○ A. A sump pump
- ○ B. An irrigation pump
- ○ C. A spa pump
- ○ D. A sprinkler pump

Question 49

A _____ system exists where multiple heating and cooling units are located throughout the residence.

- ○ A. Centralized
- ○ B. Distributed
- ○ C. Monitoring
- ○ D. Standalone

Question 50

Which of the following is added to the air handler to make it become an electric furnace? (Select all that apply.)

- ☐ A. Electric heating elements
- ☐ B. Circuit breakers
- ☐ C. A control system
- ☐ D. A solenoid

Question 51

Which of the following is a warming media used in a home heating (radiant) system?

- ○ A. Water
- ○ B. Sand
- ○ C. Gravel
- ○ D. Aluminum

Question 52

The _____ thermal collectors are used to create the heat energy required for space heating and domestic hot water.

- ○ A. Remote
- ○ B. Solar
- ○ C. Electric
- ○ D. Magnetic

Question 53

A _____ decodes the X-10 commands and places them on the 110v AC power lines.

- ○ A. Transformer
- ○ B. Router
- ○ C. Transponder
- ○ D. Calculator

Question 54

Where is the best location for a thermostat?
- ○ A. Next to the fireplace
- ○ B. In the kitchen above the stove
- ○ C. Near the warm air return grill
- ○ D. In the laundry room

Question 55

Irrigation timers are controllers that include a _____ unit capable of activating one or more valves of the irrigation system at specific times.
- ○ A. Sensor
- ○ B. Remote
- ○ C. Relay
- ○ D. Clock

Question 56

What can be used if a calm area of the sump is not found for placing the liquid level sensors?
- ○ A. A float well
- ○ B. A still well
- ○ C. A lever ball
- ○ D. A float switch

Question 57

What is used to keep criminals from reading and recording the fixed radio code for a remote control garage door opener?

- A. Rolling code
- B. A code grabber
- C. A descrambler
- D. A scrambler

Question 58

The _____ requires that all garage door operators manufactured or imported after January 1, 1993, for sale in the United States be outfitted with an external entrapment protection system.

- A. IEEE
- B. FCC
- C. CPSC
- D. NFPA

Question 59

Which of the following is a design feature that requires some type of authentication of the identity of a user attempting to gain access to a password-protected home security system from a distance?

- A. A control box
- B. A card reader
- C. A remote control
- D. Remote access

Question 60

What is the range of the control box receiver and remote transmitter?

- A. 5–25 feet
- B. 25–50 feet
- C. 50–100 feet
- D. 8–50 feet

Question 61

A reversing apparatus, such as a photoelectric sensor, is located about _____ inches off the floor on both sides of a garage door.

- ○ A. 5–6
- ○ B. 6–8
- ○ C. 8–10
- ○ D. 10–12

Question 62

Cable ratings are established by Article 800 of the _____ and also by _____. (Select two.)

- ❑ A. NEC
- ❑ B. CMG
- ❑ C. UL
- ❑ D. MPR

Question 63

A _____ is any air space between walls and under structural floors.

- ○ A. Plenum
- ○ B. CMP
- ○ C. CMR
- ○ D. Transplant

Question 64

Which cable is covered by NEC Article 725?

- ○ A. Multipurpose plenum cable
- ○ B. Multipurpose riser cable
- ○ C. Power-limited plenum cable
- ○ D. CLPT

Question 65

What is used for moving large loads between the lower and upper levels of a home?
- ○ A. A dolly
- ○ B. A dumbwaiter
- ○ C. A hoist
- ○ D. A crane

Question 66

What is the mechanism that is used to create the spark that lights a gas fireplace?
- ○ A. A lighter
- ○ B. A spark plug
- ○ C. A pilot
- ○ D. An igniter

Question 67

Automated home ceiling fans can operate with which type of controller?
- ○ A. X11
- ○ B. X10
- ○ C. X12
- ○ D. X9

Question 68

X-10 controller modules can handle _____ inductive loads and offer 16 steps in speed.
- ○ A. 1000v
- ○ B. 5000v
- ○ C. 250v
- ○ D. 500v

Question 69

Which of the following is a solenoid valve used for?

○ A. Elevators
○ B. Power converters
○ C. Electric fireplaces
○ D. Power transformers

Question 70

What is used to describe a type of network switch that uses the telephone outlets?

○ A. Star
○ B. Portal
○ C. Cable
○ D. Gateway

Answers to Practice Exam 1: Residential Systems

1. B
2. A, B, C
3. A
4. B
5. C
6. D
7. A
8. B
9. C
10. A
11. C
12. B
13. A
14. C
15. A
16. C
17. B
18. D
19. B
20. C
21. A, C
22. A
23. C
24. D
25. A
26. C
27. B
28. A
29. C
30. D
31. B
32. D
33. A, C
34. D
35. B
36. A
37. B, D
38. A
39. C
40. A
41. C
42. D
43. B
44. A
45. B
46. B
47. A
48. A
49. B
50. A, B, C
51. A
52. B
53. C
54. C
55. D
56. B
57. A
58. C
59. D
60. D
61. A
62. A, C
63. A
64. C
65. B
66. D
67. B
68. D
69. A
70. B

Question 1

Answer B is correct. Home RF is a standard specifically targeted to the residential user and uses a protocol called the Shared Wireless Access Protocol (SWAP). It is derived from the existing Digital Enhanced Cordless Telephone (DECT) standard. Answer A is incorrect because Bluetooth is designed to connect one device to another using a short-range radio link. Answer C is incorrect because IEEE 802.11 uses a wireless local area network. Answer D is incorrect because it uses the DSSS PHY layer.

Question 2

Answers A, B, and C are correct. The remote user only needs to have a Web browser, a modem, and a valid username and password to access and control the resources of residential network. Answer D is incorrect because IrDA is an international organization that creates and promotes interoperable infrared data interconnection standards for wireless networks and is not used to operate an automated subsystem using a Web page.

Question 3

Answer A is correct. It uses frequency division multiplexing (FDM) techniques to divide this bandwidth of frequencies into several communications channels. HomePNA can be used without interrupting normal voice or fax services. Answer B is incorrect because twisted-pair wiring is not required for this application. Answer C is incorrect because RJ-11 jacks are for connecting the adapter to the wall phone outlet jack and a telephone. Answer D is incorrect because an HPNA USB adapter is an external interface device that uses the computer USB port to interface to the wall telephone jack.

Question 4

Answer B is correct. The Electronic Industries Association (EIA) developed the ANSI/EIA-600 standard to specify the CEBus communications protocol. CEBus is an acronym for consumer electronics bus. Answer A is incorrect because X-10 is a home network control protocol designed to work with Web browsers for remote access control. Answer C is incorrect because routers provide Internet access with more than one computer. Answer D is incorrect because PLC stands for power-line carrier, which is a home automation product that uses power-line technology. They are referred to as PLC devices because they communicate over existing power-line wiring.

Question 5

Answer C is correct. In 1990, a major advance in Ethernet standards came with the introduction of the IEEE 802.3i 10BASE-T standard. It permitted 10Mbps Ethernet to operate over simple Category 3 UTP cable. Answer A is incorrect because 100BASE-TX operates over two pairs of Category 5 twisted-pair cable. Answer B is incorrect because 100BASE-T4 operates over four pairs of Category 3 twisted-pair cable. Answer D is incorrect because 100BASE-FX operates over two multimode fibers.

Question 6

Answer D is correct. IEEE 1394 has a data transfer rate of 100Mbps, 200Mbps, and 400Mbps over a maximum cable length of 4.5 meters. Answer A is incorrect because RS-232 has a data transfer rate of 300bps–9600bps over a maximum cable length of 50 feet. Answer B is incorrect because USB 1.1 has a data transfer rate of 12Mbps over a maximum cable length of 5 meters. Answer C is incorrect because IEEE 1284 has a data transfer rate of 2Mbps over a maximum cable length of 10 meters.

Question 7

Answer A is correct. The RJ-31x jack is a special type of jack used with home security alarm systems. Answer B is incorrect because RJ-11 is the type of connector used for terminating a telephone cable that connects the telephone to a standard wall outlet. Answer C is incorrect because RJ-45 jacks are jacks attached at both ends of a UTP cable used to connect components in a home Ethernet network. Answer D is incorrect because the RJ-14 is similar to the RJ-11, but the four wires are used for two phone lines.

Question 8

Answer B is correct. Class A addresses are reserved for large networks and use the last 24 bits (the last three octets or fields) of the address for the host address. Answer A is incorrect because Class B addresses are assigned to medium-sized networks. Answer C is incorrect because Class C addresses are normally used with smaller LANs. In a Class C address, only the last octet is used for host addresses. Answer D is incorrect because Class D is used for broadcasting.

Question 9

Answer C is correct. A dedicated audio/video system design places all audio/video equipment in one location. Answer A is incorrect because a distributed audio/video system design places the TV and audio entertainment in different areas, or zones, of the home. Answer B is incorrect because a whole home audio/video system design allows you to see and hear all the shared video and audio sources in multiple locations throughout the home. Answer D is incorrect because a remote access system is used with infrared wireless controllers.

Question 10

Answer A is correct. The sixth channel in the Dolby Digital surround sound 5.1 format dedicated for low frequency effects (LFE) is reserved for the subwoofer speaker. The low frequency effects channel gives the Dolby Digital format the ".1" designation. Answer B is incorrect because the center channel is reserved for the front of the listening position. Answer C is incorrect because the right channel is reserved for the right front speaker. Answer D is incorrect because the left channel is reserved for the left front speaker.

Question 11

Answer C is correct. The tuner portion of a stereo or A/V receiver selects radio stations in the AM and FM broadcast bands. The received signal is sent to the preamplifier and ultimately through the amplifier to the speakers. Answer A is incorrect because receivers are used to process audio, video, satellite, and AM/FM broadcast signals received from external sources and to output amplified or decoded signals to television video input jacks, speakers, or recording devices. Answer B is incorrect because the amplifier's job is to take audio inputs from a number of sources that have been selected and processed in the preamplifier and amplify them to a level suitable for driving multiple speakers. Answer D is incorrect because the volume control is used to change the amplitude of the sound heard through the speakers.

Question 12

Answer B is correct. When an amplifier is too small for the speaker system, high volume levels can cause the amplifier to automatically shut down due to clipping of the output audio signal. Answer A is incorrect because, if the amplifier power rating is too large for the speakers, you can blow them at high volume levels. Answer C is incorrect because, if the speakers are the same power rating as the amplifier, you should not see clipping. Answer D is incorrect because a monitor produces a video signal and would not affect the audio signal.

Question 13

Answer A is correct. Cathode-ray tubes (CRTs) have been used as the main display component of television receivers and desktop computer monitors for several years. CRT displays however, are not a portable technology and are not used with portable laptop computing systems. Answer B is incorrect because liquid crystal displays are used for a wide range of small portable electronic and computing devices, including laptop computers, cell phones, games, and personal digital assistants (PDAs). Answer C is incorrect because plasma displays are available for large-screen TV displays offering high resolutions and a 16:9 aspect ratio compatible with the HDTV format and are used for some types of portable computer systems. Answer D is incorrect because flat-panel displays get their name from the type of physical shape and appearance that distinguishes them from CRT displays. They are also suitable for lightweight portable laptop computers.

Question 14

Answer C is correct. RJ-45 jacks use Category 5 UTP cable. Answer A is incorrect because RCA uses coaxial cable. Answer B is incorrect because F-type uses coaxial cable. Answer D is incorrect because optical uses optical cable.

Question 15

Answer A is correct. Streaming is a popular technology for capturing audio and video files and playing them on a computer while they are being downloaded from an Internet Web server. Answer B is incorrect because the personal video recorder (PVR) is a new type of consumer electronics equipment that records television programs digitally encoded in the compressed MPEG format on a nonremovable hard disk. Answer C is incorrect because most media servers are used by commercial service providers for pay per view (PPV) and video on demand (VOD) multimedia services. Answer D is incorrect because most media servers are used by commercial service providers for pay per view (PPV) and video on demand (VOD) multimedia services.

Question 16

Answer C is correct. Direct broadcast satellite (DBS) service is a class of service defined by the Federal Communications Commission whereby home subscribers receive television programs by direct paid subscription access to a satellite rather than through an intermediate cable service provider. Answer A is incorrect because cable TV service provides television programming service to homes over coaxial cable networks. Answer B is incorrect because DTV is a new over-the-air digital television system that will be used by the nearly 1,600 local broadcast television stations in the United States. Answer D is incorrect because terrestrial television and AM/FM radio reception is the mode used with an outside antenna pointed toward commercial broadcast stations.

Question 17

Answer B is correct. The major new protocol adopted for the home entertainment industry is called the Home Audio Video Interoperability Protocol. Answers A, C, and D are incorrect because these are protocols of the World Wide Web.

Question 18

Answer D is correct. HI-8 is an 8mm tape that uses more than 400 lines. Answer A is incorrect because VHS is a standard videocassette used by VCRs and it uses 250 lines. Answer B is incorrect because 8mm has a resolution of

260 lines. Answer C is incorrect because Mini DV is a mini VHS format tape and uses 500 lines of resolution.

Question 19

Answer B is correct. The term *LNB* used in the text is an acronym for low noise block-down converter (so called because it converts a whole band, or block, of frequencies to a lower band). Answer A is incorrect because NEC stands for the National Electrical Code. Answer C is incorrect because DBS is a term for the digital broadcast system. Answer D is incorrect because HDTV is high-definition television.

Question 20

Answer C is correct. Keypads are located by most contractors on the inside wall near the door most often used by the occupants. Answer A is incorrect because the closet is not a recommended location. The keypad should be located where you can easily access it. Answer B is incorrect because you should never put a keypad in front of the house because of vandalism. Answer D is incorrect because the back of the house is not a good location because occupants do not normally use the back door as their main entry. It is recommended that you mount a keypad at the inside area near the front door.

Question 21

Answers A and C are correct. Temperature sensors are used as one component of an environmental monitoring system. They are normally used to monitor high and low temperature values in vacation homes, water pipes, furnace and heating vents, outside farm buildings, computer equipment, utility rooms, or areas that can sustain damage with extreme temperature swings. Answers B and D are incorrect because temperature monitoring is not done from right to left.

Question 22

Answer A is correct. Computer systems can be integrated with control protocols such as X-10 and CEBus to operate and manage security systems. Answers B, C, and D are all World Wide Web protocols.

Question 23

Answer C is correct. The ANSI/TIA/EIA 570A standard requires at least 1" of separation between parallel runs of security wire and AC power wiring. Answers A, B, and D are incorrect because these are not standard requirements and therefore are invalid.

Question 24

Answer D is correct. When security wire and cabling are crossed with AC power wires, the crossover must be at a 90° angle. Answers A, B, and C are not the required 90° angles and are therefore incorrect.

Question 25

Answer A is correct. The RJ-31X phone jack uses an 8-pin telephone jack that is specially designed to seize the phone line and send the alarm message. Answers B, C, and D are incorrect and do not represent the correct number of pins that the RJ-31x has.

Question 26

Answer C is correct. All alarm systems that use the telephone lines in a home to automatically dial a central monitoring station must have an RJ-31x jack installed. This enables the alarm system to seize the phone line when an alarm condition exists. Answers A, B, and D are invalid telephone jack types.

Question 27

Answer B is correct. They require two wires for 12v–16v DC and two wires for the alarm circuit. Answers A, C, and D are incorrect because the alarm circuit requires only two wires.

Question 28

Answer A is correct. Passive sensors need only two wires, and active sensors require four wires. Answers B, C, and D are not the required number of wires needed for an active sensor.

Question 29

Answer C is correct. *Hybrid* is a loosely defined generic term used in the telecommunications industry to describe any system that uses a mix of media and transmission standards. Answers A, B, and D are invalid terms that do not describe any system that uses a mix of media and transmission standards.

Question 30

Answer D is correct. Analog communications systems are those systems designed to transport voice information. Answer A is incorrect because a digital system converts analog voice signals into a digital format. Answer B is incorrect because satellite systems are used to transport audio and video, not just voice information. Answer C is incorrect because cable systems are designed to carry voice and video information.

Question 31

Answer B is correct. The Private Branch Exchange (PBX) is the name of a product line of equipment used primarily by business organizations that need their own internal telephone-switching interfaces with the external telephone network. Answer A is incorrect because CTI is for computer telephony integration technology products and is not an essential element of any PBX service or feature. Answer C is incorrect because ACD (automatic call distribution) service for a business organization is a type of added service for some PBX systems. ACD is not necessarily essential to those organizations that need their own internal telephone-switching interface with the external telephone network. Answer D is incorrect because most key systems are supported by a central processing unit called a key service unit (KSU). KSU systems typically support advanced business telephone features such as call forwarding, extension dialing, and voice mail options.

Question 32

Answer D is correct. The Point to Point Tunneling Protocol (PPTP) is used to encapsulate a message when it is transmitted over the Internet using a virtual private network (VPN). Answer A is incorrect because FTP is a file transfer protocol used with the World Wide Web. Answer B is incorrect because the Transmission Control Protocol (TCP) is a reliable protocol used to send packets over the Internet; it is not used to encapsulate other protocols. Answer C is incorrect because the User Datagram Protocol (UDP) is an unreliable protocol used to send packets over an IP network; it is not used to encapsulate other packets.

Question 33

Answers A and C are correct. The punch-down tool is also known as an impact tool. It is usually supplied with two types of blades: One is for the 66-type block, and the other is for the 110-type block. Answer B is incorrect because a hole punch is used for punching holes in paper. Answer D is incorrect because a wire cutter is used for cutting individual wires one at a time and does not use a blade-type 66 connection block.

Question 34

Answer D is correct. This restriction blocks the phone from dialing a 1 or a 0 to initiate a long-distance call. Answer A is incorrect because an 800 number block prohibits the dialing of any 800 number service. Answer B is incorrect because 011 number block blocks the use of the international prefix calling number. Answer C is incorrect because collect call blocking restricts the user's phone from accepting collect calls.

Question 35

Answer B is correct. For each cabled location, grade 2 requires two Category 5 (5E recommended by the FCC ruling) UTP cables and two 75-ohm coaxial cables to each location plus, as an option, optical fiber cabling. Answer A is incorrect because grade 1 requires a minimum of one Category 3 UTP cable and one 75-ohm coaxial cable for each location. Answers C and D are incorrect because these grades are invalid wiring standards.

Question 36

Answer A is correct. *Home run* wiring is another name for a star wiring topology in which all cabling is routed from each room to a common termination location. Answers B, C, and D are incorrect because these are invalid names and are not another name for the star wiring topology.

Question 37

Answers B and D are correct. X-10 and CEBus are the two power-line control protocols. They are used to send signals through handheld wireless keypad devices and the home power lines to remotely control the lighting scenes and illumination levels in any area of the home. Answer A is incorrect because there is no known home lighting control protocol called X-12. Answer C is incorrect because GEBus is not an identified standard for any home control protocol.

Question 38

Answer A is correct. Electrical metallic tubing (EMT), commonly called thin-wall conduit, is metallic tubing that can be used for exposed or concealed electrical installations. Answers B, C, and D are invalid names for thin-wall conduit.

Question 39

Answer C is correct. Residential lighting can be programmed for different zones. A *zone* is one group of lights controlled by one switch or a dimmer module. Answer A is incorrect because *cluster* is a term used to describe the smallest manageable piece of information in a FAT-based system. Answer B is incorrect because *PPP* is an acronym for the Point-to-Point Protocol. Answer D is incorrect because ASB is an invalid answer not relating to the question.

Question 40

Answer A is correct. The black or red (hot) wire always connects to the brass terminal located on the same side as the narrow slot. Answer B is incorrect because the white wire always connects to the silver terminal located on the same side as the wide slots. Answer C is incorrect because there is no black terminal connection. Answer D is incorrect because the green wire is always connected to the green screw terminal.

Question 41

Answer C is correct. X-10-compatible two-wire dimmer switches can be used only with incandescent lights. Answer A is incorrect because an SPST is (single pole single throw) type of switch is used for operating a single light fixture. Answer B is incorrect because DPST (double pole single throw) is another type of on-off control and is used to interrupt a two-wire circuit. Answer D is incorrect because SPDT (single pole double throw) is referred to as a three-way switch and is used for controlling a light from two locations.

Question 42

Answer D is correct. A DPST switch is another type of on-off control and is used to interrupt a two-wire circuit. Answer A is incorrect because SPST is used for operating a single light fixture. Answer B is incorrect because SPDT is referred to as a three-way switch and is used for controlling a light from two locations. Answer C is incorrect because DPST is another type of on-off control and is used to interrupt a two-wire circuit.

Question 43

Answer B is correct. A photoelectric switch contains a special sensor that operates a switch when the light intensity or darkness reaches a certain value. Answer A is incorrect because automated dimmer modules have the capability to turn the light on and off, plus dim it to any level, at the switch location or from any X-10 transmitter. Answer C is incorrect because the three-way switch circuit is frequently used in residential installations to control a light fixture or lighting array from two locations. Answer D is incorrect because clock switches are used to turn lights on and off at given intervals.

Question 44

Answer A is correct. A furnace is not designed for using plastic. Answers B, C, and D are incorrect because furnaces are designed to use these types of fuels.

Question 45

Answer B is correct. A mercury switch must be mounted completely horizontally for the liquid mercury to be capable of using the effect of gravity when making and breaking the contact between its embedded wires. Answer A is incorrect because magnetic switches do depend on gravity and can be mounted any direction. Answers C and D are incorrect because these are not valid types of switches and are not used.

Question 46

Answer B is correct. An evaporator coil is that portion of a heat pump or central air conditioning system that is located inside the home. Answer A is incorrect because the condensing unit is also known as an outdoor coil. It serves as the transfer point where heat is transferred from inside the home to the outside air. Answers C and D are incorrect because these are invalid and are not types of coil.

Question 47

Answer A is correct. The oxidation reduction potential (ORP) is a measure of the pool water's overall capability to eliminate waste. Answer B is incorrect because pH is a measure of the water alkalinity level and is not directly related to the pool water's overall capability to eliminate waste. Answer C is incorrect because CCP is not a name for any unit or process dealing with water purification or treatment. Answer D is incorrect because POR is not a term or acronym associated with pool water conditioning or measurement.

Question 48

Answer A is correct. Sump pumps are designed to remove water from basements or potential flood areas. Answer B is incorrect because most home sprinkler systems use the available pressure from the main utility water supply or, in locations near large rivers or lakes, the local irrigation service supply line. Answer C is incorrect because pumps designed for pools and spas are usually custom pumps. They fall into two general categories called in-ground and above-ground pumps. Answer D is incorrect because a sprinkler pump is used for home sprinkler systems only and not to pump water for flooding.

Question 49

Answer B is correct. Distributed systems exist where multiple heating and cooling units are located throughout the residence. Control of the various heating and cooling units is typically performed individually. Answer A is incorrect because centralized systems exist where heating and cooling units occupy a centralized location within the residence and provide heating and cooling to the entire residence, usually through metal ductwork. Answer C is incorrect because a monitoring system is a generic term that applies to a sensor system used in home security systems. A monitoring system is not unique to a distributed heating and cooling system. Answer D is incorrect because standalone systems are not associated with multiple locations. A standalone system implies a single centralized system that supplies heating and cooling to the entire home.

Question 50

Answers A, B, and C are correct. If electric heating elements, circuit breakers, and a control system are added to the air handler, it becomes an electric furnace. Answer D is incorrect because solenoids are electromechanical devices that are an integral part of an electrically operated water valve or control relay switch.

Question 51

Answer A is correct. Some home heating systems are water-based radiant heaters. In these systems, the fuel is used to heat water in a central reservoir, which is then piped into heat-radiating devices (radiators) positioned in each room. In some systems, hot water is circulated through the system, whereas in others the water is actually converted to steam prior to being forced through the radiator distribution system. Answers B, C, and D are not warming media used for radiating heat throughout a home.

Question 52

Answer B is correct. The open direct radiant heating system can be converted to solar heating. Its required electricity can be harnessed directly from the sun by using cost-effective solar thermal collectors to create the heat energy required for space heating and domestic hot water. Answers A, C, and D are invalid and not used with thermal collectors.

Question 53

Answer C is correct. A typical remote access design employs a standard telephone connection and an X-10 telephone transponder to communicate with the residential control system. The transponder receives touch-tone input signals sent to it across the telephone lines and encodes them on the residential power lines using the X-10 protocol. Answer A is incorrect because a transformer is used for transforming power levels on power lines, not over phone lines. Answer B is incorrect because a router is invalid and cannot be used for this application. Answer D is incorrect because a calculator is used for calculating numbers and not for encoding signals.

Question 54

Answer C is correct. The best area for a thermostat location is near the warm air return grill. Answers A, B, and D are incorrect because these locations can adversely affect the operation of a thermostat because they are normally warmer than the other areas.

Question 55

Answer D is correct. Irrigation timers are simple controllers that include a clock unit capable of activating one or more valves of the irrigation system at specified times. Several designs are commercially available with many features and a wide range of costs. Answer A is incorrect because sensors are not an internal part of a timing system. Answer B is incorrect because a remote is used with remote access control of a water system and does not apply to a timer unit. Answer C is incorrect because a relay is a magnetic pull of a solenoid coil that can be used to open or close an electric switch.

Question 56

Answer B is correct. The liquid level sensors should be located in a calm area of the sump. If this is not possible, you might need to use a still well, which is a smaller well or pipe by the side of the sump chamber. Answer A is incorrect because a float well is invalid and not used for sump pump operations. Answer C is incorrect because a lever ball is part of a float switch that turns the pump on after the lever ball has reached its predetermined height. Answer D is incorrect because a float switch is a sensor that turns on the pump after the water raises the lever ball to a predetermined height.

Question 57

Answer A is correct. A rolling code radio system has been designed for remote control transmitters. This feature automatically changes the signal code each time the remote control transmitter is used. Answer B is incorrect because a code grabber is used to read and record the fixed radio codes of the user's transmitter. This is what criminals use for gaining access to the garage. Answer C is incorrect because a descrambler is similar to a code grabber, except it is used to descramble cable television channels for viewing. Answer D is incorrect because a scrambler does the opposite of the descrambler and scrambles cable television channels.

Question 58

Answer C is correct. The Consumer Product Safety Commission (CPSC) requires that all garage door operators manufactured or imported after

January 1, 1993, for sale in the United States be outfitted with an external entrapment protection system. Answer A is incorrect because the IEEE has not published or been involved with standards for garage door operator safety. Answer B is incorrect because FCC rules do not apply to the operation of garage door openers. Answer D is incorrect because the NFPA does not oversee the safety requirements of remote garage door operators.

Question 59

Answer D is correct. Remote access is a design feature that requires some type of authentication of the identity of a user attempting to gain access to a locked or password-protected home security or computer system from a distant location. Answer A is incorrect because control boxes provide the intelligent adjustments required in a residential access control system. Answer B is incorrect because a card reader is for reading access codes from a card that's inserted into the reader and decoded. Answer C is incorrect because remote control is a design feature used to perform the monitoring, controlling, and supervisory functions of garage doors, gates, and residential entrance doorways at a distance.

Question 60

Answer D is correct. The transmitter is programmed at the factory with its own unique code and has a range of 8–50 feet, depending on the environmental conditions. Answers A, B, and C are incorrect because the factory set range should be 8–50 feet.

Question 61

Answer A is correct. Reverse signaling apparatus such as photoelectric sensors are located about 5–6 inches off the floor on both sides of a garage door. Answers B, C, and D are invalid and do not apply to the manufacturer's suggested height off the floor.

Question 62

Answers A and C are correct. Cable ratings are established by Article 800 of the NEC and also by Underwriter Laboratories, Inc. Answers B and D are incorrect because both answers are NEC markings. CMG is a mark that stands for communications general-purpose, and MPR stands for multipurpose riser.

Question 63

Answer A is correct. A plenum is any air space between walls, under structural floors, or above dropped ceilings that is used for environmental air, such as an air conditioning duct. Answers B and C are NEC markings that stand for communications plenum and communications riser. Answer D is incorrect because a transplant is something or someone that is taken from one environment and placed in another.

Question 64

Answer C is correct. NEC Article 725 covers CL2P, or power-limited plenum cable. Type CL2P (class 2 plenum) cables are listed as being suitable for use in ducts, plenums, and other spaces used for environmental air. Answer A is incorrect because copper multipurpose plenum (MPP) cable is suitable for residential or commercial installation in the space between a ceiling and the floor above it. Answer B is incorrect because copper multipurpose riser cable (MPR) is suitable for installation in vertical riser shafts that are designed to run from floor to floor. Answer D is an invalid cable type not used.

Question 65

Answer B is correct. The task of moving many items between the lower and upper levels of a home can be simplified with the use of a modern, multifunction dumbwaiter. A dumbwaiter can easily carry loads up to 200 lbs. and can travel approximately 30 feet per minute. Answer A is incorrect because a dolly is used for smaller loads usually at ground level. Answer C is incorrect because a hoist is typically reserved for commercial use and lifts items in the 1000-lbs. range. Answer D is incorrect because a crane is used for lifting items that are well over 1000 lbs.

Question 66

Answer D is correct. The mechanism used to create the spark that lights a gas fireplace is called the igniter. The three general classes of fireplace igniters are pilot, piezoelectric (push-button or rotary), and electronic. Answer A is incorrect because a lighter is used to light such things as a pilot light. Answer B is incorrect because a spark plug is used to ignite fuel in a combustion engine and is not used as a fireplace igniter. Answer C is incorrect because a pilot is used as a pilot light in another method of igniting a fireplace.

Question 67

Answer B is correct. Automated home ceiling fans can be operated with X10 controllers. Answers A, C, and D are invalid and are not used by ceiling fans.

Question 68

Answer D is correct. Unlike dimmer switches, X-10 controller modules can handle 500v inductive loads and offer 16 steps in speed. The dimmer control on the X-10 remote can be used to slow the fan, and the brighten control can be used to speed it up. Answers A, B, and C are incorrect and are not standard X-10 voltage levels.

Question 69

Answer A is correct. Because a solenoid is a form of electromagnet, it can be shaped into a tube and used to move a piece of metal linearly. This property can be used in many ways, especially with valves, which is why a solenoid valve is suitable for use in elevators and other lift systems. Answer B is incorrect because a power converter is used for converting power, such as AC to DC. Answer C is incorrect because electric fireplaces do not use solenoid valves. Answer D is incorrect because a power transformer uses electric solenoids.

Question 70

Answer B is correct. Home portals enable the sharing of a high-speed Internet connection throughout the home. A portal is a type of network switch that uses the telephone outlets in the home to provide additional ports for high-speed Internet access without adding new wiring. Answer A is incorrect because a star is a type of network design for a computer network. Answer C is incorrect because cable is not a valid answer used for this application. Answer D is incorrect because a gateway is not the correct term used for describing this particular network.

Practice Exam 2: Systems Infrastructure and Integration

Question 1

When installing a home theater, you should refer to _____ for the specifications.

○ A. ANSI/TIA/EIA 570-A
○ B. Home RF 2.0
○ C. IEEE 802.11b
○ D. HPNA 3.0

Question 2

What is the recommended design concept for structured wiring?

○ A. Daisy chain
○ B. Home run
○ C. Mesh
○ D. Ring

Question 3

What are the advantages of using the outer braided wire shield on STP cabling? (Select all that apply.)

- ❑ A. Maximizing the throughput
- ❑ B. Reducing the susceptibility to crosstalk
- ❑ C. Preventing signal losses
- ❑ D. Minimizing EMI radiation

Question 4

Which of the following types of cable is the preferred standard for home structured wiring?

- ○ A. ScTP
- ○ B. FTP
- ○ C. UTP
- ○ D. STP

Question 5

Which of the following describes ScTP cabling?

- ○ A. Two-pair 150-ohm STP with a foil or braided screen surrounding each pair
- ○ B. Two-pair 100-ohm UTP with a single foil or braided screen surrounding all four pairs
- ○ C. Four-pair 150-ohm STP with a foil or braided screen surrounding each pair
- ○ D. Four-pair 100-ohm UTP with a single foil or braided screen surrounding all four pairs

Question 6

For networking applications, the term UTP generally refers to _____.
- ○ A. Four-pair 100-ohm twisted wires enclosed in a common sheath
- ○ B. Four-pair 150-ohm twisted wires enclosed in a common sheath
- ○ C. Four-pair 100-ohm twisted wires enclosed in a common sheath with a foil or braided screen surrounding each pair
- ○ D. Four-pair 150-ohm twisted wires enclosed in a common sheath with a foil or braided screen surrounding each pair

Question 7

Which of the following types of cable is more susceptible to EMI?
- ○ A. ScTP cable
- ○ B. UTP cable
- ○ C. Coaxial cable
- ○ D. Fiber-optic cable

Question 8

Which standard specifies a bandwidth of no less than 250MHz?
- ○ A. Category 4
- ○ B. Category 5
- ○ C. Category 6
- ○ D. Category 5E

Question 9

When comparing the various types of cable used in low-voltage structured-wiring design, which of the following types of cable provides the greatest immunity to EMI?
- ○ A. Fiber-optic cable
- ○ B. UTP
- ○ C. STP
- ○ D. Coaxial cable

Question 10

Which of the following statements is not true regarding the use of conduits?

- A. Conduit is not required by any standard for indoor residential low-voltage circuits.
- B. Wiring and electrical equipment must be enclosed in conduit in areas that are considered to be hazardous.
- C. Conduit should be used for low-voltage wiring in outdoor locations.
- D. NEC articles 500–517 define the hazardous areas, where wiring and electrical equipment must be enclosed in conduit.

Question 11

Which topology does the home-run wiring system employ?
- A. Ring
- B. Star
- C. Bus
- D. Mesh

Question 12

Which of the following is not an advantage of using structured home-run wiring?
- A. It's accessible.
- B. It's flexible.
- C. It's economical.
- D. It's easy to troubleshoot.

Question 13

Which types of wiring are specified for grade 2 structured wiring?
- A. One four-pair Category 3 UTP cable and one 75-ohm coaxial cable
- B. Two Category 5E cables and two 75-ohm coaxial cables
- C. Two Category 5 cables and two 75-ohm coaxial cables
- D. Two Category 3 UTP cables and two 75-ohm coaxial cables

Question 14

Which component of a structured wiring system determines which services are available in a room?

- ○ A. The central distribution panel
- ○ B. The gateway device
- ○ C. The home computer network
- ○ D. The multipurpose outlet

Question 15

A professionally installed low-voltage structured-wiring system can be expected to last approximately _____ years.

- ○ A. 3
- ○ B. 5
- ○ C. 10
- ○ D. 15

Question 16

Which terms are frequently used to describe the black and white wires in residential high-voltage wiring?

- ○ A. Hot and neutral
- ○ B. Live and dead
- ○ C. Hot and cold
- ○ D. Bold and neutral

Question 17

Which wire is referred to as the equipment grounding conductor?

- ○ A. Blue
- ○ B. Green
- ○ C. Orange
- ○ D. Red

Question 18

What provides emergency AC power for the home when commercial power service is unavailable?

- ○ A. Surge suppressors
- ○ B. A ground rod
- ○ C. SMPS
- ○ D. UPS

Question 19

An emergency backup power generator is NOT fueled by _____.

- ○ A. Gasoline
- ○ B. Diesel
- ○ C. Natural gas
- ○ D. Water

Question 20

What do residential power backup systems use to transfer the electrical load to the generator and disconnect the commercial power source?

- ○ A. A voltage regulator
- ○ B. A power switching device
- ○ C. A transfer switching device
- ○ D. A power source switch

Question 21

Residential backup power systems can be sized and wired for a home in either whole house or essential power distribution designs. With whole house power distribution, the generator is sized for powering _____.

- ○ A. All devices within the home
- ○ B. Isolated essential devices
- ○ C. Isolated unessential devices
- ○ D. Limited emergency lighting

Question 22

With essential power distribution, the residential backup power system design allows for emergency backup power to _____.

- ○ A. All devices within the home
- ○ B. Isolated essential devices
- ○ C. Isolated unessential devices
- ○ D. Up to three devices

Question 23

VA ratings for UPS units are the product of volts and _____.

- ○ A. Amperes
- ○ B. Watts
- ○ C. Apparent power
- ○ D. Real power

Question 24

Power factor is a number between _____ representing the fraction of the apparent power delivered to the load that does useful work.

- ○ A. 1.0 and 2.0
- ○ B. 0.0 and 1.2
- ○ C. 0.2 and 1.0
- ○ D. 0.0 and 1.0

Question 25

The current capacity of a circuit breaker is measured in what?

- ○ A. Volts
- ○ B. Watts
- ○ C. Amperes
- ○ D. Kilowatts

Question 26

Which type of electrical outlet is used to guard against accidental electric shock?

- ○ A. Ground current fault interrupter (GCFI)
- ○ B. Ground fault circuit interrupter (GFCI)
- ○ C. Circuit ground fault interrupter (CGFI)
- ○ D. Ground fault interrupter current (GFIC)

Question 27

How does an electrical outlet that guards against accidental electric shock work?

- ○ A. It compares the amount of electric current coming into a circuit on the black wire with the amount leaving the circuit on the white wire. If more current enters than exits, it shuts off the circuit.
- ○ B. It compares the amount of electric current coming into a circuit on the white wire with the amount leaving the circuit on the black wire. If more current exits than enters, it shuts off the circuit.
- ○ C. It uses a sensor to sense when something is touching the outlet.
- ○ D. It compares the amount of electric current coming into a circuit on the red wire with the amount leaving the circuit on the blue wire. If more current enters than exits, it shuts off the circuit.

Question 28

Which of the following describes a ground fault?

- ○ A. When less current enters the circuit than leaves
- ○ B. When more current enters the circuit than leaves
- ○ C. When an equal amount of current is entering and leaving the circuit
- ○ D. When more current leaves the circuit than enters

Question 29

The service entrance does NOT include which of the following?

- ○ A. The masthead
- ○ B. The conduit
- ○ C. The wiring
- ○ D. The breaker panel

Question 30

New lighting and outlet circuits in a remodeled home require _____ American Wire Gauge wire to safely handle _____ amperes.

- ○ A. #12; 20
- ○ B. #10; 15
- ○ C. #8; 20
- ○ D. #12; 10

Question 31

A light switch used to operate a light from two locations is called what?

- ○ A. A one-way switch
- ○ B. A two-way switch
- ○ C. A three-way switch
- ○ D. A four-way switch

Question 32

Which of the following are accepted ballast technologies used with fluorescent lighting? (Select all that apply.)

- ❑ A. Magnetic ballast
- ❑ B. Hybrid magnetic ballast
- ❑ C. High-frequency electronic ballast
- ❑ D. Source energy ballast

Question 33

Which system integration interface provides wireless Internet access from any location in a home?

○ A. A Web pad
○ B. A keypad
○ C. A touchscreen
○ D. Wireless RF/IR receivers

Question 34

Which interface is often a portable or wall-mounted design with an LCD display and backlit keys?

○ A. A Web pad
○ B. A keypad
○ C. A touchscreen
○ D. Wireless RF/IR receivers

Question 35

Which interface is available in a standalone large-screen panel, wall-mounted unit, or handheld remote controller?

○ A. A Web pad
○ B. A keypad
○ C. A touchscreen
○ D. Wireless RF/IR receivers

Question 36

Which type of home networking device uses sensors?

○ A. A Web pad
○ B. A keypad
○ C. A touchscreen
○ D. Wireless RF/IR receivers

Question 37

What is a location consideration for a Web pad?
- ○ A. None; it can be operated easily from anywhere within the home.
- ○ B. It must be connected to wiring inside the wall.
- ○ C. It must be located in an area free of signal obstruction.
- ○ D. It should be located in an area of maximum access by home residents.

Question 38

What is a location consideration for a keypad?
- ○ A. None; it can be operated easily from anywhere within the home.
- ○ B. It should be located in an area of maximum access by home residents.
- ○ C. It must be connected to wiring inside the wall.
- ○ D. It must be located in an area free of signal obstruction.

Question 39

Which following statement is true with regard to the optimum location of a touchscreen in a home?
- ○ A. It should always be located near a door leading to the outside area.
- ○ B. It must be located in an area where it can be connected to wiring inside the wall.
- ○ C. It should be located in an area of maximum access by home residents.
- ○ D. It must be located in an area free of signal obstruction.

Question 40

Which gateway configuration adds full or partial Web addresses to a list to be blocked?
- ○ A. Internet content filtering
- ○ B. Port mapping
- ○ C. A demilitarized zone
- ○ D. Remote administration

Question 41

Which of the following is not a main feature of a keypad?

○ A. An LCD display that shows status information
○ B. A wall-mounted receiver
○ C. An audible feedback sound whenever a keypad button is pressed
○ D. A backlit alphanumeric keypad for low-light conditions

Question 42

Which is not a benefit of a patch/distribution panel for homeowners?

○ A. It enables people to make changes to a home network quickly and easily.
○ B. It minimizes wiring mistakes.
○ C. It can be easily replaced.
○ D. It increases system flexibility by fast and simple offline equipment bypassing.

Question 43

What is the preferred location in a home for the central distribution panel?

○ A. It should be located in any area that will make it difficult for intruders to access.
○ B. It should be located in the center of the home to maintain an even distance from all interface devices.
○ C. It should be located in an area near the circuit breaker panel.
○ D. It should be located in close proximity to the external telephone demarcation point and TV or satellite cable feed points.

Question 44

The _____ is a software-enabled device used to manage devices connected to an in-home network.

○ A. Home networking controller
○ B. Indicator
○ C. Cabinet
○ D. Distribution panel

Question 45

What is a hazard of establishing a DMZ for a home computer?
- ○ A. It severely limits Internet access.
- ○ B. It causes the entire home network to crash.
- ○ C. It leaves the machine vulnerable to hackers.
- ○ D. It causes the Internet connection to lag.

Question 46

Which is not a common controller configuration interface device?
- ○ A. A keypad
- ○ B. A touchscreen
- ○ C. An intelligent remote control
- ○ D. A Web pad

Question 47

Which is not an NEC marking used to label communications cables in automated home systems?
- ○ A. MMX
- ○ B. CMX
- ○ C. CMG
- ○ D. CMP

Question 48

Which type of cable is primarily used in home digital audio/video system interface connection?
- ○ A. Coaxial cables
- ○ B. Category 5E UTP cables
- ○ C. Fiber-optic cables
- ○ D. Wireless

Question 49

Which type of cable is standard for data and telephone communications?

- ○ A. Coaxial cables
- ○ B. Category 5E UTP cables
- ○ C. Fiber-optic cables
- ○ D. Wireless

Question 50

Which type of cable is recommended for all analog video signal connectivity?

- ○ A. Coaxial cables
- ○ B. Fiber-optic cables
- ○ C. Wireless
- ○ D. Category 5E UTP cables

Answers to Practice Exam 2

1. A
2. B
3. B, D
4. C
5. D
6. A
7. B
8. C
9. A
10. A
11. B
12. C
13. C
14. D
15. D
16. A
17. B
18. D
19. D
20. C
21. A
22. B
23. A
24. D
25. C
26. B
27. A
28. B
29. D
30. A
31. C
32. A, B, C
33. A
34. B
35. C
36. D
37. A
38. B
39. C
40. A
41. B
42. C
43. D
44. A
45. C
46. D
47. A
48. C
49. B
50. A

Question 1

Answer A is correct. ANSI/TIA/EIA 570-A Residential Telecommunications Cabling applies to telecommunications premises wiring systems installed within an individual building with residential end users. Answer B is incorrect because HomeRF is a wireless LAN standard supported by the HomeRF Working Group. Answer C is incorrect because IEEE 802.11b is the wireless LAN standard created by the Institute for Electrical and Electronic Engineers. Answer D is incorrect because HPNA 3.0, developed by the Home Phone Network Alliance, specifies the use of phone lines for networking.

Question 2

Answer B is correct. A structured wiring employs a home-run wiring plan in which all circuits are terminated at a central location. Each outlet in the home can be dedicated to a single independent source. Troubleshooting is simplified because each home-run circuit is independent from the others. Installing an outlet for all services in every room allows maximum flexibility and avoids pulling new wires into the home as floor plans change. Answer A is incorrect because daisy-chain wiring uses a single chain of low-voltage cable throughout a home. It is difficult to troubleshoot. Answers C and D are incorrect because mesh and ring are two network topologies and are not used for structured wiring.

Question 3

Answers B and D are correct because the twisted pairs in 150-ohm STP are individually wrapped in a foil shield and enclosed in an overall outer braided wire shield. The shielding is designed to minimize EMI radiation and susceptibility to crosstalk. Answers A and C are incorrect because using the outer braided wire shield on STP cabling cannot prevent signal losses or maximize the throughput.

Question 4

Answer C is correct. Unshielded twisted pair (UTP) has evolved as the preferred standard for home structured wiring and commercial networking cabling because the UTP cable and connectors cost less and are easier to install. Answers A, B, and D are incorrect because STP and ScTP (also known as FTP) are more expensive than UTP.

Question 5

Answer D is correct. ScTP can be described as four-pair 100-ohm UTP with a single foil or braided screen surrounding all four pairs to minimize EMI radiation and susceptibility to outside noise. Answers A, B, and C are invalid cabling types for ScTP.

Question 6

Answer A is correct. For networking applications, the term *UTP* generally refers to the 100-ohm, four twisted pairs of wires enclosed in a common sheath. Answers B, C, and D are invalid cabling types for UTP.

Question 7

Answer B is correct. UTP cable has no foil or braided screen surrounding the wires. It is more susceptible to EMI than ScTP and coaxial cable. Therefore, answers A and C are incorrect. Answer D is incorrect because fiber-optic cable is not made of copper and is immune from EMI.

Question 8

Answer C is correct. Category 6 standard describes a performance range for unshielded and screened twisted-pair cabling in the maximum bandwidth region of 250MHz. Answer A is incorrect because Category 4 has a rated bandwidth up to 20MHz. Answer B is incorrect because Category 5 cable has a rated bandwidth of 100MHz. Answer D is incorrect because Category 5E cable has a rated bandwidth of 100MHz also.

Question 9

Answer A is correct. When comparing the various types of low-voltage structured-wiring components and cabling, fiber-optic cable provides the greatest immunity to EMI because it is made of glass or plastic, which are not conductive. Answers B, C, and D are incorrect because UTP, STP, and coaxial cable are made of copper and are not immune from EMI.

Question 10

Answer A is correct. Conduit is not required by any standard for indoor residential low-voltage circuits except for plenum areas. Answers B, C, and D are true statements. Wiring and electrical equipment must be enclosed in conduit in areas that are considered to be hazardous areas. These areas are described in the National Electrical Code in articles 500–517. Conduit should be used for low-voltage wiring in outdoor locations.

Question 11

Answer B is correct. Home run is the recommended concept for modern home low-voltage structured wiring using a star topology. A star topology has each outlet in the home connected by its own individual home run of cabling extending back to a central distribution point where all wiring in the home is accessible on a terminal block. Answer A is incorrect because nodes in a ring system are connected into a continuous loop. Answer C is incorrect because the bus topology is used in the daisy-chain wiring scheme to connect all outlets in a single chain throughout a home. Answer D is incorrect because each node in a mesh network has a direct link with every other node.

Question 12

Answer C is correct. The daisy-chain wiring scheme that connects all the outlets in a single chain costs less than the home-run wiring scheme. Answer A is incorrect because it is true that in the home-run wiring design all wiring in a home is accessible on a terminal block. Any changes in distribution of services can be quickly and easily made at the central distribution point. Answer B is incorrect because installing an outlet for all services in every room allows maximum flexibility. Answer D is incorrect because each home-run circuit is independent from the others; therefore, the troubleshooting process is simplified.

Question 13

Answer C is correct. ANSI/TIA/EIA 570-A defines the grade 2 structured wiring using two Category 5 cables for voice and data signals and two 75-ohm coaxial cables for input and output of video signals. Answer A is incorrect because it is the specification for grade 1 structured wiring. Answer B is incorrect because two Category 5E cables and two 75-ohm coaxial cables is an invalid cabling specification for grade 2 structured wiring; however, it is the preferred structured wiring. Answer D is incorrect because it is an invalid cabling specification for grade 2 structured wiring.

Question 14

Answer D is correct. The multipurpose outlets in each room determine which services are available in that room. Each outlet can be constructed based on which services are desired in each room (cable, Internet access, telephone, audio, video, LAN, and so on). Answer A is incorrect because the function of the distribution panel is to control how all outside services enter the home. Answer B is incorrect because a gateway device is used to allow a single Internet access line, such as DSL, cable, or telephone, to be connected to the home computer network. Answer C is incorrect because it is used to connect home computers together into a network.

Question 15

Answer D is correct. A professionally installed low-voltage structured-wiring system can be expected to last approximately 15 years. It has the longest life cycle of components in a low-voltage system. Answers A, B, and C are incorrect because a professionally installed low-voltage structured wiring system can be expected to last much longer than 3, 5, or even 10 years.

Question 16

Answer A is correct. Hot and neutral are used to describe the black and white wires, respectively, in residential high-voltage wiring. Answer B is incorrect because live and dead are not correct designations for the black and white wires present in residential high-voltage wiring. Answer C is incorrect because, although hot is the correct term used to describe the black wires present in residential high-voltage wiring, cold is not used to describe the white wires. Answer D is incorrect because, although neutral is the correct designation for white wires, bold is not used in reference to the black wires present in residential high-voltage wiring.

Question 17

Answer B is correct. The equipment grounding conductor wire is green. Answer A is incorrect because blue is not the correct color of the wire. Answer C is incorrect because the equipment grounding conductor wire is not orange. Answer D is incorrect because it is not red.

Question 18

Answer D is correct. Uninterrupted power supplies (UPSs) refer to residential backup power systems, which provide emergency AC power for the home when commercial power service is unavailable. Answer A is incorrect because surge suppressors are used to protect home appliances and computers from voltage spikes. Answer B is incorrect because a ground rod is used as a safety ground in case an accidental connection occurs between the hot wire and the case. Answer C is incorrect because SMPS power supplies are widely used to power modern computers but do not provide emergency backup AC power.

Question 19

Answer D is correct. Emergency backup power generators are not fueled by water. Answer A is incorrect because an emergency backup power generator can be fueled by gasoline. Answer B is incorrect because an emergency backup power generator can be fueled by diesel, and answer C is incorrect because it can be fueled by natural gas.

Question 20

Answer C is correct. Residential power backup systems use a transfer switching device to disconnect the commercial power source and transfer the electric load to the generator. Answer A is incorrect because voltage regulators are used to maintain a constant voltage under varying load conditions. Answer B is incorrect because a power switching device is not a device used to transfer electric loads to a generator. Answer D is incorrect because a transfer switching device is not used to transfer electric loads to a generator.

Question 21

Answer A is correct. Whole house power distribution uses a generator sized to power all devices within the home. Answer B is incorrect because the whole house power distribution design describes power generation to all devices within the home as opposed to isolated ones. Answer C is incorrect because isolated unessential devices would most likely not be exclusively hooked up to a residential power backup system. Answer D is incorrect because whole house power distribution uses a generator sized to power all devices within the home.

Question 22

Answer B is correct. The essential power distribution design is used for isolated essential devices. Answer A is incorrect because, although the whole house power distribution design describes backup generator power connected to all devices within a home, the essential power distribution design does not. Answer C is incorrect because the essential power distribution design would most likely not be used for isolated unessential devices. Answer D is incorrect because no limit is designated to the number of devices that a generator in an essential power distribution design can support.

Question 23

Answer A is correct. VA ratings for UPS units are the product of volts and amperes. Answer B is incorrect because a watt measures the useful work part of apparent power, but it is not multiplied with volts to obtain a VA rating. Answer C is incorrect because apparent power is made up of real power and reactive power but is not used together with volts to make up a VA rating. Answer D is incorrect because real power is the part of apparent power that does the work.

Question 24

Answer D is correct. Numbers between 0.0 and 1.0 are used to represent the fraction of apparent power that does useful work. Answer A is incorrect because numbers between 1.0 and 2.0 are not used to represent the fraction of the apparent power that does useful work. Answer B is incorrect because, although numbers between 0.0 and 1.0 are used to represent the fraction of the apparent power that does useful work, numbers after 1.0 are not. Answer C is incorrect because, even though numbers from 0.2 and 1.0 form part of the accepted range, they do not cover all available numbers.

Question 25

Answer C is correct. The current capacity of a circuit breaker is measured in amperes. Answer A is incorrect because current capacity is not measured in volts. Answer B is incorrect because watts refer to the part of apparent power that does the useful work; they are not used to measure current capacity. Answer D is incorrect because current capacity is not measured in kilowatts.

Question 26

Answer B is correct. A ground fault circuit interrupter (GFCI) is used to guard against accidental electric shock. Answers A, B, and C are incorrect because GCFI, CFFI, and FFIC are not valid types of electrical outlets.

Question 27

Answer A is correct. A GFCI works by comparing the amount of electric current coming into a circuit through the black wire with the amount leaving a circuit through the white or neutral wire. If more current enters than leaves, it can shut down a circuit within 0.025 seconds. Answer B is incorrect because electrical current goes in through the black wire and out through the white or neutral wire. Answer C is incorrect because a GFCI is not equipped with a sensor telling it when an object is touching it. Answer D is incorrect because the wires that electric current goes through are not blue or red.

Question 28

Answer B is correct. A ground fault occurs when more current enters the circuit than leaves. Answer A is incorrect because less current entering a circuit than leaving does not describe a ground fault. Answer C is incorrect because an equal amount of current entering and leaving a circuit does not describe a ground fault. Answer D is incorrect because more current leaving a circuit than entering does not describe a ground fault.

Question 29

Answer D is correct. The breaker panel is not located in the service entrance. Answers A, B, and C are incorrect because the masthead, conduit, and wiring are all located in the service entrance.

Question 30

Answer A is correct. #12-AWG wire is recommended to safely handle 20 amperes. Answer B is incorrect because #10-AWG wire is larger than the recommended wire gauge for most lighting circuits. #10-AWG wire is also not the correct size and 15 amperes is not the maximum load that the wire must handle. Answer C is incorrect because #8-AWG wire is more than adequate to handle 20 amperes and #8-gauge wire is not normally used for lighting circuits. Answer D is incorrect because #12-AWG wire is the correct gauge of wire, but 10 amperes is not the maximum load it supports.

Question 31

Answer C is correct. A three-way switch is the name given to a light switch used to operate a light from two locations. Answer A is incorrect because a one-way switch is not the correct name for the three-way switch. Answer B is incorrect because, although this type of switch is found in two locations for the same light, it is not called a two-way switch. Answer D is incorrect because a four-way switch is not a valid name for this type of switch.

Question 32

Answers A, B, and C are correct. Magnetic ballast, hybrid magnetic ballast, and high-frequency electronic ballast are all accepted ballast technologies used with fluorescent lighting. Answer D is incorrect because source energy ballast is not a ballast technology.

Question 33

Answer A is correct. Web pads are portable, handheld computer systems that provide wireless Internet access from any location in the home. Answer B is incorrect because keypads provide the operational interface for most automated home technology systems. Answer C is incorrect because a touchscreen is an input device that enables users to operate automated systems by simply touching icons on a display screen. Answer D is incorrect because wireless RF/IR devices use sensors for receiving and processing particular events.

Question 34

Answer B is correct. Keypads use keys for their interfaces. They are often portable or wall-mounted designs with an LCD display and backlit keys. Answer A is incorrect because Web pads are completely portable and are used to provide wireless Internet access from any location in the home. Answer C is incorrect because touchscreens enable users to operate home-automated systems by simply touching icons on a display screen. Answer D is incorrect because wireless RF/IR devices use sensors for receiving and processing particular events.

Question 35

Answer C is correct. A touchscreen is an input device that enables users to operate home-automated systems by simply touching icons on a display screen. It is available in a standalone large-screen panel, wall-mounted unit, or handheld remote controller. Answer A is incorrect because Web pads are completely portable. Answer B is incorrect because keypads use keys for their interfaces. Answer D is incorrect because wireless RF/IR devices use sensors for receiving and processing particular events.

Question 36

Answer D is correct. Wireless RF/IR devices use sensors for receiving and processing particular events. Answers A, B, and C are incorrect because keypads, touchscreens, and Web pads are operated through physical contact.

Question 37

Answer A is correct. Web pads are completely portable and can be used from anywhere within the home. Answers B and C are incorrect because they are concerns for wireless RF/IR devices. Answer D is incorrect because Web pads are portable and do not need to be permanently fixed in an area of maximum access.

Question 38

Answer B is correct. Keypads are often used for security purposes, which require frequent and easy access. Answer A is incorrect because they are location concerns for keypads. Answer C is incorrect because keypads can be portable. Answer D is incorrect because keypad signals cannot be blocked like IR can.

Question 39

Answer C is correct. A touchscreen needs to be located for maximum access by the home residents. Answer A is incorrect because the location of a touchscreen should not be limited to an area near a door to the outside area. Answer B is incorrect because some types of touchscreens can be carried as handheld units. Answer D is incorrect because, although the handheld units use IR signals, they are portable and can be carried around any such obstacles.

Question 40

Answer A is correct. Specific IP ports on the gateway can be set to block particular types of traffic from the Internet entering the home network. A list of full or partial Web addresses that need to be blocked is assigned to the entire network or a defined group in the network for filtering. Answer B is incorrect because port mapping maps the IP address of the gateway to computers on the home network that access the Internet. Answer C is incorrect because establishing a demilitarized zone allocates a home computer unprotected access to the Internet, often for online gaming purposes. Answer D is incorrect because remote administration specifies the address of the remote PC for administration purposes.

Question 41

Answer B is correct. A wall-mounted receiver is a feature of an RF/IR receiver, not a keypad. Answers A, C, and D are all main features of a keypad.

Question 42

Answer C is correct. Easy replacement is not listed as one of the main benefits of using a patch/distribution panel. Answers A, B, and D are incorrect because they are all benefits of a patch/distribution panel for homeowners.

Question 43

Answer D is correct. According to the ANSI/TIA/EIA 570-A standard, the preferred location should be close to the telephone company demarcation point and near the entry point of cable TV or satellite connections. Answer A is incorrect because making it difficult for intruders to access is not a major operational concern. Answer B is incorrect because maintaining an even distance from all interface devices is not a major operational concern. The priority is to locate the central distribution panel near the area where the external telephone and cable service enters the home. The "centralized" name of the distribution panel does not imply that the panel must be located in the center of the home. Answer C is incorrect because there is no requirement for the central distribution panel to be located near the breaker panel.

Question 44

Answer A is correct. The home networking controller is the control system that manages the home network. Answer B is incorrect because the indicator is a component of the home controller. Answer C is incorrect because the cabinet is only a place to store the controller. Answer D is incorrect because the distribution panel only controls cable runs between the central controller and the system interfaces.

Question 45

Answer C is correct. A DMZ effectively turns off all security features for that particular machine. Answer A is incorrect because a DMZ does not limit Internet access. Answer B is incorrect because home network crashing is not a danger of a DMZ. Answer D is incorrect because a DMZ does not cause a lagging Internet connection.

Question 46

Answer D is correct. Web pads are not used to program controllers. Answers A, B, and C are incorrect because keypads, touchscreens, and intelligent remote controls can all be used to program controllers.

Question 47

Answer A is correct. MMX is used for video cards, not cables. Answer B is incorrect because CMX is an NEC marking indicating communications cabling for limited use. Answer C is incorrect because CMG is an NEC marking indicating communications cabling for general purpose. Answer D is incorrect because CMP marks the communications cabling used in the plenum.

Question 48

Answer C is correct. Fiber-optic cables are used for digital TV as well as digital audio. Answer A is incorrect because coaxial cables are generally used for analog video signal distribution. Answer B is incorrect because Category 5E UTP cables are standard for data and telephone communications. Answer D is incorrect because wireless technology uses RF and IR signals.

Question 49

Answer B is correct. Category 5E UTP cables are standard for data and telephone communications. Answer A is incorrect because coaxial cables are generally used for video signal distribution. Answer C is incorrect because fiber-optic cables are used in high-definition and digital TV as well as digital audio. Answer D is incorrect because wireless technology uses RF and IR signals.

Question 50

Answer A is correct. Coaxial cables are used for analog video signal distribution. Answer B is incorrect because fiber-optic cables are used in digital TV as well as digital audio. Answer C is incorrect because wireless technology uses RF and IR signals. Answer D is incorrect because Category 5E UTP cables are standard for data and telephone communications.

PART IV
Appendixes

A What's on the CD-ROM

B Using the PrepLogic Practice Exams, Preview Edition Software

Glossary

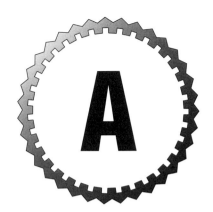

What's on the CD-ROM

This appendix provides a brief summary of what you'll find on the CD-ROM that accompanies this book. For a more detailed description of the PrepLogic Practice Exams, Preview Edition exam simulation software, see Appendix B, "Using the PrepLogic Practice Exams, Preview Edition Software." In addition to the PrepLogic Practice Exams, Preview Edition software, the CD-ROM includes an electronic version of the book—in Portable Document Format (PDF)--and the source code used in the book.

The PrepLogic Practice Exams, Preview Edition Software

PrepLogic is a leading provider of certification training tools. Trusted by certification students worldwide, PrepLogic is the best practice exam software available. In addition to providing a means of evaluating your knowledge of this book's material, PrepLogic Practice Exams, Preview Edition features several innovations that help you improve your mastery of the subject matter.

For example, the practice tests allow you to check your score by exam area or domain, to determine which topics you need to study further. Another feature allows you to obtain immediate feedback on your responses in the form of explanations for the correct and incorrect answers.

PrepLogic Practice Tests, Preview Edition exhibits all the full-test simulation functionality of the Premium Edition but offers only a fraction of the total questions. To get the complete set of practice questions, visit www.preplogic.com and order the Premium Edition for this and other challenging exam training guides.

For a more detailed description of the features of the PrepLogic Practice Exams, Preview Edition software, see Appendix B.

An Exclusive Electronic Version of the Text

As mentioned previously, the CD-ROM that accompanies this book also contains an electronic PDF version of this book. This electronic version comes complete with all figures as they appear in the book. You can use Acrobat's handy search capability for study and review purposes.

Using the PrepLogic Practice Exams, Preview Edition Software

This book includes a special version of the PrepLogic Practice Exams software, a revolutionary test engine designed to give you the best in certification exam preparation. PrepLogic offers sample and practice exams for many of today's most in-demand and challenging technical certifications. A special Preview Edition of the PrepLogic Practice Exams software is included with this book as a tool to use in assessing your knowledge of the training guide material while also providing you with the experience of taking an electronic exam.

This appendix describes in detail what PrepLogic Practice Exams, Preview Edition is; how it works; and what it can do to help you prepare for the exam. Note that although the Preview Edition includes all the test simulation functions of the complete retail version, it contains only a single practice test. The Premium Edition, available at www.preplogic.com, contains a complete set of challenging practice exams designed to optimize your learning experience.

The Exam Simulation

One of the main functions of PrepLogic Practice Exams, Preview Edition is exam simulation. To prepare you to take the actual vendor certification exam, PrepLogic is designed to offer the most effective exam simulation available.

Question Quality

The questions provided in PrepLogic Practice Exams, Preview Edition are written to the highest standards of technical accuracy. The questions tap the content of this book's chapters and help you review and assess your knowledge before you take the actual exam.

The Interface Design

The PrepLogic Practice Exams, Preview Edition exam simulation interface provides you with the experience of taking an electronic exam. This enables you to effectively prepare to take the actual exam by making the test experience familiar. Using this test simulation can help eliminate the sense of surprise or anxiety you might experience in the testing center because you will already be acquainted with computerized testing.

The Effective Learning Environment

The PrepLogic Practice Exams, Preview Edition interface provides a learning environment that not only tests you through the computer, but also teaches the material you need to know to pass the certification exam. Each question includes a detailed explanation of the correct answer, and most of these explanations provide reasons as to why the other answers are incorrect. This information helps reinforce the knowledge you already have and provides practical information you can use on the job.

Software Requirements

PrepLogic Practice Exams requires a computer with the following:

- Microsoft Windows 98, Windows Me, Windows NT 4.0, Windows 2000, or Windows XP
- A 166MHz or faster processor
- A minimum of 32MB of RAM
- 10MB of hard drive space

NOTE As with any Windows application, the more memory, the better the performance.

Installing PrepLogic Practice Exams, Preview Edition

You install PrepLogic Practice Exams, Preview Edition by following these steps:

1. Insert the CD-ROM that accompanies this book into your CD-ROM drive. The Autorun feature of Windows should launch the software. If you have Autorun disabled, select Start, Run. Go to the root directory of the CD-ROM and select <u>setup.exe</u>. Click Open, and then click OK.

2. The Installation Wizard copies the PrepLogic Practice Exams, Preview Edition files to your hard drive. It then adds PrepLogic Practice Exams, Preview Edition to your desktop and the Program menu. Finally, it installs test engine components to the appropriate system folders.

Removing PrepLogic Practice Exams, Preview Edition from Your Computer

If you elect to remove the PrepLogic Practice Exams, Preview Edition, you can use the included uninstallation process to ensure that it is removed from your system safely and completely. Follow these instructions to remove PrepLogic Practice Exams, Preview Edition from your computer:

1. Select Start, Settings, Control Panel.

2. Double-click the Add/Remove Programs icon. You are presented with a list of software installed on your computer.

3. Select the PrepLogic Practice Exams, Preview Edition title you want to remove. Click the Add/Remove button. The software is removed from your computer.

How to Use the Software

PrepLogic is designed to be user-friendly and intuitive. Because the software has a smooth learning curve, your time is maximized because you start practicing with it almost immediately. PrepLogic Practice Exams, Preview Edition has two major modes of study: Practice Exam and Flash Review.

Using Practice Exam mode, you can develop your test-taking abilities as well as your knowledge through the use of the Show Answer option. While you are taking the test, you can expose the answers along with detailed explanations of why answers are right or wrong. This helps you better understand the material presented.

Flash Review mode is designed to reinforce exam topics rather than quiz you. In this mode, you are shown a series of questions but no answer choices. You can click a button that reveals the correct answer to each question and a full explanation for that answer.

Starting a Practice Exam Mode Session

Practice Exam mode enables you to control the exam experience in ways that actual certification exams do not allow. To begin studying in Practice Exam mode, you click the Practice Exam radio button from the main exam customization screen. This enables the following options:

- *The Enable Show Answer button*—Clicking this button activates the Show Answer button, which allows you to view the correct answer(s) and full explanation for each question during the exam. When this option is not enabled, you must wait until after your exam has been graded to view the correct answer(s) and explanation for each question.

- *The Enable Item Review button*—Clicking this button activates the Item Review button, which allows you to view your answer choices. This option also facilitates navigation between questions.

- *The Randomize Choices option*—You can randomize answer choices from one exam session to the next. This makes memorizing question choices more difficult, thereby keeping questions fresh and challenging longer.

On the left side of the main exam customization screen, you are presented with the option of selecting the preconfigured practice test or creating your own custom test. The preconfigured test has a fixed time limit and number of questions. Custom tests allow you to configure the time limit and the number of questions in your exam.

The Preview Edition on this book's CD-ROM includes a single preconfigured practice test. You can get the complete set of challenging PrepLogic Practice Exams at www.preplogic.com to make certain you're ready for the big exam.

You click the Begin Exam button to begin your exam.

Starting a Flash Review Mode Session

Flash Review mode provides an easy way to reinforce topics covered in the practice questions. To begin studying in Flash Review mode, click the Flash Review radio button from the main exam customization screen. Then you either select the preconfigured practice test or create your own custom test.

You click the Best Exam button to begin a Flash Review mode session.

Standard PrepLogic Practice Exams, Preview Edition Options

The following list describes the function of each of the buttons you see across the bottom of the screen:

Depending on the options, some of the buttons will be grayed out and inaccessible—or they might be missing completely. Buttons that are appropriate are active.

➤ *Exhibit*—This button is visible if an exhibit is provided to support the question. An *exhibit* is an image that provides supplemental information necessary to answer a question.

➤ *Item Review*—This button leaves the question window and opens the Item Review screen, from which you can see all the questions, your answers, and your marked items. You can also see correct answers listed here, when appropriate.

➤ *Show Answer*—This option displays the correct answer, with an explanation about why it is correct. If you select this option, the current question is not scored.

➤ *Mark Item*—You can check this box to flag a question you need to review further. You can view and navigate your marked items by clicking the Item Review button (if it is enabled). When your exam is being graded, you are notified if you have any marked items remaining.

- *Previous Item*—You can use this option to view the previous question.
- *Next Item*—You can use this option to view the next question.
- *Grade Exam*—When you have completed your exam, you can click Grade Exam to end your exam and view your detailed score report. If you have unanswered or marked items remaining, you are asked whether you want to continue taking your exam or view the exam report.

Seeing Time Remaining

If your practice test is timed, the time remaining is displayed on the upper-right corner of the application screen. It counts down the minutes and seconds remaining to complete the test. If you run out of time, you are asked whether you want to continue taking the test or if you want to end your exam.

Getting Your Examination Score Report

The Examination Score Report screen appears when the Practice Exam mode ends—as a result of time expiration, completion of all questions, or your decision to terminate early.

This screen provides a graphical display of your test score, with a breakdown of scores by topic domain. The graphical display at the top of the screen compares your overall score with the PrepLogic Exam Competency Score. The PrepLogic Exam Competency Score reflects the level of subject competency required to pass the particular vendor's exam. Although this score does not directly translate to a passing score, consistently matching or exceeding this score does suggest that you possess the knowledge needed to pass the actual vendor exam.

Reviewing Your Exam

From the Your Score Report screen, you can review the exam you just completed by clicking the View Items button. You can navigate through the items, viewing the questions, your answers, the correct answers, and the explanations for those questions. You can return to your score report by clicking the View Items button.

Contacting PrepLogic

If you would like to contact PrepLogic for any reason, including to get information about its extensive line of certification practice tests, you can do so online at www.preplogic.com.

Customer Service

If you have a damaged product and need to contact customer service, please call

800-858-7674.

Product Suggestions and Comments

PrepLogic values your input! Please email your suggestions and comments to feedback@preplogic.com.

License Agreement

YOU MUST AGREE TO THE TERMS AND CONDITIONS OUTLINED IN THE END USER LICENSE AGREEMENT ("EULA") PRESENTED TO YOU DURING THE INSTALLATION PROCESS. IF YOU DO NOT AGREE TO THESE TERMS, DO NOT INSTALL THE SOFTWARE.

Glossary

10BASE-T
An Ethernet specification defined by the IEEE 802.3 committee officially as specification 802.3i. The specification describes a 10Mbps wire speed using baseband information. The specification requires a star topology using Category 3, 4, or 5 unshielded twisted-pair (UTP) cable.

100BASE-T
An Ethernet IEEE 802.3u specification utilizing a wire speed of 100Mbps using Category 5 UTP cable.

AC cable
AC (armored cable) is identified by the code name AC cable. AC cable is often referred to as *BX*, although this is the trademark of a specific manufacturer. Armor clad and metal clad (MC) might sound like the same cable, but some important differences exist. AC cable uses the interior bond wire in combination with the exterior interlocked metal armor as the equipment grounding means of the cable. MC cable, on the other hand, is manufactured with a green insulated grounding conductor, and this conductor, in combination with the metallic armor, comprises the equipment ground. AC cable can have up to four insulated conductors only; a fifth insulated conductor is allowed by UL if it is a grounding conductor. Each conductor in AC cable is paper wrapped.

AC-3 (Audio Coding-3)
Dolby's digital audio data compression algorithm adopted for HDTV transmission and used in laserdiscs and CDs for 5.1 surround sound multichannel home theater use.

ADSL terminal unit (ATU)
The engineering term used for describing an ADSL modem. The ADSL terminal unit is also known as the ADSL transceiver unit, ADSL transmission unit, and ADSL termination unit. The

ATU-R (remote) unit is installed in the home, and the ATU-C (central) office unit is installed in the local telephone company central office.

Audio Engineering Society (AES)

Founded in 1949, the largest professional organization for electronic engineers and all others actively involved in audio engineering. It's primarily concerned with education and standardization.

air handler

The subassembly contained in an HVAC air distribution system that typically houses items such as the supply fan(s), return fan(s), heating coil, cooling coil, and filters. Its purpose is to provide the pressure force and conditioning necessary to get the airflow through the entire system and at the design temperature and humidity condition.

alarm system

A dedicated electronic system using a central controller and distributed sensors that detects unauthorized intrusion into a home interior or protected external areas. Typically used with door and window sensors, glass break sensors, motion detectors, and occasionally closed circuit television. When armed by the user, the system detects intrusion from the outside and can react with responses that vary from the activation of interior sirens, bells, and lighting, to the notification of a central alarm monitoring system.

ampere (A)

A unit of measure of the rate of electron flow or current in an electrical conductor. One ampere of current represents one coulomb of electrical charge (6.24×10^{18} charge carriers) moving past a specific point in 1 second. The ampere is named after Andre Marie Ampere, a French physicist (1775–1836).

amplitude

The magnitude of a signal in voltage or current. It's frequently expressed in terms of peak, peak-to-peak, or RMS.

amplifier bridging

Bridging an amplifier is a technique used to configure a two-channel stereo amplifier to drive a single load (speaker) with more power than the sum of the two original channels.

analog

A term used in telecommunications technology to describe a continuously variable waveform that is analogous to the electrical signal produced by the human voice when processed by a microphone and audio amplifier. Also, it's a signal in which a base carrier's alternating current frequency is modified in some way, such as by amplifying the strength of the signal or varying the frequency to add information to the signal (also called *amplitude modulation*). Broadcast and telephone transmissions have conventionally used analog technology. An analog signal can be represented as a series of sine waves. The term originated because the modulation of the carrier wave is analogous to the fluctuations of the human voice or other sound that is being transmitted.

Advanced Television Systems Committee (ATSC)
A group that developed voluntary national standards for high-definition television (HDTV), standard-definition television (SDTV), data broadcasting, multichannel surround-sound audio, and satellite direct-to-home broadcasting.

apparent power
A value of electrical power obtained by multiplying the root mean square (RMS) volts and RMS amps. The term *RMS* is used to describe what is obtained by measuring AC voltage with a standard volt meter and amp meter. The result is expressed as volt-amperes (VA) and not watts. Apparent power is not a measure of the true power in a circuit. See also true power.

asymmetric digital subscriber line (ADSL)
Asymmetric DSL is a class of DSL service described in ANSI T1.413-1998 that offers higher download speeds with a slower upstream speed. The actual network bandwidth a customer receives from a DSL service in the home depends on the total length of the telephone local loop distance to the local office.

backwash valve
A valve used to reverse the flow of water in a swimming pool filter during the filter cleaning cycle. Modern backwash valves are molded of temperature- and chemical-resistant chlorinated polyvinyl chloride (CPVC) material.

bandwidth
The transmission capacity of a medium in terms of a range of frequencies expressed in hertz (Hz). A greater bandwidth indicates the capability to transmit a greater amount of data over a given period of time.

Bayonet-Neil-Concelman (BNC) connector
A coaxial cable connector that, in its male form, has a center pin (bayonet) connected to the center conductor and a metal tube connected to the cable shield. A rotating ring surrounds the tube and contains small holes that mate with projections on the female connector to make the connection. The connector was developed in the late 1940s and is named after the creators—Amphenol engineer Carl Concelman and Bell Labs engineer Paul Neill. Neill designed the N-type connector, and Concelman designed the C-type connector. The BNC is a hybrid N/C-type with a mechanical extra appropriately called a *bayonet*. BNC connectors are often identified using other terms of unknown origin, such as British naval connectors, bayonet navy connectors, or bayonet nut connectors.

bimetal elements

A type of temperature sensor used in older thermostats containing two strips of different metals that are attached together. As the room temperature changes, the lengths of the metals change. Because the metals are different and can change their lengths at different rates, they are used to operate thermostat switches that react to temperature changes.

bit (binary digit)

The smallest component of information in a binary notation system. A bit is a single 1 or 0.

bit rate

The number of binary digits transmitted over a media per unit of time. Bit rate is specified as the number of bits transmitted in one second (bps).

Bluetooth

Bluetooth is a wireless technology standard designed to connect one device to another with a short-range radio link. This standard has evolved from early work and engineering studies performed by Ericsson Mobile Communications in 1994. The IEEE Project 802.15.1 has derived a Wireless Personal Area Network (WPAN) standard based on the Bluetooth v1.1 Foundation specifications.

bundled cabling

A type of cable available from vendors described in addendum 3 to the ANSI/TIA/EIA-568-A standard designed to make installation of grade 2 structured wiring easier. The addendum describes four-pair cable assemblies that are not covered by an overall sheath (as specified for hybrid cables) but by any binding method, such as speed-wrap or cable ties. A standard bundled cable for grade 2 consists of two Category 5E cables and two RG-6 quad-shield cables.

byte

A group of bits generally accepted in the computer industry as 8 bits but can be 9 bits on 36-bit machines. Some older computer architectures used *byte* for quantities of 6 or 7 bits, but these usages are now obsolete. An octet always contains 8 bits and is used to avoid confusion with the term *byte*.

cable tester

A type of network cable test equipment used to check cables for opens, shorts, and continuity.

call waiting

A local exchange telephone service offering that provides an indication to a subscriber conducting a call that one or more other phones are trying to make a connection to a number. With call waiting service, the second call is announced by a soft beep. The user can ask the first caller to wait while she answers the second call. She then has the option to return to the first call or disconnect from the first to talk to the second caller.

caller line identity (CLI)
A local exchange telephone system service used to identify the number of the caller to the receiving (called) party. This type of service is one of many listed under the title of caller line identification or calling line identity (CLI) services. Features such as caller display and call return use a feature of modern telephone networks that transmits the number of the caller as each call is set up. Called customers using these CLI services can have access to this information. With caller display, a customer can see the number on his phone before answering the call.

Category 5
A type of UTP communications cable described in ANSI/TIA/EIA 568-A that has four twisted pairs of wires. Category 5 cable is tested to 100MHz. Category 5 cable was previously used in residential structures wiring and 100BASE-T Ethernet LAN installations but has been replaced with Category 5E as the preferred structured wiring cable for new installations. Cat 5E provides improved performance at approximately the same cost.

Category 5E
An enhanced version of Category 5 cable described in specification ANSI/TIA/EIA 568A-addendum 5. Category 5E cable was officially ratified as a standard on December 13, 1999, and was updated with the release of ANSI/TIA/EIA 568B.1-2000. Although it has a rated bandwidth of 100MHz, it has improved specifications for near-end crosstalk (NEXT), power sum equal level far-end crosstalk (PSELFEXT), and attenuation. The specification also enforces several attributes that were optional in the original Category 5 specification. Category 5E cable is usually tested to a bandwidth of 350MHz, despite its 100MHz specified bandwidth.

Category 6
A type of data communications cable that complies with the transmission requirements in the TIA/EIA 568-B.2-1 Commercial Building Telecommunications Category Standard. Cable carrying the Category 6 rating has a rated bandwidth of 250MHz.

Category 7
A type of high-performance data communications twisted-pair cable that uses a braided shield surrounding all four foil-shielded pairs to reduce noise and interference. Cat 7 cable, connecting hardware, and patch cords are rated to a maximum frequency bandwidth of 600MHz.

cathode-ray tube (CRT)
A type of vacuum tube in which images are produced when an electron beam strikes a phosphorescent surface. CRTs are used as the primary display device in computer monitors and television receivers to display video information.

CEBus (consumer electronics bus)

CEBus is a communications protocol developed by the Electronic Industries Alliance (EIA) as the ANSI/EIA-600 standard. The CEBus is designed to control devices on a power line, but it also works on other media. The CEBus standard involves device addresses that are set in hardware at the factory and includes four billion possibilities.

cladding

The portion of a fiber-optic cable encircling the core material.

composite video

A single video signal that contains luminance, color, and synchronization information. NTSC, PAL, and SECAM are all examples of composite video systems.

condenser

A heat exchanger (also known as an outdoor coil) assembly used with residential air conditioning or heat pump systems. As an integral component of a heat pump or a central air conditioning system, it is located in an outdoor casing that is vented where a large fan passes air over the coils to exhaust heat. It serves as the transfer point where heat is transferred from inside the home to the outside air.

conduit

A type of enclosure for wiring installed in locations that are classified as hazardous due to potentially explosive air environments, underground wiring, or outdoor and moist locations. It is available in a variety of types for many environments. They range from rigid metal conduit to intermediate metal conduit to thin-walled conduit called electrical metallic tubing (EMT) to nonmetallic types called polyvinyl chloride (PVC).

copper communications riser cable (CMR)

A fire rating for riser cable established by the National Electrical Code (NEC) ANSI/NFPA 70 standard. The NEC assigns fire ratings to communications transmission cables and discusses suitable applications for each cable type. Copper communications riser cable is suitable for installation in vertical riser shafts that are designed to run from floor to floor. It cannot be used in any environmental air spaces, such as air conditioning ducts, unless the local code allows it. CMR cable might also be labeled as flame test (FT)-4 and is not as expensive as plenum cable.

crossover cable

A cable used to directly connect two computers, or two hubs, in which the receive and transmit wires of a twisted-pair cable have been transposed at each end of the cable connector to perform the crossover function normally performed by a single hub.

crossover networks

A device used with multiple speaker systems to separate portions of the audio spectrum for routing to individual speakers for optimum performance.

daisy chain
Name for a type of residential wiring design used in home lighting systems in which load requirements for individual lighting systems do not require a dedicated home-run connection to the main power distribution panel. Some lighting systems include multiple light fixtures controlled with wall switches that share a 15 ampere daisy-chain service from the main power distribution panel.

digital
Circuitry or telecommunications transmission systems in which information carrying signals are restricted to either of two states, corresponding to logic 1 or 0.

digital subscriber line (DSL)
A high-speed Internet access service offered by local telephone companies using regular telephone lines. DSL has the capability to move data over the phone lines at speeds up to 6Mbps, or 140 times faster than the fastest analog modems (56,000 bits per second). In addition to its very high speed, DSL has many benefits over analog connections. Unlike dial-up connections that require analog modems to dial in to the Internet service provider every time the user wants to retrieve email or obtain access to the Internet, DSL connections are always on. Another benefit is the capability to use the telephone at the same time the user is accessing the Internet.

digital versatile disc (DVD)
A high-volume storage media with the same physical appearance and dimensions as a compact disc. Initially called digital video discs by the creators, DVDs were named digital versatile discs in 1996. DVDs are standardized with five physical formats and four media storage versions.

DVD-ROM is a high-capacity data storage medium. DVD-Video is a digital storage medium for feature-length motion pictures, and DVD-Audio is an audio-only storage format similar to CD-Audio. DVD-R offers a write-once, read-many storage format akin to CD-R. DVD-RAM was the first rewritable (erasable) flavor of DVD to come to market and has subsequently found competition in the rival DVD-RW and DVD+RW formats. With the same overall size as a standard 120mm diameter, 1.2mm thick CD, DVD discs provide up to 17GB of storage with transfer rates that are higher than CD-ROMs and access times similar to CD-ROMs. DVDs come in four versions. DVD-5 is a single-sided, single-layered disc boosting capacity sevenfold to 4.7GB. DVD-9 is a single-sided, double-layered disc offering 8.5GB, whereas DVD-10 is a 9.4GB dual-sided, single-layered disc. Finally, DVD-18 increases capacity to a huge 17GB on a dual-sided, dual-layered disc. DVDs are similar in appearance to CDs—they are both 4 3/4" in diameter. DVD tracks, however, are closer together and the compression ratio is much higher. The result is DVDs can hold 4.7GB per side compared to approximately 660MB for a CD.

dual-tone multifrequency (DTMF)

A method of tone dialing using a 12-button keypad and dual tones representing each of the 12 buttons.

electromagnetic interference (EMI)

A type of disturbance occurring in communications circuits when unwanted electrical currents are induced into the wiring or components of an electrical system. EMI can also degrade the performance of wireless communications channels.

electromagnetic lock

A type of lock used to control access through a door where maximum security is required. It consists of an electromagnet mounted in the door frame and a matching plate installed in the door. The electromagnet can be energized from a switch by the person controlling the access. When energized, the electromagnet holds the plate with a strong magnetic field. The lock has no moving parts. The amount of pressure an electromagnetic lock can withstand is somewhat determined by its integrity rating. The Builders Hardware Manufacturers Association (BHMA) together with the American National Standards Institute (ANSI) have determined three integrity grades for electromagnetic locks. A grade 3 integrity rating provides a locking strength of 500–900 lbs. An integrity rating of grade 2 means that the electromagnetic lock must be strong enough to withstand a pressure of 1,000–1,400 lbs. And an integrity rating of grade 1 means that the electromagnetic lock is designed to withstand pressures of 1,500–2,700 lbs.

electromechanical controllers

A type of irrigation system controller used to operate the valve stations at prescribed times. Electromechanical controllers use clocks in addition to stored irrigation programs. They are respected as a reliable solution for simple irrigation systems. They have few sophisticated electronic components and are driven by electric motors and gears. Turning dials or flipping switches programs the controller to select watering times, how long each zone will be watered, and on which days the watering will occur.

electrostatic speaker

A special type of speaker designed to avoid the problems inherent in cone speakers. They use a graphite-coated plastic membrane suspended between two perforated metal sheets. A high voltage of several thousand volts is applied to the membrane. The input signal is also raised to a high voltage by a transformer and is applied to the perforated metal sheets.

encapsulation
A communications process in which a packet or frame is enclosed within another packet for the purpose of hiding the header of the encapsulated packet during transit through a network. At the destination, the receiving node recognizes the encapsulation header and recovers the original enclosed packet.

Ethernet
A popular local area network technology that is standardized by the IEEE as the 802.3 Carrier Sense Multiple Access with Collision Detection (CSMA/CD) standard. Ethernet was originally developed by two engineers at the Xerox Corporation and then developed further by a partnership of Xerox, DEC, and Intel. An Ethernet LAN typically uses coaxial cable or special grades of UTP cable as the transmission media.

European Broadcasting Union (EBU)
A professional society that helps establish standards in the audio and broadcast industry in Europe as well as other nations.

evaporators
A component of a heat pump or central air conditioning system located inside the home. It is also known as an *indoor coil* and functions as the heat transfer point for warming or cooling indoor air. It consists of a series of pipes connected to a furnace or air handler, which blows the indoor air across the evaporator coil, causing the coil to absorb heat from the air. The cooled air is then delivered to the house through the ducting.

fast Ethernet
The common name for the IEEE 802.3u Ethernet 100BASE-T specification that transmits information at a wire speed of 100Mbps.

fax
An abbreviation for facsimile machines used in home offices or businesses to send and receive documents using standard telephone lines or digital dedicated lines. Faxing is a less expensive alternative to overnight package delivery and can respond to situations in which a document is needed the same day for an urgent business transaction.

firewall
A system designed to prevent unauthorized access to or from a private network. Firewalls can be implemented in both hardware and software, or a combination of both. Normally, a firewall is deployed between a trusted, protected private network and a public network. For example, the trusted network might be a corporate network and the public network might be the Internet.

FireWire
The product name for Apple Computer's version of the standard IEEE 1394 High Performance

Serial Bus. It is used for connecting peripheral devices to personal computers. FireWire provides a single plug-and-socket connection on which up to 63 devices can be attached with data transfer speeds up to 400Mbps.

flat-panel display

A common name used to describe plasma or LCD displays. Flat-panel displays get their name from the type of construction that sets them apart from CRT displays. They are suitable for laptop computers and provide portable computing with the technology to make them truly portable and lightweight. Flat screens are a special type of CRT display.

frequency

Within the context of wireless communications terminology, a measure of the number of cycles or reversals of an electromagnetic field that occur in 1 second. It's also used to express the quantity of units of information transmitted in a given time period expressed in hertz.

gateway

A special interface device similar to a router that allows a single Internet access line such as DSL, cable, or telephone dial-up access to be connected to the home computer network. The residential gateway serves as the communications hub for the entire suite of home networking devices deployed throughout the home. Gateways provide integrators with two functions—one is to connect the home network to the Internet and the other is to act as a central communications interface resource for all in-home digital devices, ranging from lighting, security, and HVAC to phones, home entertainment, and personal digital assistants. Various types of residential gateways are available based on the gradients of functionality required. Most provide capability for advanced data, voice, and video routing within the home. Most gateway products also include a firewall. The gateway is normally located near the distribution panel to simplify the connections from the gateway to the home network patch panel and the incoming Internet access port.

G.DMT (G.Discrete Multitone)

A type of ADSL service defined in ITU Recommendation 992.1. The main difference between G.Lite and G.DMT is bandwidth. Sometimes called *full-rate ADSL*, the G.DMT variety can download data at up to 8Mbps and send data upstream at up to 1.5Mbps.

G.Lite

A type of Universal DSL described in ITU Recommendation G.992.2. Sometimes called *splitterless DSL*, it is a form of ADSL that does not require a splitter installation at the subscriber location. However, it does so at the expense of lower data rates. G.Lite can be used with a distributed splitter configuration but with added complexity and limited benefit. It represents the most consumer-friendly version of DSL. G.Lite equipment and service costs less than other varieties, and it reportedly has a do-it-yourself installation. G.Lite supports a maximum of 1544Kbps downstream and 384Kbps upstream.

Gigabit Ethernet
A term used to describe the IEEE 802.3z 1000BASE-X Ethernet specification. Gigabit Ethernet, a transmission technology based on the Ethernet frame format and protocol used in LANs, provides a data rate of 1 billion bits per second (1 gigabit). Gigabit Ethernet is carried primarily on optical fiber cable (with very short distances possible on copper media).

grade 1
The grade level described in ANSI/TIA/EIA 570-A (Residential Telecommunications Cabling Standard) that provides a generic cabling system that meets the minimum requirements for telecommunications services. It specifies a minimum of one twisted-pair cable and associated connecting hardware (Category 3) and one coaxial cable (75-ohm) configured in a star topology. This grade provides for telephone, TV (digital or analog), satellite, CATV, and low-speed data services.

grade 2
The grade level described in ANSI/TIA/EIA 570-A (Residential Telecommunications Cabling Standard) in which each cabled location requires two Category 5 UTP cables and two 75-ohm coaxial cables plus, as an option, optical fiber cabling configured in a star topology. This grade provides a broader range of residential services than grade 1.

ground fault circuit interrupter (GFCI)
A type of electrical outlet used to protect people from electrical shock due to faulty extension cords, appliances, or tools. A GFCI detects any minor imbalance in the current flowing in the hot and neutral conductors that might be caused by a ground fault in the connected appliance and disconnects the outlet automatically.

high-definition television (HDTV)
An advanced television broadcast standard adopted by the ATSC that provides higher resolution, a 16×9 aspect ratio, and 5.1 surround sound channels.

home audio/video interoperability (HAVi)
A vendor-neutral audio-video standard aimed specifically at the home entertainment environment. HAVi allows different home entertainment and communications devices (such as VCRs, TVs, stereos, security systems, and video monitors) to be networked together and controlled from one primary device, such as a PC.

HomePNA (HPNA)

The HomePNA abbreviation stands for Home Phone Network Alliance. It is a standard based on a process that enables voice and data transmissions to share the available bandwidth of the telephone cabling in a home without mutual interference. It uses frequency division multiplexing (FDM) techniques to divide this bandwidth of frequencies into several communications channels. The acronyms for phone-line networking and power-line networking standards organizations can be confusing. The standard adopted for using the phone lines for networking is derived from the standards group that developed it (Home Phone Network Alliance). The published standard is often referenced as HPNA 2.0. See also HPLA.

HomeRF

A wireless LAN home technology standard supported by the HomeRF Working Group. It is not compatible or interoperable with the IEEE 802.11 family of wireless LANs. The founding members of the HomeRF consortium include among others, Microsoft, Intel, HP, Motorola, and Compaq.

home-run wiring

A method used in modern automated home design in which all the wiring is routed from a central point in the house, usually in a wiring closet. At the central point, the wires terminate at various types of patch panels or cross-connect terminals. This point becomes the interface between home wiring and external telecommunications wiring. Home-run wiring is the basic design concept used in structured wiring.

horizontal resolution

Chrominance and luminance resolution (detail) expressed horizontally across a picture tube. This is usually expressed as a number of black-to-white transitions or lines that can be differentiated. Horizontal resolution is limited by the bandwidth of the video signal or equipment.

HPLA

An acronym for a standard called the HomePlug Power-line Alliance. The name is derived from a group of companies that developed the standard for power-line networking. The standard is referenced as HomePlug 1.0 or sometimes HPLA 1.0.

hybrid fiber coaxial (HFC)

A network that includes both fiber-optic cable and coaxial cable as the transmission media. HFC networks are used by the cable TV industry for the distribution of digital video channels and high-speed Internet access services.

Hypertext Markup Language (HTML)

Authoring software used on the World Wide Web. HTML is basically ASCII text surrounded by HTML commands.

IEEE 802 standards

A series of standards approved by the IEEE, as follows:

- 802.1 (High-level Interface)
- 802.2 (Logical Link Interface)
- 802.3 (CSMA/CD)
- 802.4 (Token Bus)
- 802.5 (Token Ring)
- 802.6 (Metropolitan Area Networks)
- 802.7 (Broadband LANs)
- 802.8 (Fiber Optics)
- 802.9 (Integrated Voice Data)
- 802.10 (LAN Security)
- 802.11 (Wireless Networks)
- 802.12 (Demand Priority Access (100VG AnyLAN))
- 802.13 (Not Used)
- 802.14 (Cable TV-based Broadband)
- 802.15 (Wireless Personal Area Network)
- 802.16 (Broadband Packet)

IEEE-1394

The IEEE 1394 Port Standard, also known as the FireWire port, is a very fast external serial bus standard that supports data transfer rates of up to 400Mbps. Products supporting the IEEE-1394 standard have different names, depending on the company. Apple, which originally developed the technology, uses the trademarked name FireWire. Other companies use other names, such as i.Link (Sony) and Lynx (TI), to describe their IEEE-1394 standard products. A single 1394 port can be used to connect up to 63 external devices.

insulation displacement connection (IDC)

IDC is the recommended method of telecommunications copper wire termination recognized by the ANSI/TIA/EIA-568-A standard. These termination devices are the popular choice for home network star topology connection points. Commonly called punch-down connections, these connections require the use of a small punch-down tool to properly secure the cable to terminal block. Punch-down connections remove or displace the conductor's insulation as it is seated in the connector. During termination, the cable is pressed between two edges of a metal clip, which displaces the insulation and exposes the copper conductor. This ensures a solid connection between the copper conductor and terminating clip.

integrated services digital network (ISDN)

A type of broadband, all-digital access service offered by local exchange carriers. ISDN provides circuit-switched access to the public network for voice, data, and video transmission. Basic rate interface (BRI) and primary rate interface (PRI) are two types of ISDN service.

interlaced scanning
A type of scan used in television broadcasting in which half the screen image is developed on successive frames of odd and even lines.

intermodulation distortion (IM)
Intermodulation distortion is a measured performance feature of an amplifier listed in the specifications. It is caused when two or more frequencies become mixed in a nonlinear device. New frequencies that are not part of the source information are generated and produce undesirable effects in the amplifier output.

Internet
A worldwide collection of networks that make up an international Internet where almost any client can access any server. The Internet was initially a U.S. government network called the ARPAnet, which was named from the developing organization, the Advanced Research Projects Agency.

Internet Assigned Number Authority (IANA)
An organization that previously performed services for IP address assignments and related engineering services, which are now performed by ICANN.

Internet Corporation for Assigned Names and Numbers (ICANN)
A nonprofit corporation that was formed to assume responsibility for the IP address space allocation, protocol parameter assignment, domain name system management, and root server system management functions previously performed under U.S. government contract by IANA and other entities.

Internet Engineering Task Force (IETF)
An international community of network designers, operators, vendors, and researchers concerned with the evolution of the Internet architecture and the operation of the Internet. The actual technical work of the IETF is done in its working groups.

Internet Protocol (IP)
A primary protocol included in the TCP/IP suite used for encapsulating other protocols, such as TCP and UDP. The IP header includes the source and destination addresses of the packet and other routing information.

Internet service provider (ISP)
A company that provides individuals and other companies access to the Internet and other related services such as Web site building and virtual hosting. An ISP has the equipment and the telecommunication line access required to have a point-of-presence on the Internet for the geographic area served.

ionization detectors
A type of sensor used in home smoke detectors. They work on the principle related to the disturbance caused by smoke in a small ionization chamber.

IrDA
IrDA stands for Infrared Data Association. It is an international organization that creates and promotes interoperable infrared data interconnection standards for wireless networks. The name is associated with products that conform to the interconnection standard.

key systems
A type of privately owned telephone system used by small business firms and home offices. It is less expensive than a PBX and is designed for organizations that need as few as 3 or 4 phones to as many as 250 phones. Key system phones have buttons that select outside lines and have the capability to call other phones internally.

Most key systems include a controller-type cabinet called a key service unit (KSU) and a set of proprietary telephones. The KSU holds all the system's switching components, system power supply, outside lines, and internal station cards.

kilowatt-hour meter
A device installed by the local electrical power utility to measure the power consumed in a home. The meter records the cumulative power consumed in kilowatt-hours. A single-phase watt-hour meter is essentially an induction motor whose speed is directly proportional to the voltage applied and the amount of current flowing through it. The phase displacement of the current, as well as the magnitude of the current is automatically taken into account by the meter. In other words, the power factor influences the speed and the moving element (disk) rotates with a speed proportional to true power. The read-out dials are simply a means for counting the motor revolutions, and by proper gearing, it is arranged to read directly in kilowatt-hours.

LAN
Acronym for local area network. Also used to describe wireless local area networks (WLANs).

laserdisc
An older analog video optical disc format first introduced in 1974. Each side of a disc could hold about an hour of video, so feature films required a pause in the middle to flip the disc over. Some films are on two discs. The introduction of DVDs made laserdiscs obsolete for the video market.

liquid crystal display (LCD)
A type of flat display technology used in laptop computer design, calculators, PDAs, and the majority of flat-screen displays. LCD displays are available as active matrix, dual-scan, or passive matrix display types. An LCD uses the fact that certain organic molecules (liquid crystals) can be reoriented by an electric field. As these materials are optically active, their natural twisted structures can be used to alter the polarization of light on a flat screen.

load

A term used to describe the maximum amount of electrical current (in amperes) required for all the electrical devices in a home. Load analysis is usually performed by an electrical contractor for new home or remodeling construction. Load is also used to determine the power consumed by household electrical appliances. The actual power used by the load is called *true* power, or just *power*, and is measured in watts. (Even though watts = volts × amps, apparent power is measured in VA to differentiate it from true power.)

local exchange carrier (LEC)

Any public telephone company in the United States that provides local telephone service. Some of the largest LECs are the Bell operating companies (BOCs), which were grouped into holding companies known collectively as the regional Bell operating companies (RBOCs) when the Bell System was broken up by a 1983 consent decree. LEC companies are also sometimes referred to as *telcos* or *telephone companies*. A *local exchange* is the local central office of an LEC. When you pick up a phone to make a call, the dial tone you hear is coming from the central office of your local exchange carrier. Telephone lines from homes and businesses terminate at a local exchange facility. Local exchanges connect to other local exchanges within a local access and transport area (LATA).

LonWorks

A multipurpose networking protocol developed by the Echelon Corporation that can be implemented over any medium, including power lines, twisted-pair cable, wireless (RF), infrared (IR), coaxial cable, and fiber optics. It has no central controller but uses a system of intelligent nodes that communicate with each other. The LonWorks protocol was published as ANSI/TIA/EIA-709 in 1998.

low-voltage wiring

A generic term used to describe all residential cabling that is not associated with 240v/120v high-voltage alternating current (AC) primary power wiring used in all U.S. residential construction. Although no exact definition of low-voltage exists in residential cabling standards, it is generally accepted in the industry to include those circuits that are connected to low-energy current-limited sources described in Article 725 of the NEC. Low-voltage wiring includes all wiring for audio/video components, network data cabling, security sensor wiring, phone lines, telecommunications systems, and low-voltage landscape lighting. This encompasses all residential wiring with the exception of 120v or higher voltage power wiring.

magnetic stripe card

A type of security card used to control access to restricted areas or residential gated communities. It consists of a plastic card with a narrow strip of magnetic material fused to the back. Data stored on the strip as narrow bars form the basis of a binary code, which is used by the access control system reader.

MC cable
A type of metal clad (MC) cable that includes its own flexible metal covering. MC cable is composed of THHN soft-drawn, copper wire, conductors, and an insulated grounding conductor. It is suitable for branch, feeder, and service power distribution in commercial and industrial applications as well as in multifamily buildings, theaters, and other populated structures. The acronym *THHN* refers to heat-resistant thermoplastic (90° C) for dry locations. This is a reference to the type of insulation on the wiring, coded per the NEC.

motorized dampers
A component of an HVAC system used to regulate the amount of warm or cold air that enters a room or an area.

Moving Picture Experts Group (MPEG)
A set of video and audio compression standards. The MPEG format employs compression algorithms to remove redundant picture information from successive video scenes. The MPEG methodology compresses only key objects within a frame every 15th frame. Between the key frames, only the information that changes from frame to frame is recorded. The MPEG standard includes compression specifications for both video and audio signals.

multiplexer
An electronic high-speed switching device that combines the information contained on several digital video or data channels into a single high-speed channel. Multiplexers are used with home surveillance systems for recording several channels simultaneously.

National Electrical Code (NEC)
A document sponsored by the National Fire Protection Association. The National Electrical Code covers the requirements for electric conductors and equipment installed within or on public and private buildings or other structures. This includes mobile homes and recreational vehicles; floating buildings; and other premises such as yards, carnivals, parking and other lots, and industrial substations. It also covers conductors that connect the installations to a supply of electricity and other outside conductors and equipment on the premises, including optical fiber cable. It covers buildings used by electric utilities, such as office buildings; warehouses; garages; machine shops; and recreational buildings that are not an integral part of a generating plant, substation, or control center.

Most states require a permit and an inspection for this type of work. Even though following the NEC cannot guarantee safe electrical installations, it is the best guide available. However, every state might differ slightly in its requirements for inspection and code compliance.

National Electrical Manufacturers Association (NEMA)

An organization of manufacturers of electrical equipment, including, but not limited to, wiring devices, wire and cable, conduit, load centers, pressure wire connectors, circuit breakers, and fuses. NEMA is the voice of the electrical industry, and through it, standards for electrical products are formulated. Generally, these standards promote interchangeability between products of one manufacturer with like products made by another manufacturer. In some cases, standards relating to product performance are also formulated by NEMA, but these are the exception rather than the rule.

National Fire Protection Association (NFPA)

A U.S. standards organization that develops, publishes, and disseminates timely consensus codes and standards intended to minimize the possibility and effects of fire and other risks. Virtually every building, process, service, design, and installation in society is affected by NFPA documents.

National Television System Committee (NTSC)

The organization that developed the U.S. broadcast television standards. NTSC has also become the name of the standard. Other countries, such as Canada, Mexico, and Japan, have also adopted the NTSC broadcast standard.

network interface card (NIC)

A network interface card is a printed circuit board with a connector that plugs in to a PC expansion slot. A NIC contains a transceiver for sending and receiving data frames on a network as well as the Data Link layer hardware needed to format the sending bits and to decipher received frames.

NH cable

A type of cable used in electrical wiring that is halogen-free (NH stands for nonhalogen.) The noncorrosive material chosen enables this cable to be used for fixed installations in public buildings and in governmental installations where halogen-free products are demanded. The jacket is made of flame-retardant polyolefin material.

NM cable

Abbreviation for nonmetallic (NM) cable, it is the most widely used electrical power cable for indoor wiring. Romex is a brand name for a type of plastic insulated wire. It is often called nonmetallic sheath; however, the formal code name is NM cable. This type of wire is suitable for stud walls, on the sides of joists, and so on that are not subject to mechanical damage or excessive heat. Newer homes are wired almost exclusively with NM wire.

octet
Term used to describe 8 bits in an IP address. An octet is thus an 8-bit byte. Because a byte is not 8 bits in all computer systems, *octet* is used to describe a precise reference to 8 bits. Technically, an octet represents any 8-bit quantity. By definition, an octet ranges in mathematical value from 0 to 255. Typically, an octet is also a byte, but the term *octet* came into existence because historically some computer systems did not represent a byte as 8 bits.

optoelectronic
Any device that functions as an electrical-to-optical or optical-to-electrical transducer, or an instrument that uses such a device in its operation.

oxidation reduction potential (ORP)
A measure for determining a swimming pool water's overall capability to eliminate wastes. High oxidation is present when pollution is low and the water is of high quality. ORP is rapidly becoming the standard means of testing and regulating pool water sanitation.

packet
A unit of information formatted for transmission over a network. A packet contains control and header information that corresponds to the OSI Network layer, or layer 3. *Packet* is often used interchangeably (and incorrectly) with *frame*.

passive infrared (PIR)
A type of motion detector sensor used in home security systems. The term *infrared* comes from the sensor's capability to see areas of the optical spectrum outside the range of light frequencies discernable by human vision. PIR motion detectors operate on the principle that the heat or infrared emissions emanating from a human entering an area will disturb an otherwise stable infrared background and trigger an alarm. Any area has a normal infrared background temperature that does not change abruptly over time. A PIR is referred to as passive because it does not release any energy of its own but instead causes an alarm when the surrounding infrared background environment changes abruptly.

peer network
A peer network (or peer-to-peer network), as the name suggests, is where all computers on a LAN are peers and have equal status concerning sharing rights and capabilities. Any individual machine can share data or peripherals with any other machine on the same local area network.

Personal Communications Service (PCS)
A second-generation U.S. digital cellular telephone service established by the Federal Communications Commission. PCS services use the 1900MHz band. All the licenses for the use of

the PCS frequency bands by various service providers were awarded through public auctions for services in the major metropolitan areas. The FCC has also allocated spectrum for unlicensed PCS services in the 1910MHz–1930MHZ band. The unlicensed PCS service accommodates a wide range of services for small in-building areas such as data networking within office buildings, wireless private branch exchanges (PBXs), personal digital assistants, laptop computers, portable facsimile machines, wireless replacements for portions of the wire-line telephone network, and other types of short-range communications.

phase alternate lines (PAL)
A color television broadcast standard used in Europe. Broadcasts in NTSC format can have problems with color hues varying in unwanted ways because of phase shifts in the color subcarrier. PAL fixed that by reversing the phase on every other line, hence the name. Because TV came later to Europe than to the United States, this superior format was adopted there in contrast to NTSC.

phone-line splitter
A device installed on a telephone line that is connected to both high-speed broadband digital service and analog voice devices such as telephones and fax machines. The splitter routes the broadband and analog signals on the telephone line to the correct device. Most splitters must be installed by the telephone company; however, some can be installed by the customer.

photoelectric switch
A special type of switch used for controlling lights depending on the light intensity (lux value). This type of switch contains a special sensor that operates a switch when the light intensity or darkness reaches a certain value. Photoelectric switches are typically used for outdoor locations for landscape and security lighting. Most photoelectric switches are integrated in a single fixture that includes the sensor and lamp fixture and have adjustments to turn off lights after a number of hours where it is unnecessary to have lighting from dusk to dawn.

plasma display panel (PDP)
A type of advanced large-screen display technology that uses hundreds of thousands of tiny cells (pixels) containing minute amounts of an inert gas sandwiched between two sheets of glass. Electrodes are placed in pairs on the inner side of the front plate. When electrically charged, they produce an ultraviolet beam, which activates the phosphorous coating of the cell transmitting light through the glass surface. The brightness of the emission depends on the current passing through the ionized gas. Large and brilliant color images can be produced with different colors of phosphorous. Plasma displays provide a totally flat screen design

with excellent off-center viewing capability and superior resolution.

plenum
A compartment or chamber to which one or more air ducts are connected and that forms part of the air distribution system.

Point to Point Tunneling Protocol (PPTP)
A network protocol that encapsulates Point to Point Protocol (PPP) packets into IP datagrams for transmission over the Internet or other public TCP/IP-based networks. PPTP can also be used in private intranets.

pop-up sprinkler
A popular type of sprinkler head used for home lawn and garden irrigation systems. They are installed below the ground, and the sprinkler head remains out of sight while inactive. When the sprinkler system is turned on, a small portion of the head emerges above the surface to disperse water to the irrigation area. Spray head sprinkler pop-ups are designed to spray a small fixed area, whereas rotor types are normally used for large, unobstructed areas.

power-line carrier (PLC)
A term used to describe the capability for the power-line wiring used in the electrical system of a home to carry a command signal for controlling lighting and appliances. The X10 control protocol that uses PLC technology was invented to exploit this capability. The home automation products that use power-line technology are also referred to as power-line carrier (PLC) devices because they communicate over existing power-line wiring. See also HomePNA, HPLC, and X10.

private branch exchange (PBX)
A telephone switch owned by a private organization or home office user. The system is usually located at the owner's facility. The PBX provides phone services including internal calling and access to the public switched telephone network. It allows a small number of outside lines to be shared among all the people of the organization. Advanced PBX phone switches sometimes provide auto-attendant, voice-mail, and ACD (automatic call distribution) services for the organization.

programmable thermostat
A special type of thermostat designed to enable temperature adjustments to be scheduled according to the day of the week and the time of day. It controls both heating and air conditioning equipment with a single integrated control processor.

progressive scan
A high-definition TV (HDTV) format. The progressive scan system scans the total number of lines, 60 times per second. This means you see the complete image displayed on your TV screen two

times more often than in the interlaced scan method. This results in smoother motion in moving images, fewer motion artifacts, and no visible flicker. A progressive scan system with 720 lines of resolution is written as 720p. Progressive scan has been used for many years for computer display monitors. It is also employed for DVD players that play back DVD movies in the progressive scan mode for viewing on advanced digital television receivers.

protocol
A formal set of rules that establishes the method of handling data transmissions in both wired and wireless networks.

public safety answering point (PSAP)
A location where emergency response dispatchers answer 911 calls. The dispatch operator at the PSAP attempts to obtain as much information as possible from the caller and send the appropriate help to the caller's location. The caller might be in distress and unable to provide location information. The caller's phone number, however, is available to the dispatch operator through a system called automatic number identification (ANI). ANI is a service that provides the receiver of a telephone call with the number of the calling phone. The method of providing this information is determined by the local exchange carrier.

pull station
A device typically mounted on a wall that is used to manually activate a fire alarm system.

punch-down tool
A tool used to connect wires to an insulation displacement connection (IDC) block. This type of termination is the recommended method of copper termination recognized by the ANSI/TIA/EIA-568-A standard for UTP cable terminations. These termination devices are the popular choice for home network star topology connection points. Commonly called punch-down connections, these connections require the use of a small punch-down tool to properly secure the cable to the terminal block. Punch-down connections remove or displace the conductor's insulation as it is seated in the connector. During termination, the tool is used to press the cable between two edges of a metal clip, which displaces the insulation and exposes the copper conductor. This ensures a solid connection between the copper conductor and terminating clip.

registered jack (RJ)
In the United States, telephone jacks are also known as *registered* jacks, sometimes described as RJ-XX. This series of telephone connection interfaces (receptacle and plug) is registered with the FCC. They originate in name from interfaces that were part of AT&T's Universal Service Order Codes (USOC) and were adopted as part

of FCC regulations (specifically Part 68, Subpart F. Section 68.502). The term *jack* sometimes means both receptacle and plug and sometimes just the receptacle.

remote access
A type of telecommunications service that provides user access to a computer or network from a location away from the home or work location. It's often used by employees on business travel who need access to the corporate private network to read email or access shared files. Remote access is also a valuable tool for home network users who need to access automated home features from a remote location.

Remote Authentication Dial-In User Service (RADIUS)
A client/server protocol that enables remote access servers to communicate with a central server on a LAN to authenticate dial-in users and authorize their access to the requested system or service. RADIUS enables a company to maintain user profiles in a central database that all remote servers can share. It provides better security, allowing a company to set up a policy that can be applied at a single administered network point.

reversing valve
A type of valve used by heat pumps to redirect the refrigerant flow. It enables the inside coil and outside coil to act as either an evaporator or a condenser.

ribbon speakers
A special type of speaker where the input signal is applied to a foil ribbon suspended between magnets or metal sheets. Ribbon speakers do not use a cone and voice coil; the input signal from the amplifier is applied to the foil ribbon. The varying electrical charge caused by the input signal forces it to be repelled or attracted to the magnets, thereby moving air and producing sound. This design overcomes some of the deficiencies of cone speakers in the mid and upper range of the audio spectrum.

riser
A type of duct for holding wire and cable that connects one floor to another, usually in a commercial building, and penetrates fire-rated walls or floors. Cable installed in risers must meet specific fire standards published by the NFPA. This type of cable is referred to as riser cable.

RJ-11
A registered jack (RJ) type of connector used for terminating a telephone cable that connects the phone to a standard wall outlet. It is used by many home automation devices and is the standard used for all residential phones lines.

RJ-31x connector
A type of registered jack used for attaching alarm systems to telephone lines. The connector is used to take control of the phone line during a security alarm condition.

RJ-45 connector
A type of registered jack (RJ) that uses an eight-conductor connector with a small locking pin. It's used to connect the network interface card installed in networked computers onto a local area network.

rotor sprinkler
Rotor sprinkler heads are pop-up sprinklers designed to disperse water over a large circular area. Small rotors can cover radii of up to 50 feet, whereas large rotors are able to cover radii up to 100 feet. The two types of rotor sprinklers are called impact and gear driven. They differ only in the mechanism used to move the head in a circular motion.

router
A component consisting of hardware and software that provides intelligent connections between networks. Routers operate at the Network layer (layer 3) of the OSI model and are responsible for making decisions about which paths through a network the data packets will use.

S/PDIF
Sony/Philips digital interface format is a consumer version of the AES/EBU digital audio interconnection standard based on coaxial cable and RCA connectors.

scenes
The effect created by a lighting control system when a number of lights dim and brighten to different intensities.

screened twisted-pair (ScTP) cabling
A new type of 100-ohm twisted-pair cabling that evolved from the ANSI/TIA/EIA standard IS-729—the *IS* indicates it is an interim standard. ScTP cable types are also being developed to meet new more rigorous criteria for Category 6E and 7E cable.

SECAM (Sequential Couleur Aver Memoire)
Translated as sequential color and memory, SECAM is a composite color television transmission system that potentially eliminates the need for both the color and hue controls. It has a 625-line vertical resolution and 25-frame-per-second frame rate. One of the color difference signals is transmitted on one line, and the second is transmitted on the second line. Memory is required to obtain both color difference signals for color decoding. This standard is used in France, Eastern Europe, Africa, and Asia.

set points
Temperature settings used on residential programmable and communicating thermostats that indicate how high or low the user wants the temperature to be in a home.

shielded twisted-pair (STP) cable
A type of network cable use in the design of LANs originally developed as an IBM cabling system. It was used to support Token-Ring LANs. The original specification

for STP was released in 1984 and defined the 150-ohm STP cable types 1, 2, 6, 8, and 9 for support of frequencies up to 16MHz. It also defined the 100-ohm type 3 UTP cable, as well as type 5 and 5J fiber-optic cables. Later, an enhanced IBM cabling system defined STP-A cable types 1A, 2A, 6A, and 9A for support of a network standard called Fiber Distributed Data Interface (FDDI). The *A* suffix denotes the enhanced IBM cabling system. The original IBM cabling system was defined in IBM publication GA27-3773. The enhanced, or STP-A, cabling is defined in the /ANSI/TIA/EIA 568-A standard.

Simple Network Management Protocol (SNMP)
A protocol used for network management and monitoring of network devices and their functions.

single-room system
A lighting control system designed specifically for the control of lights in one room, such as a home theater.

small office home office (SOHO)
A generic term used in the home networking industry to refer to products and services primarily intended for the small business community and residential computer networks.

solenoid
An electromechanical device used with irrigation system control valves. Solenoids consist of an electromagnet that, when energized, moves a valve to the open position. A spring retainer moves the valve to the closed position when the voltage is removed from the solenoid.

spread spectrum
A digital radio transmission technology in which the bandwidth of the radio carrier signal is considerably greater than that normally required to convey the digital baseband information. Spread spectrum technology provides excellent resistance to interference from other radio signals and intentional radio frequency jamming sources at the expense of wider spectrum utilization.

standalone
A term used to describe a home technology component capability to operate independent of another component.

Shared Wireless Access Protocol (SWAP)
A protocol derived from the existing Digital Enhanced Cordless Telephone (DECT) standard. The SWAP specification describes wireless transmission devices and protocols for interconnecting computers, peripherals, and electronic appliances in a home wireless (HomeRF) network environment. The Shared Wireless Access Protocol referenced in the HomeRF standard should not be confused with the Wireless Applications Protocol (WAP), which is used primarily for wireless Internet access on cellular networks.

structured wiring

The term *structured cabling/wiring* refers to all the cabling and components installed in a logical, hierarchical way with a central distribution box serving as a common termination point for all residential cabling. It's designed to be relatively independent of the computer or telecommunications network that uses it, so either can be updated with a minimum of rework to the home. Structured wiring uses a wiring/cable design in which each room in a residence is wired for dual coaxial cable and UTP cable outlets.

subwoofer

A type of speaker designed for use with surround sound multiple-speaker systems. These speakers are optimized to reproduce sound only in the lower frequencies of the audio spectrum region of 20Hz–80Hz.

switcher

A video controller used with security systems to provide the capability to connect several surveillance cameras to a single monitor. The switcher can be programmed to cycle through all the cameras in a surveillance system or dwell on each camera for a specified length of time.

Telecommunications Industry Association (TIA)

An organization accredited by ANSI to develop voluntary industry and home technology standards. The TIA's Standards and Technology Department is composed of five divisions, which sponsor more than 70 standards-setting groups.

Telnet

A TCP/IP terminal emulation protocol that enables remote login (the capability to log in to another computer) from your local computer. Telnet is a common way to remotely control Web servers. telnet is also a Unix command that starts the Telnet program on a Unix computer. The Telnet program is the software (based on the Telnet protocol) that enables the user to log in to a remote computer and use its resources. The most common reason for using a Telnet program is to log in to a remote computer to use email or run programs not available on the user's local computer. Remote login to a more powerful computer can provide access to a full range of Internet services and sophisticated programs that often cannot be run on a low-end personal computer. Telnet can also be used to access various library catalogs and databases. Telnet programs (or clients) are available for a wide variety of computer systems.

three-way calling

A telephone subscriber service offered by local exchange carriers, long-distance carriers, and mobile wireless cellular carriers. It enables a subscriber to talk to two people in two different locations at the same time.

Three-way calling is initiated by calling the first party, pressing and releasing the switch hook to get a second dial tone, and dialing a second number. When the second party answers, the switch hook is pressed and released again. The three-way connection is then in place and the call can proceed with all three parties connected.

three-way switch
A type of switch used to control a light fixture from two locations. An example of this type of installation is where a light needs to be controlled at the top and bottom of a stairway. A three-way switch is required at both locations.

tie wrap
A plastic tie for holding cables together or holding cable in place, usually in the interior of a wall or in a cable tray.

Transmission Control Protocol/Internet Protocol (TCP/IP)
The name applied to a suite of protocols developed by the Advanced Research Projects Agency (ARPA).

TRIAC (thyrister for AC applications)
A type of solid-state device used in home lighting dimmer controls. TRIACs switch the AC voltage on and off using a variable phase control principle permitting a gradual increase or decrease in the percentage of the phase that is applied to the load.

TOSLINK
A popular consumer equipment fiber-optic interface based on the Sony/Philips digital interface format (SPDIF) protocol, using an implementation first developed by Toshiba.

touchscreen
A type of technology that uses a catode-ray tube (CRT) or a liquid crystal display (LCD) screen as both a keypad and a display. Touchscreens are popular with kiosks, point-of-sale cash registers, library search terminals, and computer-aided design (CAD) and drafting applications.

true power
The actual power used by the load of an electrical circuit measured in watts. (Even though watts = volts × amps, apparent power is measured in VA to differentiate it from true power.) The relationship between true power and VA is represented by the cosine (the trigonometric function) of the phase angle between the voltage and current. The relationship between real power (in watts) and apparent power (in VA) is expressed as power (watts) = cosine (phase) × apparent power (VA).

twisted pair
A communication medium consisting of two copper conductors insulated from one another and twisted together to reduce the effects of electromagnetic interference. Twisted-pair cables comprise the

majority of installed telephone and home LAN cables.

uninterrupted power supply (UPS)
The most widely used type of backup power system for servers and other computers in home networking. A UPS is designed to provide power to client computers and servers in a business location or home office for a brief period after a power failure to allow a gradual planned shutdown of computer systems. UPS specifications indicate the length of time as well as the amount of electrical power the unit can deliver when commercial power is lost. The most important rating to be aware of is the volt-ampere (VA) rating, which is the product of the volts and amperes delivered to the load.

universal serial bus (USB)
A computer standard for a serial interface designed to connect peripheral devices to personal computers. The USB Specification, Revision 2.0, covers three speeds: 480Mbps, 12Mbps, and 1.5Mbps. The term *Hi-Speed USB* refers to just the 480Mbps portion of the USB Specification; the term *USB* refers to the 12Mbps and 1.5Mbps speeds.

unshielded twisted-pair (UTP) cable
A type of cable widely used in LANs. It has insulated conductors twisted together, leading to better electrical performance, lower crosstalk, and significantly higher bit rates than untwisted cable. The grading system divides UTP cable into separate categories. The spacing of the twists in a given length along with fire code and installation criteria dictate the official category rating to which the cable belongs and the bandwidth that can be supported.

video home system (VHS)
A standard for video tape systems developed by JVC in 1976. With VHS processing, the intensity information (luminance) and color information (chrominance) are filtered separately to reduce the bandwidth and are then remixed when played back. Therefore, both this filtering and the fact that the playback format is composite video serve to reduce the quality of the image.

Voice over IP (VoIP)
A voice communications system using the Internet Protocol. VoIP systems send voice information in digital form as discrete packets rather than in the traditional analog circuit-switched protocols of the public switched telephone network. It's also a term used in IP telephony for a set of facilities for managing the delivery of voice information using the Internet Protocol. A major advantage of VoIP and Internet telephony is that they avoid the tolls charged by ordinary telephone service.

Web pad
A wireless, portable, tablet-shaped, consumer-focused digital device

usually with a touchscreen user interface. It has a browser-based interface to simplify and enhance the home control experience. The two components of a Web pad are a portable LCD built into a tablet-shaped device and a base station that is plugged in to the home network, which sends and receives wireless (RF) transmissions. Transmissions between the Web pad and the base station use RF or wireless home networking technologies, such as HomeRF, Bluetooth, or IEEE 802.11b, and consequently have a limited range of 150 feet.

wide area network (WAN)
A telecommunications network encompassing a large geographical area. Usually, WANs are considered to cover distances beyond a metropolitan area.

Wi-Fi
An abbreviation for Wireless Fidelity. Wireless networking equipment that meets the 802.11 standard is approved to use the Wi-Fi label. It is meant to be used generically when referring to any type of wireless 802.11 network device, whether 802.11b, 802.11a, dual-band, and so on. The term is promulgated by an organization called the Wi-Fi Alliance. Any products tested and approved as Wi-Fi Certified (a registered trademark) by the Wi-Fi Alliance are certified as interoperable with each other, even if they are from different manufacturers. A user with a Wi-Fi Certified product can use any brand of access point with any other brand of client hardware that also is certified. Typically, however, any Wi-Fi product using the same radio frequency (for example, 2.4GHz for 802.11b or 11g, and 5GHz for 802.11a) will work with any other product, even if it's not Wi-Fi Certified.

World Wide Web Consortium (W3C)
An organization created to promote interoperability and establish an open forum for discussion for the World Wide Web (WWW). In October 1994, Tim Berners-Lee, inventor of the Web, founded the World Wide Web Consortium (W3C) at the Massachusetts Institute of Technology, Laboratory for Computer Science in collaboration with the European Organisation for Nuclear Research (CERN), where the Web originated, with support from the Defense Research Projects Agency (DARPA) and the European Commission.

X10
A power-line carrier (PLC) protocol that enables compatible devices throughout the home to communicate with each other via the existing 120v AC wiring in the house. Using X10, you can control lights and virtually any other electrical device from anywhere in the house with no additional wiring. The protocol name is derived from *Experiment 10*, which was a

successful communications protocol perfected by Pico Electronics of Scotland in the 1970s and was originally intended to control audio turntables. The original X10 patent expired in December 1997. X10 is now an open standard and many manufacturers are developing new and improved X10 products.

XDSL
A term used to refer generically to any of the digital subscriber line variants, such as HDSL, SDSL, RADSL, ISDL, or VDSL.

zone
A designated area established in an automated home design for programming lighting scenes or audio source material from a remote source or control point.

Index

Symbols

8mm video format, 112
10BASE-T specification (Ethernet), 21-22
66/110 connection blocks, punch-down tool, 173-174
100BASE-FX specification (Ethernet), 22
100BASE-T specification (Ethernet), 22
100BASE-T4 specification (Ethernet), 22
100BASE-TX specification (Ethernet), 22
720p progressive format (HDTV), 116
802.3i 10BASE-T standard (Ethernet), 21-22
802.3u 100BASE-T standard (Ethernet), 22
802.11a standard (IEEE)
    compatibility with 802.11b standard, 8
    data rates, 10
    Physical Layer (PHY), 8
802.11b standard (IEEE)
    ad hoc mode, 8-9
    complementary code keying (CCK), 8
    data rates, 10
    Direct Sequence Spread Spectrum (DSSS), 8
    distribution system (DS), 9
    infrastructure mode, 8-9
    versus 802.11g standard, 8
    Wi-Fi certification, 8
802.11g standard (IEEE)
    transmission rates, 10
    versus 802.11b standard, 8
911 emergency systems
    enhanced (E911), 188
    external services, 187-188
1080i interlaced transmission standard (HDTV), 115

A

A/V cabinetry, home automation features, 342-343
A/V controller keypads, 349-350
A/V equipment in high-voltage systems
    component placement, 425
    locating, 423-424
A/V receivers, 90
AC-to-DC power converters, home automation systems, 352-353
access points, designing for device connectivity, 54-55
acoustical windows, 136
ad hoc mode (IEEE 802.11b standard), 8-9
ADSL terminal unit (DSL modem), 65
Advanced Television System Committee (ATSC), 109
air conditioning systems (HVAC), troubleshooting and maintenance, 270-271
    air handler problems, 271-272
air handlers
    air conditioning systems (HVAC), troubleshooting and maintenance, 271-272
    HVAC systems, 238-239, 249
        attic fans, 250
        whole house fans, 250
alarms, sounder types, 149
ambient lighting, 215
American Wire Gauge (AWG), telephone wiring, 192

amplifiers
    bridging, 91-92
    low-voltage wiring, 392
    power ratings, 91
analog audio, audio/video device
    connectivity, 104
analog telecommunications systems, 162-163
ANSI/EIA-600 standard (CEBus), 19, 73
ANSI/TIA/EIA computer networking
    standards
        568 specification, 72, 182
        570-A specification, 72-73, 151, 180,
            190-191, 467
Apple QuickTime Player, 106-107
armored cable (AC)
    device connectivity in high-voltage
        systems, 438
    home lighting wiring, 220-221
ARP utility (TCP/IP), 45
ARPANET (DoD), IP addressing
    origins, 42
asymmetric DSL (ADSL)
    bandwidth allocation, 63
    local loop cable length performance , 63
ATSC TV standard, 114
attic ducting, 268-269
attic fans for HVAC systems, 250
audio distribution system
    connection components, 84
    features, 84-85
    typical equipment, 85
    zone design, 84
audio wire, low-voltage device connectivity,
    399-400
Audio/Video Cable Installer's Pocket Guide, 126
audio/video systems
    component settings, 100
        distribution channels, 100
        equalization, 101
        internal broadcasting, 101
        volume, 100
    designs
        dedicated type, 82
        distributed type, 82
        whole type, 83-85
        zoning distribution, 85
    device connectivity
        analog audio, 104
        component video, 102-103
        composite video, 103
        digital audio, 105
        low-voltage cabling and connectors,
            105
        S-video, 104
        termination points, 105-106
        video signal formats, 102
    exam prep questions, 121-125
    externally provided audio/video
        services
        cable TV services, 108
        digital TV services, 109-110
        external terrestrial off-the-air
            broadcast services, 109
        satellite services, 108
    in-house services
        media servers, 107
        personal video recorders, 107
        streaming audio/video, 106-107
    installations
        satellite service, 117-118
        speakers, 119
    location considerations, 86
        keypads, 88
        speakers, 86-87
        television, 87
        touchscreens, 88
        volume controls, 88
    maintenance, 120
    physical products
        amplifiers, 91-92
        keypads, 94
        receivers, 89-90
        source equipment, 99-100
        speakers, 92-93
        video displays, 95-97
        video projection systems, 98-99
    reference resources, 126
    standards
        IEEE, 116
        image resolution, 111
        protocols, 111
        surround sound, 113
        television, 114-116
        video recording, 111-113
        wiring, 110
automated dimming modules (home
    lighting systems), 212
automated homes
    component location
        keypads, 349
        sensors, 349
    configuration and settings
        keypad programming, 357
        sensors, 357-360
        time-of-day scheduling, 360
    control processors, 456-458

device connectivity
    Category 5E UTP cables, 465
    coaxial cables, 464
    communications cable, 360-361, 464
    fiber-optic cables, 465
    infrared transmission, 465
    low-voltage wiring, 361
    radio frequency, 466
    termination points, 361, 466
distribution panels, 455
exam prep questions, 364-368, 475-480
features
    A/V cabinetry, 342-343
    dumbwaiters, 347-348
    elevator systems, 346
    fireplace igniters, 347
    home fans, 348
    life systems, 343
    skylights, 348
    stair lift designs, 346
    window shading designs, 345
installation plans
    central home controllers, 469
    user interfaces, 469
interfaces, location of, 455
keypads
    core component functions, 454
    location considerations, 453
maintenance plans
    testing stage, 470-471
    troubleshooting tools, 471-474
patch panels, 458
products
    control boxes, 351
    distribution panels, 355
    keypads, 349-350
    lift systems, 356
    portals, 357
    power supplies, 352-353
    solenoids, 354-355
standards
    ANSI/TIA/EIA 570-A, 467
    EIA 600 CEBus, 362
    IEEE, 468
    National Electrical Code (NEC), 362, 467
    National Electrical Contractors Association (NECA), 468
    TIA/EIA 570-A, 362
    Underwriters Laboratories (UL), 363, 468
systems integration design, 450
    keypads, 451
    touchscreens, 451-452
    Web pads, 450
    wireless RF/IR receivers, 452
touchscreens, location considerations, 453
user interface programming, 459
    controller systems, 461-462
    intelligent remote controls, 463
    keypad-based controllers, 462
    residential gateways, 459-461
    touchscreens, 462
Web pads, location considerations, 453
wireless RF/IR receivers
    core component functions, 454
    location considerations, 453
automated window treatments, 217
automobile lift systems, 356
auxiliary disconnect outlet (ADO), 191

B

backup systems, residential gateways, 460
ballasts (fluorescent lighting), 217
basic rate interface (BRI) service, 67
Betacam video format, 113
binding post terminals (amplifiers), 393
Bluetooth wireless standard
    channel configurations, 6
    connections
        point-to-multipoint, 7
        point-to-point, 7
    development of, 6
    function of, 6
    master devices, 7
    piconets, 7
    slave devices, 7
BNC (Bayonet Neill Concelman)
    connectors, 391-392
    coaxial cable, 24
    low-voltage wiring, 391-392
broadband ISDN service, 67
Builders Hardware Manufacturers Association (BHMA), home access control standards, 331
bundled cabling in low-voltage wiring, 380

C

cable TV service
    development of, 109
    program transmission methods, 108

How can we make this index more useful? Email us at indexes@quepublishing.com

cabling
  home networks, 11-12
    HomePNA standard, 12-13
    interference, 27
    power-line control protocols, 16-20
    power-line networking, 14-16
    selection considerations, 27-28
    STP, 27-28
    UTP, 27-28
  hybrid fiber/coax (HFC) networks, 67
  low-voltage wiring, placement of, 387
  telecommunications systems, installation of, 192
call blocking (telecommunications), 186-187
call conferencing (telecommunications), 184
call restrictions (telecommunications), 177-178
call waiting (telecommunications), 187
caller ID (telecommunications), 185
caller line identification (CLI), 172
cameras for home security and surveillance systems, 141, 146
  installation plans, 153
  lux ratings, 141
  resolution, 141
Category 5 UTP cable
  low-voltage wiring, device connectivity, 397
  termination points, 58
Category 5E UTP cables, home automation systems, device connectivity, 465
cathode-ray tubes (CRTs), 95
CEBus (consumer electronics bus) protocol, 19-20
  development of, 19
  home lighting systems, 206-207
  home networks, 385
  media types, 20
  performance parameters, 20
  spread spectrum modulations, 20
central home controllers, home automation, 469
centralized HVAC systems, design considerations, 236-237
  residential cooling systems, 238
  residential heating systems, 237-238
circuit breaker boxes, 433
  high-voltage systems, configuring, 434
  reactivating, 442
Class A addresses (IP), 43
Class B addresses (IP), 43
Class C addresses (IP), 43
clock lighting switches, 215

closed circuit TV (CCTV) monitors
  home security and surveillance systems, 141
  quad camera switchers, 142
closed condition monitoring, 315
coaxial cable
  Bayonet-Neill-Concelman (BNC) connector, 24
  composition of, 23
  home automation systems, device connectivity, 464
  home security and surveillance systems, 148
  low-voltage wiring, 379
    device connectivity, 397
    installing, 407
    layer shield construction, 380
    manufacturer specifications, 379
  RG-6, 23
  RG-58, 23
  RG-59, 23
  television uses, 23
  twisted-pair, 23
  video connectors, 394
code grabbers (garage door operators), 312
communications cable
  home access systems
    device connectivity, 327-329
    fire ratings, 328
    limited use, 328
    multipurpose plenum, 329
    multipurpose riser, 329
    NEC markings, 327
    plenum, 328
    power-limited plenum, 329
    riser, 328
  home automation systems, connecting, 360-361, 464
  HVAC system types, 260-261
Complete Introductory Networking Course, 79
component video, audio/video device connectivity, 102-103
components
  high-voltage systems, 425
    circuit breaker boxes, 433
    dimming modules, 427-428
    fixtures, 431-433
    light switches, 429
    outlets, 426
    three-way light switches, 430-431
  low-voltage wiring
    amplifiers, 392
    binding posts, 393

connecting **599**

BNC connectors, 391-392
coaxial video connectors, 394
distribution panels, 389
low-pass filters, 392
placement considerations, 387-388
RG-6 coaxial cable, 391
RJ-11 jack connectors, 390
RJ-45 jack connectors, 389
composite video, audio/video device connectivity, 103
computers
  power backup systems, 420-422
  typical components, 37
condensers (HVAC systems)
  evaporators, 254
  heat pumps, 254-255
  reversing valves, 255
condition monitoring (doors/gates), 314
  closed, 315
  locked, 314
  opened, 315
  unlocked, 315
conduits
  home lighting systems
    selecting, 208
    types, 207
  low-voltage wiring, design considerations, 382
cone speakers, 92
configuring
  controller systems, home automation systems, 461-462
  high-voltage systems
    circuit breaker panels, 434
    distribution panels, 434
  home access controls, 324
    sensors, 325-326
    software programming, 325
    time-of-day settings, 326
    user access, 324
  home lighting systems, 217
    scenes, 218
    security, 218-219
    zones, 219
  home security and surveillance systems, 143
    camera locations, 146
    exterior lighting locations, 146
    keypad locations, 144
    passwords, 143-144
    sensor locations, 144-145
    zone layout, 143

HVAC systems
  seasonal presets, 259
  system programming, 257-258
  time-of-day programming, 258-259
  zone programming, 256-257
intelligent remote controls (home automation systems), 463
keypad-based controllers (home automation systems), 462
low-voltage wiring
  distribution panels, 395-396
  termination points, 395
residential gateways (home automation systems), 459-461
telecommunications systems, 175
  call restrictions, 177-178
  intercom features, 178
  phone-line extensions, 175
  phone-line splitters, 175
  voice mail, 177
touchscreens (home automation systems), 462
connecting
  Category 5E UTP cables for home automation systems, 465
  coaxial cables for home automation systems, 464
  communications cable in home automation systems, 464
  fiber-optic cables in home automation systems, 465
  home access controls, 327
    communications cable, 327-329
    low-voltage wiring, 330
    phone lines, 330
    termination points, 330
    wireless intercoms, 329-330
  home lighting systems, 219
    armored cable (AC), 220-221
    low-voltage wiring, 221
    metal clad (MC) cable, 219-220
    nonmetallic (NM) cable, 219
    termination points, 221
  HVAC systems
    communications cable, 260-261
    termination points, 261
    thermostat communication, 260
  infrared technology in home automation systems, 465
  low-voltage wiring
    audio wire, 399-400
    Category 5 UTP cable, 397

How can we make this index more useful? Email us at indexes@quepublishing.com

connecting

coaxial cable, 397
fiber-optic cable, 398
plenum cable, 399
security wire, 400
termination points, 400
radio frequency technology in home automation systems, 466
telecommunications systems
  physical jacks, 179
  physical wiring types, 180
  RJ-11 connectors, 181-182
  RJ-45 termination standards, 182
  termination points, 183
  wireless, 179
termination points in home automation systems, 466
connection blocks (66/110), 173-174
connectors
  coaxial video (low-voltage wiring), 394
  fiber-optic cable
    patch cable, 26-27
    SC, 25
    ST, 25
Consumer Product Safety Commission (CPSC)
  garage door regulations, 313
  home access control standards, 332
control boxes
  home access controls, 319-320
  home automation systems
    IR Xpander command center, 351
    X10 IR command center, 351
control panels, home security and surveillance systems, 137
control points in sump pumps, 290
control processors
  central, 457
  configuring, 461-462
  gateway, 457
  home automation systems, 456-458
  indicator lights, 457
cooling systems, centralized residential systems, 238
core components (systems integration)
  control processors, 456-458
  distribution panels, 455
  interface location, 455
  keypads, 454
  patch panels, 458
  wireless RF/IR receivers, 454
crossover network patch cable, termination points, 56

D

daisy-chain wiring topology
  home lighting systems, 206
  low-voltage wiring, design considerations, 382
damper controls (HVAC systems), 250-251
  duct planning, 251
  heating adjustments, 251
  static pressure measurements, 252
data ports
  cable specifications, 32
  IEEE 1394 Port Standard, 33
  parallel
    DB-25, 33
    EPP, 33
  USB Version 2.0, 33
DB-25 parallel port connector, 33
DBS satellite services
  decoder boxes, 108
  installing, 117-118
  providers, 108
dedicated audio/video systems, design of, 82
demilitarized zones, residential gateways, 460
designing
  garden water systems
    timed systems, 282-283
    zoned systems, 284
  high-voltage systems, 416-417
    ground connections, 417-419
    load requirements, 417
    power backups (UPS), 419-422
    safety issues, 422
    surge protection, 419
  home access controls, 312
    condition monitoring, 314-315
    remote access systems, 313-314
    remote control systems, 312-313
  home lighting, 202
    conduits, 207-208
    daisy-chain wiring, 206
    grounding connections, 204-205
    home-run wiring, 205
    load requirements, 202-204
    location criteria, 209
    power-line controls, 206-207
    remote access, 208
    wire runs, 205
    wireless runs, 205
    zones, 208

home security and surveillance systems
  control panel specifications, 137
  door specifications, 136
  keypad specifications, 137
  windows specifications, 136
home water systems, 282
  remote access, 284
  timed systems, 283
  zoned systems, 284
HVAC systems, 234
  air handlers, 238-239
  centralized systems, 236-238
  distributed systems, 236-237
  non-zoned, 234
  remote access, 241-242
  water-based (radiant) heating, 239-241
  zoned, 234-236
security/fire alarm systems, 133
  cohesion with existing systems, 134
  existing home environments, 134
  new home environments, 135
  safety and code regulations, 134
  utility outlet specifications, 133
systems integration, 450
  keypads, 451
  touchscreens, 451-452
  Web pads, 450
  wireless RF/IR receivers, 452
telecommunications systems, 162
  analog systems, 162-163
  digital systems, 163-164
  home environmental factors, 168
  hybrid systems, 162
  key systems, 165-166
  PBX systems, 165
devices
  connectivity
    access points design, 54-55
    home security and surveillance
      systems, 147-148
    interfaces with legacy systems, 53-54
    systems integration, 465-466
    termination points, 56-59
  home access controls
    communications cable, 327-329
    connecting, 327-330
    low-voltage wiring, 330
    phone lines, 330
    termination points, 330
    wireless intercoms, 329-330
  home automation systems, connecting,
    360-361

home lighting systems connectivity,
  219-221
home/garden water systems connectivity,
  300-302
HVAC connectivity
  communications cable, 260-261
  thermostats, 260
  low-voltage wiring connectivity, 397-400
DHCP (Dynamic Host Configuration
  Protocol), IP address allocation, 48
digital audio, 105
Digital Enhanced Cordless Telephone
  (DECT), 6
Digital Subscriber Line. *See* DSL
digital telecommunications systems,
  163-164
digital TV services
  Advanced Television System Committee
    (ATSC), 109
  HDTV, 110
Digital8 video format, 112
dimming modules
  high-voltage systems, variable phase
    controls, 427-428
  home lighting systems
    automated, 212
    variable phase control, 211-212
    variable resistors, 211-212
direct inward dialing (DID), 184-185
Direct Sequence Spread Spectrum
  (DSSS), 8
distributed audio/video systems,
  design of, 82
distributed HVAC systems, design
  considerations, 236-237
distributed splitter DSL service, 65
distribution channels for audio/video
  component settings, 100
distribution device (DD), 191
distribution panels
  high-voltage systems, 434
  home access controls, 322-323
  home automation systems
    automated skylights, 355
    component functions, 455
  home/garden water systems, 291
  HVAC systems, 256
  low-voltage wiring, 389
    configuring, 395-396
    placement of, 387
DOCSIS (Data Over Cable Service
  Interface Specification), 68

How can we make this index more useful? Email us at indexes@quepublishing.com

Dolby Digital surround format, 113
domain name service (DNS), top-level codes, 46-47
doors/gates
- condition monitoring, 314
  - closed, 315
  - locked, 314
  - opened, 315
  - unlocked, 315
- home security and surveillance systems, 136
- remote access systems, 313-314
- switches, security sensors, 139

double pole double throw (DPDT) switches, 213, 429
double pole single throw (DPST) switches, 213, 429
DSL (digital subscriber line), 62
- asymmetric, 63
  - bandwidth allocation, 63
  - local loop cable length performance, 63
- distributed splitters, 65
- home services
  - G.DMT, 65-66
  - G.Lite, 65-66
- modems (ADSL terminal unit), 64-65
- splitterless, 64
- splitters, 64-65
- symmetric, 63

DTS 5.1 surround format, 114
DTV TV standard, 114
ducts (HVAC systems)
- air leakage, 267-269
- attic, 268-269
- installing, 266-269
- location of, 266-269
- residential thermal distribution, 266

dumbwaiters, home automation features, 347-348
Dynamic Host Configuration Protocol. *See* DHCP
dynamic speakers, 92

E

E911 (enhanced 911 systems), 188
earth ground rods for high-voltage systems, 418-419
EIA 600 CEBus standard, home automation systems, 362
EIA/TIA standards, high-voltage systems, 440

electric furnaces, 253
electrical metallic tubing (EMT) conduit, 207
electromagnetic locks (solenoid-operated), 321-322, 331
Electronic Industries Alliance (EIA), 222
- CEBus standard, 19
- HVAC system standards, 262-263

electronic irrigation controllers, 290
electronic thermostats, 249
electromechanical irrigation controllers, 289
elevators, home automation systems, 346
- hydraulic, 354
- solenoid applications, 354

email
- client programs, 70
- POP3 protocol, 71
- required configuration information, 71
- SMTP protocol, 71

emergency response systems, 132-133
ENERGY STAR program, 249
enhanced 911 (E911), 188
environmental monitoring systems, 132
EPP parallel port connector, 33
equalization, audio/video component settings, 101
Essential Guide to Home Networking Technologies, 79, 199, 231
essential power distribution for power backups, 420
Ethernet
- LAN standards, 21
  - coaxial cable, 23
  - connectors, 24-25
  - fiber-optic cable, 24-25
  - fiber-optic patch cable, 26-27
  - IEEE 802.3i 10BASE-T, 21-22
  - IEEE 802.3u 100BASE-T, 22
- network adapters, 21

evaporators (HVAC systems), 254
exam
- practice #1
  - answers, 504-524
  - questions, 485-503
- practice #2
  - answers, 539-552
  - questions, 525-538
- prep questions
  - audio/video systems, 121-125
  - home access control systems, 335-339
  - home automation systems, 364-368
  - home lighting systems, 226-230

home networking, 74-78
home security and surveillance
  systems, 154-158
home water systems, 305-309
HVAC systems, 273-278
systems integration, 475-480
telecommunications systems,
  194-198
Residential Systems content
  audio/video component settings,
    100-101
  audio/video device connectivity,
    102-106
  audio/video equipment location,
    86-88
  audio/video in-house services,
    106-107
  audio/video standards, 110-116
  audio/video system designs, 82-85
  audio/video system installations,
    117-119
  audio/video system maintenance,
    120
  device connectivity, 53-59
  equipment components, 32-35
  equipment location, 28-31
  Ethernet LAN standards, 21-27
  externally provided audio/video
    services, 108-110
  externally provided data services,
    62-71
  hardware configuration, 41-52
  home network cabling, 11-20
  network media considerations,
    27-28
  network physical devices, 36-41
  objectives overview, 4
  physical audio/video products,
    89-100
  remote access methods, 11
  shared in-house services, 60-61
  standards, 71-73
  wireless network protocols/standards,
    5-10
extension dialing (telecommunications),
  184-185
exterior lighting, home security and
  surveillance systems, 146
external services
  home security and surveillance systems,
    149
    professional monitoring, 150
    remote access, 150

telecommunications systems
  911 emergency systems, 187-188
  call blocking, 186-187
  call waiting, 187
  caller ID, 185
  personal emergency systems, 189
  three-way calling, 187
  voice mail, 185-186
external terrestrial off-the-air broadcast
  services
  UHF channels, 109
  VHF channels, 109
externally provided audio/video services
  cable TV services, 108
  digital TV services, 109-110
  external terrestrial off-the-air broadcast
    services, 109
  satellite services, 108
externally provided data services
  cable
    DOCSIS standard, 68
    hybrid fiber/coax (HFC) networks, 67
    transmission rates, 68
  DSL, 62
    asymmetric, 63
    G.DMT, 65-66
    G.Lite, 65-66
    modems, 64-65
    splitters, 64-65
    symmetric, 63
  email
    client programs, 70
    POP3 protocol, 71
    required configuration information,
      71
    SMTP protocol, 71
  ISDN, 66
    basic rate interface (BRI) service, 67
    broadband service, 67
    primary rate interface (PRI) service,
      67
  PPP, 68
  PPTP, 68-69
  RAS, 69-70
extranets, firewall implementation, 51-52

F

fan sensors, home automation systems, 359
Fast Ethernet
  100BASE-FX standard, 22
  100BASE-T4 standard, 22
  100BASE-TX standard, 22

fax machines
  multifunction, 171
  transmission speeds, 172
  transmission standards/protocols,
    ITU-T Group 3 & 4, 189-190
FCC (Federal Communications
  Commission)
  HDTV decree, 116
  low-voltage wiring standards, 406
fiber-optic cable, 24-25
  connectors
    SC, 25
    ST, 25
  home automation systems, device
    connectivity, 465
  low-voltage wiring, 381
    device connectivity, 398
    interconnection design, 381
    multimode transmission, 381
    single-mode transmission, 381
  multimode, 24-25
  single mode, 24-25
  TOSLINK, 26-27
file servers, shared in-house services, 61
fire alarm systems
  design considerations, 133
    cohesion with existing systems,
      134
    existing home environments, 134
    new home environments, 135
    safety and code regulations, 134
    utility outlet specifications, 133
  smoke detectors, location requirements,
    134
fire detection systems, 131
  heat sensors, 131-132
  smoke detectors, 131-132
fireplaces, home automation systems
  controller keypads, 350
  igniters, 347, 358
  solenoids, 355
firewalls
  configuration, 51
    extranets, 51-52
    intranets, 51
  filtering, 51
    extranets, 51-52
    intranets, 51
  home network connections, 31
fixtures for high-voltage systems, 431
  fluorescent, 432-433
  incandescent, 431-432

flat-panel video displays, 96
  LCD, 96-97
  plasma, 97
fluorescent lighting, 216
  ballasts, 217
  fixtures, 432-433
fountains, timed systems, 283
frequency division multiplexing (FDM),
  HomePNA standard, 12
front video projection, 98
furnaces (HVAC systems), 252-253
  electric type, 253
  gas type, 253

G

G.DMT service (DSL), 65-66
G.Lite service (DSL), 65-66
garage door openers
  code resets, 334
  photoelectric sensors, 326
  remote controls
    code grabbers, 312
    safety regulations, 313
  reverse sensors, 333
garden water systems
  configuration and settings
    seasonal preset programming, 300
    time-of-day programming,
      299-300
    zone programming, 298-299
  design considerations, 282
    remote access, 284
    timed systems, 282-283
    zoned systems, 284
  device connectivity, 300
    communication cable, 300-301
    low-voltage wiring, 301-302
    termination points, 302
  equipment locations, 284
    backflow prevention control valves,
      285
    heaters, 286-287
    irrigation control valves, 285
    keypads, 286
    pumps, 287-288
    relays, 286
    solenoids, 288
  exam prep questions, 305-309
  industry standards, 302-303
  installation plans, 303-304
  maintenance plans, 304

products, 289
　control points, 290
　distribution panels, 291
　interface locations, 291
　irrigation controllers, 289-290
　keypads, 292
　power supplies, 292-293
　sensors, 293
　solenoids, 295-296
　sprinklers, 296-298
　sump pumps, 293-295
　reference resources, 310
gas furnaces, 253
gates, home access controls
　sliding-type, 323
　swinging-type, 323
gateway controllers, 457
geothermal heat pumps, 241
glass breaker detectors, 139, 145
Grade 1 cabling, 190
Grade 2 cabling, 190
graphic equalizers, 101
ground connections in high-voltage systems
　designing, 417-419
　earth ground rods, 418-419
　home lighting overview, 204-205
　wire colors, 204-205
ground fault circuit interrupters (GFCIs), 422
　circuit breakers, maintaining, 442
　power outlets, 211
A Guide to Security System Design and Equipment Selection and Installation, 159

H

hard-wired security systems
　advantages, 129-130
　disadvantages, 130
hardware
　home lighting systems
　　automated window treatments, 217
　　dimming modules, 211-212
　　light fixtures, 215-217
　　lighting switches, 213-215
　　power outlets, 209-211
　home networks
　　firewalls, 31
　　network interface cards (NICs), 28-29
　　printers, 30
　　routers, 31

servers, 30
switches, 31
wireless access points, 31
home/garden water systems
　backflow prevention control valves, 285
　heaters, 286-287
　irrigation control valves, 285
　keypads, 286
　location of, 284
　pumps, 287-288
　relays, 286
　solenoids, 288
HVAC systems
　air handlers, 249-250
　condensers, 254-255
　damper controls, 250-252
　distribution panels, 256
　furnaces, 252-253
　thermostats, 246-249
network configuration, 41
　DHCP, 48
　firewalls, 51-52
　Internet domain names, 46-47
　network address translation (NAT), 51
　network interface cards (NICs), 52
　operating systems, 41-45
　port numbers, 49-50
　TCP/IP utility programs, 45-46
　UDP, 49
　uniform resource locators (URLs), 47
network equipment components, 32
　data ports, 32-33
　jack types, 34-35
systems integration, location considerations, 452-453
hazardous locations, low-voltage wiring standards (NEC), 406
HDTV (high-definition television), 110
　720p transmission format, 116
　1080i interlaced transmission standard, 115
　aspect ratio, 115
　FCC decree, 116
　standards, 115-116
heat pumps (HVAC systems), 254-255
heaters for pools/spas, 286-287
heating systems in centralized residential systems, 237-238

How can we make this index more useful? Email us at indexes@quepublishing.com

heating, ventilating, and air conditioning systems. *See* HVAC systems
HI-8 video format, 112
high-voltage wiring, 416
   A/V equipment location, 423
      new construction, 424
      remodeling upgrades, 423-424
   configuration and settings
      circuit breaker panels, 434
      distribution panels, 434
   design considerations, 416-417
      ground connections, 417-419
      load requirements, 417
      power backups (UPS), 419-422
      safety, 422
      surge protection, 419
   device connectivity
      armored cable (AC), 438
      metal clad (MC) cable, 437-438
      nonmetallic (NM) cable, 437
      termination points, 439
   equipment components, 425
      circuit breaker boxes, 433
      dimming modules, 427-428
      fixtures, 431-433
      light switches, 429
      outlets, 426
      placement, 425
      three-way light switches, 430-431
   exam prep questions, 443-447
   installation plans, 441
   maintenance plans, 441
      breaker reactivation, 442
      GFCI breaker maintenance, 442
   power consumption, appliance listing, 435-436
   reference resources, 448
   standards
      EIA/TIA, 440
      IEEE, 440
      National Electrical Code (NEC), 439-440
      Underwriters Laboratories (UL), 441
   termination points, 435-436
home access controls
   design considerations, 312
      condition monitoring, 314-315
      remote access systems, 313-314
      remote control systems, 312-313
   device connectivity, 327
      communications cable, 327-330
      low-voltage wiring, 330
      phone lines, 330
      termination points, 330
   exam prep questions, 335-339
   installation plans, reverse garage door sensors, 333
   location considerations
      keypads, 315
      relays, 316
      sensors, 317
   physical products, 318
      control boxes, 319-320
      distribution panels, 322-323
      gates, 323
      keypads, 318-319
      power supplies, 320
      solenoid-operated locks, 321-322
   reference resources, 340
   standard configuration and settings, 324
      sensors, 325-326
      software programming, 325
      time-of-day settings, 326
      user access, 324
   standards
      Builders Hardware Manufacturers Association (BHMA), 331
      Consumer Product Safety Commissions (CPSC), 332
      National Electrical Code (NEC), 332
      Underwriters Laboratories (UL), 332
   troubleshooting wireless garage door openers, 334
Home Audio Video Interoperability (HAVi), 111
home automation
   A/V cabinetry, 342-343
   component location
      keypads, 349
      sensors, 349
   configuration and settings
      keypad programming, 357
      sensors, 357-360
      time-of-day scheduling, 360
   control processors, 456-458
   device connectivity
      Category 5E UTP cables, 465
      coaxial cables, 464
      communications cable, 360-361, 464
      fiber-optic cables, 465
      infrared transmission, 465
      low-voltage wiring, 361
      radio frequency, 466
      termination points, 361, 466

home lighting

distribution panels, 455
dumbwaiters, 347-348
elevator systems, 346
exam prep questions, 364-368, 475-480
fireplace igniters, 347
home fans, 348
installation plans
    central home controllers, 469
    user interfaces, 469
interfaces, location of, 455
keypads
    core component functions, 454
    location considerations, 453
life systems, 343
maintenance plans
    testing stage, 470-471
    troubleshooting tools, 471-474
patch panels, 458
products
    control boxes, 351
    distribution panels, 355
    keypads, 349-350
    lift systems, 356
    portals, 357
    power supplies, 352-353
    solenoids, 354-355
reference resources, 369
skylights, 348
stair lift designs, 346
standards
    ANSI/TIA/EIA 570-A, 467
    EIA 600 CEBus, 362
    IEEE, 468
    National Electrical Code (NEC), 362, 467
    National Electrical Contractors Association (NECA), 468
    TIA/EIA 570-A, 362
    Underwriters Laboratories (UL), 363, 468
systems integration design, 450
touchscreens, location considerations, 453
user interface programming, 459
    controller systems, 461-462
    intelligent remote controls, 463
    keypad-based controllers, 462
    residential gateways, 459-461
    touchsceens, 462
Web pads, location considerations, 453
window shading designs, 345
wireless RF/IR receivers
    core component functions, 454
    location considerations, 453
Home Automation and Wiring, 231, 310, 369
home fans, 348
home lighting
    configurations and settings, 217
        scenes, 218
        security, 218-219
        zones, 219
    design consideration
        conduits, 207-208
        daisy-chain wiring, 206
        grounding connections, 204-205
        home-run wiring, 205
        power-line controls, 206-207
        remote access, 208
        wire runs, 205
        wireless runs, 205
        zones, 208
    design considerations, load requirements, 202-204
    device connectivity, 219
        armored cable (AC), 220-221
        low-voltage wiring, 221
        metal clad (MC) cable, 219-220
        nonmetallic (NM) cable, 219
        termination points, 221
    exam prep questions, 226-230
    industry standards
        Electronic Industries Alliance (EIA), 222
        National Electrical Code (NEC), 222-223
        National Electrical Manufacturers Association (NEMA), 223
        National Fire Protection Association (NFPA), 224
        Underwriters Laboratories (UL), 223
    installation plans, single-pole switches, 224
    location considerations, 209
    maintenance plans, 225
    products
        automated window treatments, 217
        dimming modules, 211-212
        light fixtures, 215-217
        lighting switches, 213-215
        power outlets, 209-211
    reference resources, 231

How can we make this index more useful? Email us at indexes@quepublishing.com

home networks

home networks
  cabling, 11-12
    HomePNA standard, 12-13
    media considerations, 27-28
    power-line control protocols, 16-20
    power-line networking, 14-16
  device connectivity
    access points design, 54-55
    interfaces with legacy systems, 53-54
    termination points, 56-59
  equipment components, 32
    data ports, 32-33
    jack types, 34-35
  equipment location, 28
    firewalls, 31
    network interface cards (NICs), 28-29
    printers, 30
    routers, 31
    servers, 30
    switches, 31
    wireless access points, 31
  exam prep questions, 74-78
  externally provided data services
    cable, 67-68
    DSL, 62-66
    email, 70-71
    ISDN, 66-67
    PPP, 68
    PPTP, 68-69
    RAS, 69-70
  hardware configuration, 41
    DHCP, 48
    firewalls, 51-52
    Internet domain names, 46-47
    network address translation (NAT), 51
    network interface cards (NICs), 52
    operating systems, 41-45
    port numbers, 49-50
    TCP/IP utility programs, 45-46
    UDP, 49
    uniform resource locators (URLs), 47
  physical devices
    computer systems, 37
    network-enabled devices/appliances, 40-41
    printers, 38-39
    residential gateways, 39-40
    wireless access points, 36-37
  reference resources, 79
  remote access methods, 11
  shared in-house services
    file and print services, 61
    media services, 61
    video surveillance, 60
  standards
    ANSI/EIA, 73
    ANSI/TIA/EIA, 72-73
    IEEE, 71
home security and surveillance systems, 128
  cohesion with existing systems, 134
  configuration and settings, 143
    camera locations, 146
    exterior lighting location, 146
    keypad locations, 144
    passwords, 143-144
    sensor locations, 144-145
    zone layout, 143
  control panels, 137
  design considerations, 128
  device connectivity
    coaxial cabling, 148
    low-voltage wiring, 147
    telephone lines, 147
    termination points, 148
    wireless, 147
  doors, design specifications, 136
  emergency response, 132-133
  environmental monitoring, 132
  exam prep questions, 154-158
  external services, 149
    professional monitoring, 150
    remote access, 150
  fire detection
    heat sensors, 131-132
    smoke detectors, 131-132
  hard-wired
    advantages, 129-130
    disadvantages, 130
  in-house services
    alarm types, 149
    video surveillance, 148
  installation plans
    cameras, 153
    monitors, 153
    motion detectors, 153
    RJ-31x telephone jacks, 152-153
  keypads, design specifications, 137
  location considerations, designing, 133-135
  maintenance plans, 153

physical devices, 138
    cameras, 141
    closed circuit TV (CCTV) monitors, 141
    keypads, 138
    security panels, 140
    sensors, 139-140
    switchers, 142
reference resources, 159
remote access, 130-131
safety and code regulations, 134
standards
    ANSI/TIA/EIA 570-A, 151
    National Electrical Code (NEC), 150
    Underwriters Laboratory (UL), 152
temperature sensors, 133
utility outlet specifications, 133
windows, design specifications, 136
wireless
    advantages, 128-129
    disadvantages, 129
home telecommunications
    configuration and settings, 175
        call restrictions, 177-178
        intercom features, 178
        phone-line extensions, 175
        phone-line splitters, 175
        voice mail, 177
    connection methods
        physical jack connections, 179
        physical wiring types, 180
        RJ-11 connectors, 181-182
        RJ-45 termination standards, 182
        termination points, 183
        wireless, 179
    design methods, 162
        analog systems, 162-163
        digital systems, 163-164
        hybrid systems, 162
        key systems, 165-166
        PBX systems, 165
    environmental factors, 168
    equipment location considerations
        home environmental factors, 169
        industry standards, 169
        restrictions, 169
    exam prep questions, 194-198
    external services
        911 emergency systems, 187-188
        call blocking, 186-187
        call waiting, 187
        caller ID, 185
        personal emergency systems, 189
        three-way calling, 187
        voice mail, 185-186
    home-run wiring, 180
    in-house services
        call conferencing, 184
        extension dialing, 184-185
        intercoms, 183-184
        voice mail, 183
    installation plans
        cable tips, 192
        termination wall outlets, 191-192
    remote access
        L2TP (Layer 2 Tunneling Protocol), 167
        PPP (Point-to-Point Protocol), 168
        PPTP (Point-to-Point Tunneling Protocol), 168
    standards/protocols
        cabling/wiring, 190-191
        fax transmissions, 189-190
    troubleshooting and maintenance guidelines, 193
    Voice over IP, 167
home theaters. *See* audio/video systems
home water systems
    configuration and settings
        seasonal preset programming, 300
        time-of-day programming, 299-300
        zone programming, 298-299
    design considerations, 282
        remote access, 284
        timed systems, 282-283
        zoned systems, 284
    device connectivity, 300
        communication cable, 300-301
        low-voltage wiring, 301-302
        termination points, 302
    equipment locations, 284
        backflow prevention control valves, 285
        heaters, 286-287
        irrigation control valves, 285
        keypads, 286
        pumps, 287-288
        relays, 286
        solenoids, 288
    exam prep questions, 305-309
    industry standards, 302-303

How can we make this index more useful? Email us at indexes@quepublishing.com

installation plans, 303-304
maintenance plans, 304
products, 289
    control points, 290
    distribution panels, 291
    interface locations, 291
    irrigation controllers, 289-290
    keypads, 292
    power supplies, 292-293
    sensors, 293
    solenoids, 295-296
    sprinklers, 296-298
    sump pumps, 293-295
reference resources, 310
home-run wiring topology
    home lighting systems, 205
    low-voltage wiring, design considerations, 383-384
    telecommunications, 180
HomePlug Powerline Alliance (HPLA) standard, 14-16
    devices, 16
    network adapters, 14
    required network resources, 15-16
HomePNA (Home Phone Network Alliance) standard, 12
    Ethernet speeds, 13
    frequency division multiplexing (FDM), 12
    home benefits, 13
    interface adapters
        network interface card (NIC), 12
        USB adapters, 13
    number of devices, 13
    official Web site, 12
    transmission rates, 12
    version specifications, 12
HomeRF standard
    corporate support, 6
    device range, 5
    founding members, 5
    future applications of, 6
    spectrum frequency, 5
    SWAP (Shared Wireless Access Protocol), 6
    transmission rates, 5
HVAC Instant Answers, 279
HVAC (heating, ventilating, and air conditioning systems), 234
    air handlers, 249
        attic fans, 250
        whole house fans, 250
    condensers, 254
    evaporators, 254
    heat pumps, 254-255
    reversing valves, 255
    configuration and settings
        seasonal presets, 259
        system programming, 257-258
        time-of-day programming, 258-259
        zone programming, 256-257
    damper controls, 250-251
    duct planning, 251
    heating adjustments, 251
    static pressure measurements, 252
    design considerations, 234
        air handlers, 238-239
        centralized systems, 236-238
        distributed systems, 236-237
        non-zoned, 234
        remote access, 241-242
        water-based (radiant) heating, 239-241
        zoned, 234-236
    device connectivity
        communications cable, 260-261
        termination points, 261
        thermostat communication, 260
    distribution panels, 256
    ENERGY STAR program, 249
    equipment location considerations, 242
        sensors, 245
        thermostats, 242-244
    exam prep questions, 273-278
    furnaces, 252-253
        electric type, 253
        gas type, 253
    installation plans
        ducts, 266-269
        thermostats, 264-265
    maintenance issues, 270
        air conditioning systems, 270-272
        heating systems, 271
    reference resources, 279
    standards and organizations, 262
        IEEE, 263
        National Electrical Code (NEC), 262
        National Fire Protection Association (NFPA), 264
        TIA/EIA, 262-263
        Underwriters Laboratories (UL), 263

thermostats
  electronic, 249
  localized control, 246-247
  mercury, 247-248
  overall control, 246-247
  staging, 248-249
hybrid fiber/coax (HFC) networks (cable), 67
hybrid irrigation controllers, 290
hybrid telecommunications systems, 162
hydraulic elevators, 354

I

IEEE (Institute of Electrical and Electronic Engineers)
  computer networking standards, 71
  high-voltage systems standards, 440
  home automation standards, 468
  HVAC standards, 263
IEEE 802.11 standard
  802.11a standard
    compatibility with 802.11b standard, 8
    data ranges, 8
    data rates, 10
  802.11b standard
    ad hoc mode, 8-9
    complementary code keying (CCK), 8
    data ranges, 8
    data rates, 10
    distribution system (DS), 9
    infrastructure mode, 8-9
    versus 802.11g standard, 8
    Wi-Fi certification, 8
  802.11g standard
    transmission rates, 10
    versus 802.11b standard, 8
  development of, 7
IEEE 802.XX Ethernet standards
  802.3i 10BASE-T, 21-22
  802.3u 100BASE-T, 22
IEEE 1394 External Bus Standard, 116
IEEE 1394 Serial Port Standard, 33
image resolution in audio/video systems, 111
impedance (speakers), 93
in-house services
  audio/video systems
    media servers, 107
    personal video recorders, 107
    streaming audio/video, 106-107

home security and surveillance systems, 148
  alarm types, 149
  video surveillance, 148
telecommunications systems
  call conferencing, 184
  extension dialing, 184-185
  intercoms, 183-184
  voice mail, 183
incandescent lighting, 215-216, 431-432
infrared controls, home lighting, 205
infrared transmission, home automation systems, 358, 465
infrastructure mode, IEEE 802.11b standard, 8-9
ink-jet printers, 38
installing
  high-voltage systems, wall outlets, 441
  home access controls, reverse garage door sensors, 333
  home lighting systems, 224
  home security and surveillance systems
    cameras, 153
    maintenance plans, 153
    monitors, 153
    motion detectors, 153
    RJ-31x telephone jacks, 152-153
  home/garden water systems, 303-304
  HVAC systems
    ducts, 266-269
    thermostats, 264-265
  low-voltage wiring
    coaxial cable, 407
    structured outlets, 407
    UTP cable, 407
  satellite systems, 117-118
  speakers, 119
  systems integration
    central home controllers, 469
    user interfaces, 469
  telecommunications systems
    cabling, 192
    termination wall outlets, 191-192
Institute for Electrical and Electronics Engineers. *See* IEEE
insulation displacement connection (IDC), 57, 173
integrated cable testers, troubleshooting tool, 472
Integrated Services Digital Network. *See* ISDN

intelligent remote controls, home
   automation systems, 463
intelligent thermostats
   installation guidelines, 265
   location of, 244
intercoms, 183-184
   configuration options, 178
interfaces
   home automation systems, location of, 455
   home/garden water systems, location of, 291
internal broadcasting, audio/video component settings, 101
International Corporation of Assigned Names and Numbers (ICANN), 42
International Standards Organization (ISO), 402-403
Internet content filtering, residential gateways, 460
Internet telephony, Voice over IP, 167
Internetworking Packet Exchange/Sequential Packet Exchange (IPX/SPX), 52
intranets, firewall implementation, 51
IP addressing
   allocation via DHCP, 48
   ARPANET (DoD) origins, 42
   classes, 43
   dotted-decimal notation, 43
   International Corporation of Assigned Names and Numbers (ICANN), 42
   network address translation (NAT), 51
   octets, 43
   packet delivery via TCP, 48
   packet delivery via UDP, 49
   port numbers, 49-50
   reserved, 43
   reserved private, 44
   subnets, 44-45
IP Xpander command center (home automation), 351
IPCONFIG utility (TCP/IP), 45
IPX/SPX (Internetworking Packet Exchange/Sequential Packet Exchange), 52
IrDA (Infrared Data Association), 10
irrigation controllers
   electromechanical, 289
   electronic, 290
   hybrid, 290

irrigation systems
   backflow prevention valves, location of, 285
   control valves, location of, 285
   home/garden water systems, 287-288
   keypads, 286
   relays, 286
   timers, 283
ISDN (Integrated Services Digital Network), 66
   basic rate interface (BRI) service, 67
   broadband service, 67
   primary rate interface (PRI) service, 67
   terminal adapters, 66
   versus DSL, popularity of, 66
ISO (International Standards Organization), 402-403
ITU-T Group 3 & 4 standards, fax transmissions, 189-190

J - K

jacks
   FCC-registered, 34
   RJ-11, 34
   RJ-14, 35
   RJ-31x, 35
   RJ-45, 35
   Uniform Service Ordering Code (USOC), 181-182

key telecommunications systems, 165-166
keypads
   audio/video systems, location of, 88
   home access controls, 318
      location of, 315
      portable, 319
      wall-mounted, 318
   home automation systems
      A/V controllers, 349-350
      component functions, 454
      configuring, 462
      fireplace controllers, 350
      location considerations, 349, 453
      programming of, 357
      types, 349
   home security and surveillance systems, 144
      design specifications, 137
      function of, 138
   home/garden water systems, 292
   irrigation systems, 286

remote-control, 94
systems integration design, 451
wall-mounted, 94
kilowatt usage, calculating, 202-204
KSU telecommunications system, 170-171
KSU-less telecommunications system, 170-171

L

laser printers, 39
LCD flat-panel video displays, 96
    active matrix, 97
    passive matrix, 97
legacy systems, device connectivity, 53-54
legacy thermostats
    installation guidelines, 265
    location of, 242-244
lift systems, home automation, 343, 356
light fixtures
    ambient source, 215
    fluorescent source, 216-217
    incandescent source, 215-216
    task source, 215
light switches, high-voltage systems
    double pole double throw (DPDT), 429
    double pole single throw (DPST), 429
    home lighting systems
        clock, 215
        photoelectric, 215
        standard configurations, 213
        three-way, 213-214, 430-431
    single pole double throw (SPDT), 429
    single pole single throw (SPST), 429
lighting (home)
    configurations and settings, 217
        scenes, 218
        security, 218-219
        zones, 219
    design considerations, 202
        conduits, 207-208
        daisy-chain wiring, 206
        grounding connections, 204-205
        home-run wiring, 205
        load requirements, 202-204
        power-line controls, 206-207
        remote access, 208
        wire runs, 205
        wireless runs, 205
        zones, 208

device connectivity, 219
    armored cable (AC), 220-221
    low-voltage wiring, 221
    metal clad (MC) cable, 219-220
    nonmetallic (NM) cable, 219
    termination points, 221
exam prep questions, 226-230
industry standards
    Electronic Industries Alliance (EIA), 222
    National Electrical Code (NEC), 222-223
    National Electrical Manufacturers Association (NEMA), 223
    National Fire Protection Association (NFPA), 224
    Underwriters Laboratories (UL), 223
installation plans, 224
location considerations, 209
maintenance plans, 225
products
    automated window treatments, 217
    dimming modules, 211-212
    light fixtures, 215-217
    lighting switches, 213-215
    power outlets, 209-211
reference resources, 231
limited use communications cable, 328
load requirements
    high-voltage systems, designing, 417
    home lighting, calculating, 202-204
locating
    A/V equipment, 423-424
    home access controls, 315
        keypads, 315
        relays, 316
        sensors, 317
    home lighting, design criteria, 209
    keypads in home automation systems, 349
    low-voltage wiring
        new construction, 385-387
        remodeling upgrades, 384-385
    sensors for home automation systems, 349
locked condition monitoring, 314
locks, electromagnetic, 331
log files, residential gateways, 460
LonWorks protocol (home networks), 385
Low Voltage Wiring – Security/Fire Alarm Systems, 159
low-pass filters, 392

How can we make this index more useful? Email us at indexes@quepublishing.com

614 low-voltage wiring

low-voltage wiring
   Category 5E UTP cables, 465
   coaxial cables, 464
   components
      amplifiers, 392
      binding posts, 393
      BNC connectors, 391-392
      cabling, 387
      coaxial video connectors, 394
      distribution panels, 387-389
      gateways, 388
      low-pass filters, 392
      multipurpose outlets, 387
      RG-6 coaxial cable, 391
      RJ-11 jack connectors, 390
      RJ-45 jack connectors, 389
   configuration and settings
      distribution panels, 395-396
      termination points, 395
   design considerations, 374
      bundled cabling, 380
      coaxial cable, 379-380
      conduits, 382
      daisy-chain topology, 382
      fiber-optic cable, 381
      home-run topology, 383-384
      screen twisted pair (ScTP), 375-376
      shielded twisted pair (STP), 375
      unshielded twisted pair (UTP), 376-379
      wire types, 374-381
   device connectivity
      audio wire, 399-400
      Category 5 UTP cable, 397
      coaxial cable, 397
      fiber-optic cable, 398
      plenum cable, 399
      security wire, 400
      termination points, 400
   exam prep questions, 409-413
   fiber-optic cables, 465
   home access systems, device connectivity, 330
   home automation systems, connecting, 361
   home lighting systems, 221
   home security and surveillance systems, 147
   home/garden water systems, 301-302
   installation plans
      coaxial cable, 407
      structured outlets, 407
      UTP cable, 407
   location considerations
      new construction, 385-387
      remodeling upgrades, 384-385
   maintenance plans, 408
   reference resources, 414
   standards organizations, 401-406

M

maintenance
   low-voltage wiring, 408
   systems integration
      testing stage, 470-471
      troubleshooting tools, 471-474
master devices, Bluetooth standard, 7
media servers
   pay per view (PPV), 107
   video on demand (VOD), 107
media services, streaming audio/video, 61
mercury thermostats, 247-248
metal clad (MC) cable
   high-voltage systems, 437-438
   home lighting wiring, 219-220
Microsoft Media Player, 106-107
Mini DV video format, 112
modems
   cable, DOCSIS standard, 68
   DSL, 64-65
monitors
   home security and surveillance systems, installation plans, 153
   quad camera switchers, 142
motion detectors
   home security and surveillance systems, installation plans, 153
   location of, 144
   security lighting, 218-219
   security sensors, 140
multimode fiber-optic cable transmission, 381
multipurpose outlets, low-voltage wiring, 387
multipurpose plenum communications cable, 329
multipurpose riser communications cable, 329

N - O

National Electrical Code (NEC), 150, 204, 223
   high-voltage systems
      standards, 439-440
      wiring types, 440
   home access control standards, 332
   home automation standards, 362, 467

HVAC systems, 262
low-voltage wiring standards
  hazardous locations, 406
  power-limited circuits, 403-406
wiring types, 222
National Electrical Contractors Association (NECA), 468
National Electrical Manufacturers Association (NEMA), 223
National Fire Protection Association (NFPA), 224, 264
NBTSTAT utility (TCP/IP), 45
NET VIEW utility (TCP/IP), 45
NETSTAT utility (TCP/IP), 45
network adapters (Ethernet), 21
network address translation (NAT), 51
network interface cards (NICs)
  configuration/installation, 52
  HomePNA standard, 12
  installation of, 28-29
  PCMCIA expansion slots, 28
  protocol drivers, 52
network interface device (NID), 190
network sniffers, troubleshooting tool, 472
network-enabled devices/appliances, Universal Plug and Play (UPnP), 40-41
new home construction
  A/V equipment location, 424
  low-wiring design options, 385-387
non-zoned HVAC systems, 234
nonmetallic (NM) cable
  high-voltage systems, 437
  home lighting wiring, 219
NTSC TV standard, 114

ohmmeters, troubleshooting tool, 471
opened condition monitoring, 315
operating systems
  IP addressing
    classes, 43
    dotted-decimal notation, 43
    octets, 43
    origins of, 42
    reserved, 43
    reserved private, 44
    subnets, 44-45
  software
    configuration of, 41-42
    user settings, 41-42
outlets for high-voltage systems, 426
oxidation reduction potential (ORP), pool pollution, 284

P

parallel ports
  DB-25, 33
  EPP, 33
  printers, 38
passive infrared (PIR) detectors, 140
passwords
  home security and surveillance systems, 143-144
  residential gateways, 461
patch panels, home automation systems, 458
pay per view (PPV) media servers, 107
PBX (private branch exchange) telecommunications system, 165, 170-172
pedestal sump pumps, 293-294
personal emergency response systems (PERS), 189
personal video recorders (PVRs), 107
phone lines. *See also* HomePNA standard
  home access systems, 330
  RJ-11 jack type, 34
  RJ-14 jack type, 35
  RJ-31x jack type, 35
  RJ-45 jack type, 35
photoelectric sensors
  garage door openers, 326
  lighting switches, 215
physical devices, home security and surveillance systems, 138
  cameras, 141
  closed circuit TV (CCTV) monitors, 141
  keypads, 138
  security panels, 140
  sensors, 139-140
  switchers, 142
piconets, Bluetooth wireless standard, 7
PING utility (TCP/IP), 46
plasma flat-panel video displays, 97
plenum-rated communications cable, 328, 399
point-to-multipoint connections, Bluetooth standard, 7
point-to-point connections, Bluetooth standard, 7
Point-to-Point Protocol (PPP), 68, 168
Point-to-Point Tunneling Protocol (PPTP), 68-69, 168
polyvinyl chloride (PVC) conduit, 207
pools
  heaters, 286-287
  oxidation reduction potential (ORP), 284
  pumps, 288
  timed systems, 283

How can we make this index more useful? Email us at indexes@quepublishing.com

pop-up sprinklers, 296
POP3 protocol, 71
port mapping, residential gateways, 460
portable keypads (home access systems), 319
power backups (UPS) for high-voltage systems
   computer systems, 420-422
   designing, 419-422
   essential distribution, 420
   whole house distribution, 420
power consumption, appliance listing, 435-436
power load requirements, home lighting, 202-204
power outlets
   ground fault circuit interrupter (GFCI), 211
   National Electrical Manufacturers Association (NEMA), 209-211
   wire configuration, 209-211
power supplies
   home access controls, 320
   home automation systems, AC-to-DC power converters, 352-353
   home/garden water systems, 292-293
power-limited circuits, 403-406
power-limited plenum communications cable, 329
power-line controls, home lighting systems
   CEBus protocol, 206-207
   X10 protocol, 206-207
power-line networking
   control protocols, 16-17
      CEBus standard, 19-20
      X10, 17-19
   HomePlug Powerline Alliance (HPLA) standard, 14-16
PPP (Point-to-Point Protocol), 68
PPTP (Point-to-Point Tunneling Protocol), 68-69
Practical Guide to Today's Home Entertainment Systems, 126
Practical Home Theater – A Guide to Video and Audio Systems, 126
practice exam #1
   answers, 504-524
   questions, 485-503
practice exam #2
   answers, 539-552
   questions, 525-538
primary rate interface (PRI) service, 67
print servers, 61

printers
   home network connections, 30
   ink-jet, 38
   laser, 39
   parallel port, 38
   universal serial bus (USB) port, 38
private network reserved IP addresses, 44
programming home access controls, 325
protocols
   audio/video system standards, 111
   telecommunications
      L2TP (Layer 2 Tunneling Protocol), 167
      PPP (Point-to-Point Protocol), 168
      PPTP (Point-to-Point Tunneling Protocol), 168
public safety answering points (PSAP), 911 emergency systems, 187-188
pumps, home/garden water systems
   irrigation type, 287-288
   pool/spa type, 288
punch-down connections, 57
punch-down tools
   66-type block, 173-174
   110-type block, 173-174

R

radio frequency, home automation systems, 466
radio transmitter controls, 205
RADIUS (remote authentication dial-in user service), 69-70
rapid start ballasts (fluorescent lighting), 217
RAS (Remote Access Service), 69-70
RealNetwork RealOne Player, 106-107
rear video projection, 98
receivers (audio/video systems), 89-90
   A/V, 90
   satellite, 90
   stereo, 90
   tuners, 90
reference resources
   audio/video systems, 126
   exam prep questions, 369
   home access control systems, 340
   home automation systems, 369
   home lighting systems, 231
   home networks, 79
   home security and surveillance systems, 159
   home water systems, 310

HVAC systems, 279
low-voltage wiring, 414
telecommunications systems, 199
relays
    home access controls, location of, 316
    irrigation systems, 286
remodeling upgrades
    A/V equipment location, 423-424
    low-wiring design options, 384-385
remote access
    devices, 11
    doors/gates, 313-314
    garage doors
        code grabbers, 312
        safety regulations, 313
    home lighting systems, 208
    home/garden water systems, 284
    HVAC systems, 241-242
    Remote Access Service (RAS), 11
    security systems, 130-131
    virtual private networks (VPNs)
        L2TP (Layer 2 Tunneling Protocol), 167
        PPP (Point-to-Point Protocol), 168
        PPTP (Point-to-Point Tunneling Protocol), 168
    Web page controls, 11
Remote Access Service. *See* RAS
remote-control keypads, 94
reserved IP addresses, 43
residential gateways
    backup systems, 460
    demilitarized zones, 460
    function of, 39-40
    home automation systems, configuring, 459-461
    Internet content filtering, 460
    log files, 460
    password length, 461
    port mapping, 460
    remote administration, 460
Residential Systems exam section
    audio/video component settings, 100
        distribution channels, 100
        equalization, 101
        internal broadcasting, 101
        volume, 100
    audio/video device connectivity
        analog audio, 104
        component video, 102-103
        composite video, 103
        digital audio, 105

low-voltage cabling and connectors, 105
        S-video, 104
        termination points, 105-106
        video signal formats, 102
    audio/video equipment location, 86
        keypads, 88
        speakers, 86-87
        television, 87
        touchscreens, 88
        volume controls, 88
    audio/video in-house services
        media servers, 107
        personal video recorders, 107
        streaming audio/video, 106-107
    audio/video standards
        IEEE, 116
        image resolution, 111
        protocols, 111
        surround sound, 113
        television, 114-116
        video recording, 111-113
        wiring, 110
    audio/video system designs, 82
        dedicated type, 82
        distributed type, 82
        whole type, 83-85
        zoning distribution, 85
    audio/video system installations
        satellite service, 117-118
        speakers, 119
    audio/video system maintenance, 120
    device connectivity
        access points design, 54-55
        interfaces with legacy systems, 53-54
        termination points, 56-59
    equipment components, 32
        data ports, 32-33
        jack types, 34-35
        location considerations, 28-31
    Ethernet LAN standards, 21
        coaxial cable, 23-24
        fiber-optic cable, 24-25
        fiber-optic patch cable, 26-27
        IEEE 802.3i 10BASE-T, 21-22
        IEEE 802.3u 100BASE-T, 22
    externally provided audio/video services
        cable TV services, 108
        digital TV services, 109-110
        external terrestrial off-the-air broadcast services, 109
        satellite services, 108

How can we make this index more useful? Email us at indexes@quepublishing.com

externally provided data services
cable, 67-68
DSL, 62-66
email, 70-71
ISDN, 66-67
PPP, 68
PPTP, 68-69
RAS, 69-70
hardware configuration, 41-51
firewalls, 51-52
network interface cards (NICs), 52
network media considerations, 27-28
network physical devices
computer systems, 37
network-enabled devices/appliances, 40-41
printers, 38-39
residential gateways, 39-40
wireless access points, 36-37
physical audio/video products
amplifiers, 91-92
keypads, 94
receivers, 89-90
source equipment, 99-100
speakers, 92-93
video displays, 95-97
video projection systems, 98-99
remote access methods, 11
shared in-house services
file and print services, 61
media services, 61
video surveillance, 60
standards
ANSI/EIA, 73
ANSI/TIA/EIA, 72-73
IEEE, 71
wireless networks
protocols, 5-10
standards, 5-10
reverse garage door sensors, installing, 333
reversing valves (HVAC systems), 255
RF spectrum analyzers, troubleshooting tool, 473-474
RG-6 coaxial cable, 23, 391
RG-58 coaxial cable, 23
RG-59 coaxial cable, 23
rigid metal conduit, 207
riser communications cable, 328
RJ-11 jack connectors, 34, 179-182, 390
RJ-14 jack type, 35
RJ-31x jack type, 35, 152-153
RJ-45 connectors, 21-22, 35, 59, 182, 389

Romex wire, 205, 219
rotor sprinklers, 297
routers for home network connections, 31

S

S-video, audio/video device connectivity, 104
safety of high-voltage systems
designing, 422
ground fault circuit interrupters (GFCIs), 422
satellite receivers, 90, 108
SC connectors (fiber-optic), 25
screen twisted pair (ScTP) cable, 375-376
seasonal presets
home/garden water systems, 300
HVAC systems, 259
security and surveillance systems, 128
cohesion with existing systems, 134
configuration and settings
camera locations, 146
exterior lighting locations, 146
keypad locations, 144
passwords, 143-144
sensor locations, 144-145
zone layout, 143
control panels, design specifications, 137
design considerations, 128
device connectivity
coaxial cabling, 148
low-voltage wiring, 147
telephone lines, 147
termination points, 148
wireless, 147
doors, design specifications, 136
emergency response, 132-133
environmental monitoring, 132
exam prep questions, 154-158
external services, 149
professional monitoring, 150
remote access, 150
fire detection, 131
heat sensors, 131-132
smoke detectors, 131-132
hard-wired
advantages, 129-130
disadvantages, 130
home lighting systems
designing, 218
motion detectors, 218-219
zones, 219

in-house services
  alarm types, 149
  video surveillance, 148
installation plans
  cameras, 153
  monitors, 153
  motion detectors, 153
  RJ-31x telephone jacks, 152-153
keypads, design specifications, 137
location considerations, designing, 133-135
maintenance plans, 153
physical devices, 138
  cameras, 141
  closed circuit TV (CCTV) monitors, 141
  keypads, 138
  security panels, 140
  sensors, 139-140
  switchers, 142
remote access, 130-131
safety and code regulations, 134
standards
  ANSI/TIA/EIA 570-A, 151
  National Electrical Code (NEC), 150
  Underwriters Laboratory (UL), 152
temperature sensors, 133
utility outlet specifications, 133
windows, design specifications, 136
wireless
  advantages, 128-129
  disadvantages, 129
sensors
  home access controls
    configuration of, 325-326
    garage doors, 326
    location of, 317
  home automation systems, 357
    fans, 359
    fireplace igniters, 358
    infrared, 358
    lights, 360
    location of, 349
  home security and surveillance systems, 144-145
    door switches, 139
    function of, 139
    motion detectors, 140
    window detectors, 139
  home/garden water systems, 293
  HVAC systems, 245

motion detectors, location of, 144
windows, location of, 145
serial ports
  IEEE 1394 Port Standard, 33
  USB Version 2.0, 33
servers, physical location of, 30
service set identifiers (SSIDs), 36-37
shared in-house services
  file and print services, 61
  media services, 61
  video surveillance, 60
Shared Wireless Access Protocol (SWAP), 6
shielded twisted pair (STP) cable, 375
signal bridges, X10 power-line protocol, 19
single pole double throw (SPDT) switches, 213, 429
single pole single throw (SPST) switches, 213, 429
single-mode fiber-optic cable transmission, 381
skylights, 348, 355
slave devices, Bluetooth standard, 7
sliding gates (home access systems), 323
smart thermostats, 249
smoke alarms, installing, 135
smoke detectors
  ionization type, 131-132
  location requirements, 134
  photoelectric type, 131-132
SMTP (Simple Mail Transfer Protocol), 71
software programming, home access controls, 325
solenoid-operated locks, 321-322
solenoids, home automation systems
  elevator applications, 354
  fireplaces, 355
  home/garden water systems, 288, 295-296
  locks, 321-322
spas
  heaters, 286-287
  pumps, 288
  timed systems, 283
speakers
  cone, 92
  dynamic, 92
  impedance, 93
  installing, 119
  location overview, 86
  power ratings, 93
  surround sound, 93
    typical layout, 86-87
  wire gauge guidelines, 119

How can we make this index more useful? Email us at indexes@quepublishing.com

spikes, surge protection, 419
splitter-based DSL, 64-65
splitterless DSL, 64
spray sprinklers, 298
spring clip terminals (speakers), 393
sprinkler water systems, 296
    pop-up, 296
    rotor, 297
    spray, 298
ST connectors (fiber-optic), 25
staging thermostats, automatic changeover features, 248-249
stair lifts, 346
standards
    ANSI/EIA, 600 CEBus specification, 73
    ANSI/TIA/EIA, 72
        568 specification, 72
        570-A specification, 72-73
    audio/video systems
        IEEE, 116
        image resolution, 111
        protocols, 111
        surround sound, 113
        television, 114-116
        video recording, 111-113
        wiring, 110
    high-voltage systems
        EIA/TIA, 440
        IEEE, 440
        National Electrical Code (NEC), 439-440
        Underwriters Laboratories (UL), 441
    home access controls
        Builders Hardware Manufacturers Association (BHMA), 331
        Consumer Product Safety Commission (CPSC), 332
        National Electrical Code (NEC), 332
        Underwriters Laboratories (UL), 332
    home automation systems
        EIA 600 CEBus, 362
        National Electrical Code (NEC), 362
        TIA/EIA 570-A, 362
        Underwriters Laboratories (UL), 363
    home lighting systems
        Electronic Industries Alliance (EIA), 222
        National Electrical Code (NEC), 222-223
        National Electrical Manufacturers Association (NEMA), 223
        National Fire Protection Association (NFPA), 224
        Underwriters Laboratories (UL), 223
    home security and surveillance systems
        ANSI/TIA/EIA 570-A, 151
        National Electrical Code (NEC), 150
        Underwriters Laboratory (UL), 152
    home/garden water systems, 302-303
    HVAC systems
        IEEE, 263
        National Electrical Code (NEC), 262
        National Fire Protection Association (NFPA), 264
        TIA/EIA, 262-263
        Underwriters Laboratories (UL), 263
    IEEE, 71
    low-voltage wiring, 401-406
    systems integration
        ANSI/TIA/EIA 570-A, 467
        IEEE, 468
        National Electrical Code (NEC), 467
        National Electrical Contractors Association (NECA), 468
        UL, 468
    telecommunications systems
        ANSI/TIA/EIA 570-A, 180
        cabling/wiring, 190-191
        equipment location characteristics, 169
        fax transmissions, 189-190
star extension wiring (telecommunications), 175
star topology, termination points, 56
static pressure, damper controls (HVAC systems), 252
stereo receivers, 90
streaming audio/video
    media services, 61
    players, 106-107
structured wiring (high-voltage), 416
    A/V equipment location, 423
        new construction, 424
        remodeling upgrades, 423-424
    configuration and settings
        circuit breaker panels, 434
        distribution panels, 434
    design considerations, 416-417
        ground connections, 417-419
        load requirements, 417
        power backup (UPS), 419-422
        safety, 422
        surge protection, 419

device connectivity
    armored cable (AC), 438
    metal clad (MC) cable, 437-438
    nonmetallic (NM) cable, 437
    termination points, 439
equipment components, 425
    circuit breaker boxes, 433
    dimming modules, 427-428
    fixtures, 431-433
    light switches, 429
    outlets, 426
    three-way light switches, 430-431
exam prep questions, 443-447
installation plans, 441
maintenance plans, 441
    breaker reactivation, 442
    GFCI breaker maintenance, 442
power consumption, appliance listing, 435-436
reference resources, 448
standards
    EIA/TIA, 440
    IEEE, 440
    National Electrical Code (NEC), 439-440
    Underwriters Laboratories (UL), 441
termination points, 435-436
structured wiring (low-voltage)
    components
        amplifiers, 392
        binding posts, 393
        BNC connectors, 391-392
        cabling, 387
        coaxial video connectors, 394
        distribution panels, 387-389
        gateways, 388
        low-pass filters, 392
        multipurpose outlets, 387
        RG-6 coaxial cable, 391
        RJ-11 jack connectors, 390
        RJ-45 jack connectors, 389
    configuration and settings
        distribution panels, 395-396
        termination points, 395
    design considerations, 374
        bundled cabling, 380
        coaxial cable, 379-380
        conduits, 382
        daisy-chain topology, 382
        fiber-optic cable, 381
        home-run topology, 383-384
        screen twisted pair (ScTP), 375-376
        shielded twisted pair (STP), 375
        unshielded twisted pair (UTP), 376-379
        wire types, 374-381
    device connectivity
        audio wire, 399-400
        Category 5 UTP cable, 397
        coaxial cable, 397
        fiber-optic cable, 398
        plenum cable, 399
        security wire, 400
        termination points, 400
    exam prep questions, 409-413
    installation plans
        coaxial cable, 407
        structured outlets, 407
        UTP cable, 407
    location considerations
        new construction, 385-387
        remodeling upgrades, 384-385
    maintenance plans, 408
    reference resources, 414
    standards organizations, 401-406
submersible sump pumps, 293-294
subnets (IP addresses), 44-45
sump pumps
    control points, 290
    home/garden water systems
        pedestal, 293-294
        sensors, 294-295
        submersible, 293-294
Super VHS video format, 112
surge protection, high-voltage systems, designing, 419
surround sound standards
    Dolby Digital, 113
    DTS 5.1, 114
    speakers, 93
        typical layout, 86-87
    THX Surround EX, 113
SWAP (Shared Wireless Access Protocol)
    Digital Enhanced Cordless Telephone (DECT), 6
    HomeRF standard, 6
    versus WAP (Wireless Application Protocol), 6
swinging gates (home access systems), 323
switchers, home security and surveillance systems, 142
systems integration
    configuration and settings, user interface programming, 459-463

How can we make this index more useful? Email us at indexes@quepublishing.com

core components
    control processors, 456-458
    distribution panels, 455
    interface location, 455
    keypads, 454
    patch panels, 458
    wireless RF/IR receivers, 454
design considerations, 450
    keypads, 451
    touchscreens, 451-452
    Web pads, 450
    wireless RF/IR receivers, 452
device connectivity
    Category 5E UTP cables, 465
    coaxial cables, 464
    communications cables, 464
    fiber-optic cables, 465
    infrared transmission, 465
    radio frequency, 466
    termination points, 466
exam prep questions, 475-480
installation plans
    central home controllers, 469
    user interfaces, 469
location considerations, 452
    keypads, 453
    touchscreens, 453
    Web pads, 453
    wireless RF/IR receivers, 453
maintenance plans
    testing stage, 470-471
    troubleshooting tools, 471-474
standards
    ANSI/TIA/EIA 570-A, 467
    IEEE, 468
    National Electrical Code (NEC), 467
    National Electrical Contractors Association (NECA), 468
    UL, 468

T

tablet PCs (Web pads), 450
task lighting, 215
TCP (Transmission Control Protocol), 48
    IP address, guaranteed packet delivery, 48
    port numbers, 49-50
TCP/IP (Transmission Control Protocol/Internet Protocol) utility programs
    ARP, 45
    Internet domain names, 46-47
    IPCONFIG, 45
    launching, 46
    NBTSTAT, 45
    NET VIEW, 45
    NETSTAT, 45
    PING, 46
    TRACERT, 46
    WINIPCFG, 45
telecommunications
    cabling, 190
    configurations and settings, 175
        call restrictions, 177-178
        intercom features, 178
        phone-line extensions, 175
        phone-line splitters, 175
        voice mail, 177
    connection methods
        physical jack connections, 179
        physical writing types, 180
        RJ-11 connectors, 181-182
        RJ-45 termination standards, 182
        termination points, 183
        wireless, 179
    design methods, 162
        analog systems, 162-163
        digital systems, 163-164
        hybrid systems, 162
        key systems, 165-166
        PBX systems, 165
    equipment location considerations
        home environmental factors, 169
        industry standards, 169
        restrictions, 169
    exam prep questions, 194-198
    external services
        911 emergency systems, 187-188
        call blocking, 186-187
        call waiting, 187
        caller ID, 185
        personal emergency systems, 189
        three-way calling, 187
        voice mail, 185-186
    home environmental factors, 168
    in-house services
        call conferencing, 184
        extension dialing, 184-185
        intercoms, 183-184
        voice mail, 183
    installation plans
        cable tips, 192
        termination wall outlets, 191-192
    products and services
        66/110 connection blocks, 173-174
        caller line identification, 172
        fax machines, 171-172

KSU, 170-171
KSU-less, 170-171
PBX, 170-172
video conferencing, 172
reference resources, 199
remote access, 167
   L2TP (Layer 2 Tunneling Protocol), 167
   PPP (Point-to-Point Protocol), 168
   PPTP (Point-to-Point Tunneling Protocol), 168
standards/protocols
   cabling/wiring, 190-191
   fax transmissions, 189-190
troubleshooting and maintenance guidelines, 193
Voice over IP, 167
Telecommunications Industry Alliance (TIA), HVAC system standards, 262-263
telephone lines, home security and surveillance systems, 147
television. *See also* audio/video systems
   location of, 87
   satellite systems, installing, 117-118
   standards
      ATSC (Advanced Television Systems Committee), 114
      DTV (Advanced Television Systems Committee), 114
      HDTV, 115-116
      NTSC (National Television Systems Committee), 114
   viewing distances, 87
temperature sensors, 133, 245
terminal adapters (ISDN), 66
termination points
   audio/video device connectivity, 105-106
   device connectivity, 56
      Category 5 twisted-pair cable termination points, 58
      crossover network patch cable, 56
      insulation displacement connection type termination blocks, 57
      RJ-45 tip and ring designations, 59
      star topology, 56
      Type 66 punch-down termination blocks, 57
      Type 110 punch-down termination blocks, 58
   high-voltage systems, 435-436, 439
   home access systems, device connectivity, 330

home automation systems
   connecting, 361
   device connectivity, 466
home lighting systems, 221
home security and surveillance systems, 148
home/garden water systems, 302
HVAC systems, 261
low-voltage wiring
   configuring, 395
   device connectivity, 400
telecommunications systems, 183
termination wall outlets, installation of, 191-192
testing home automation, maintenance plans, 470-471
thermostats (HVAC systems), 247
   connections, 260
   electronic, 249
   installing, 264-265
   intelligent, 244, 265
   legacy, 242-244, 265
   localized control, 246-247
   location of, 242-244
   mercury, 247-248
   operation of, 247
   overall control, 246-247
   staging, 248-249
   troubleshooting and maintenance, 271
   wireless, 244
thin-wall conduit, 208
three-way calling (telecommunications), 187
three-way lighting switches, 213-214, 430-431
THX Surround EX surround format, 113
TIA/EIA 570-A standard (HVAC systems), 263, 362
TIA/EIA 570-A-1 standard (HVAC systems), 263
time domain reflectometers (TDRs), 193, 472
time-of-day programming
   home/garden water systems, 299-300
   home access controls, 326
   home automation systems, 360
   HVAC systems, configuring, 258-259
TOSLINK cable (fiber-optic patch), 26-27
touchscreens
   audio/video systems, location of, 88
   home automation systems
      configuring interfaces, 462
      location considerations, 453
      systems integration design, 451-452
   video displays, 99

How can we make this index more useful? Email us at indexes@quepublishing.com

TRACERT utility (TCP/IP), 46
troubleshooting
   high-voltage systems, 441
      breaker reactivation, 442
      GFCI breaker maintenance, 442
   home access controls, wireless garage door openers, 334
   home automation, maintenance plans, 471-474
   home lighting systems, 225
   home/garden water systems, 304
   HVAC systems, 270
      air conditioning, 270-272
      heating, 271
   telecommunications systems guidelines, 193
tuners (audio/video systems), 90
twisted-pair coaxial cable, 23
Type 66 connection blocks, 174
Type 66 punch-down termination blocks, 57
Type 110 connection blocks, 174
Type 110 punch-down termination blocks, 58

U

UDP (User Datagram Protocol), 49
   IP address, non-guaranteed packet delivery, 49-50
   port numbers, 49-50
Understanding and Servicing Alarm Systems, 159
Underwriters Laboratories (UL), 223
   fire detection standard, 134
   high-voltage system standards, 441
   home access control standards, 332
   home automation system standards, 363, 468
   home security and surveillance systems, 152
   HVAC standards, 263
uniform resource locators (URLs), 47
Uniform Service Ordering Code (USOC), telephone jacks, 181-182
uninterruptible power supplies (UPSs), high-voltage systems, 419
   computer backups, 420-422
   volt-ampere (VA) ratings, 420-422
Universal Plug and Play (UPnP), 40-41
unlocked condition monitoring, 315
unshielded twisted pair (UTP) cable, 376-377
   Category designations, 377-378
   new Category development, 378-379

unused channels, 101
USB adapters, HomePNA standard, 13
USB port printers, 38
USB Version 2.0, data ports, 33
User Datagram Protocol (UDP), 49-50
user interface programming, 459
   controller systems, 461-462
   installation plans for home automation, 469
   intelligent remote controls, 463
   keypad-based controllers, 462
   residential gateways, 459-461
   touchscreens, 462
UTP cable (Ethernet)
   installing, 407
   RJ-45 connectors, 21-22

V

VHS video format, 111
vibration windows, 136
video displays
   cathode-ray tubes (CRTs), 95
   flat-panel, 96
      LCD, 96-97
      plasma, 97
   projection systems, 98
      front projection method, 98
      rear projection method, 98
      touchscreens, 99
video distribution systems, 83-84
video on demand (VOD) media servers, 107
video projection systems
   front projection method, 98
   rear projection method, 98
   touchscreens, 99
video recording standards, 111
   8mm, 112
   Betacam, 113
   Betacam SP, 113
   Digital8, 112
   HI-8, 112
   Mini DV, 112
   Super VHS, 112
   VHS, 111
video signal formats, audio/video device connectivity, 102
video surveillance systems, 60
videoconferencing, 172
virtual private networks (VPNs)
   L2TP (Layer 2 Tunneling Protocol), 167

PPP (Point-to-Point Protocol), 168
PPTP (Point-to-Point Tunneling Protocol), 168
Voice and Data Internetworking, 199
voice mail
    configuration options, 177
    telecommunications
        external services, 185-186
        in-house services, 183
Voice Over IP (VoIP), 167
volt-ampere (VA) ratings (UPSs), 420-421
    converting to watts, 421-422
voltmeters, troubleshooting tool, 471
volume controls (A/V systems)
    component settings, 100
    location of, 88

W - Z

wall outlets, installing, 441
wall-mounted keypads (home access systems), 94, 318
WAP (Wireless Application Protocol) versus SWAP, 6
water-based (radiant) heating (HVAC systems), 239
    geothermal heat pumps, 241
    solar alternatives, 240
    traditional, 239
Web pads, home automation systems
    location considerations, 453
    systems integration design, 450
    tablet PCs, 450
whole audio/video systems
    audio distribution system
        connection components, 84
        features, 84-85
        typical equipment, 85
        zone design, 84
    design of, 83-85
    video distribution systems, 83-84
whole house fans (HVAC systems), 250
whole house power distribution, power backups, 420
Wi-Fi certification (IEEE 802.11b standard), 8
windows
    acoustic detectors, 139
    glass protection systems
        acoustical, 136
        vibrational, 136

home security and surveillance systems, design specifications, 136
shading, 345
vibration detectors, security sensors, 139
WINIPCFG utility (TCP/IP), 45
wireless access points, 31
    service set identifiers (SSIDs), 36-37
Wireless Ethernet Compatibility Alliance (WECA), 8
wireless garage door openers, troubleshooting, 334
wireless intercoms, home access systems, 329-330
wireless networks
    infrared transmission, 465
    protocols, 5-10
    radio frequency, 466
    standards, 5
        Bluetooth, 6
        HomeRF, 5-6
        IEEE 802.11 group, 7-10
        IrDA, 10
wireless RF/IR receivers
    home automation systems
        component functions, 454
        location considerations, 453
        systems integration design, 452
wireless runs
    home lighting overview, 205
    infrared, 205
    radio transmitter, 205
wireless security systems
    advantages, 128-129
    connectivity, 147
    disadvantages, 129
wireless telecommunications
    cordless phones for home consumer market, 179
    E911 (enhanced 911) connectivity, 188
    wireless LANs, 179
    wireless PBX equipment, 179
wireless thermostats, 244
wiring
    armored cable (AC), 438
    home lighting (Romex), 205
    low-voltage
        new construction options, 385-387
        remodeling options, 384-385
    metal clad (MC), high-voltage systems, 437-438

How can we make this index more useful? Email us at indexes@quepublishing.com

National Electrical Code (NEC)
  gauge, 222
  type abbreviations, 222
nonmetallic (NM), 437
telecommunications systems
  ANSI/TIA/EIA 570-A standard, 180
  Grade 1, 180
  Grade 2, 180
transmission standards/protocols, ANSI/TIA/EIA 570-A, 190-191

X10 power-line protocol
  addresses, 17
  command transmission rate, 17
  development of, 17
  home lighting systems, 206-207
  home networks, 385
  HVAC systems, 241
  IR command center (home automation), 351
  potential problems, 18-19
  receiver module, 17
  signal bridges, 19
  signal bursts, 17
  transmission rates, 17
  transmitters, 17

zones
  distribution of audio/video systems, 85
  programming
    home/garden water systems, 298-299
    home security and surveillance systems, 143
    home water management, 284
    HVAC systems, configuring, 234-236, 256-257
    lighting systems, 208, 219

What if Que

joined forces to deliver the best technology books in a common digital reference platform?

We have. Introducing
**InformIT Online Books
powered by Safari.**

- **Specific answers to specific questions.**
InformIT Online Books' powerful search engine gives you relevance-ranked results in a matter of seconds.

- **Immediate results.**
With InformIt Online Books, you can select the book you want and view the chapter or section you need immediately.

- **Cut, paste, and annotate.**
Paste code to save time and eliminate typographical errors. Make notes on the material you find useful and choose whether or not to share them with your workgroup.

- **Customized for your enterprise.**
Customize a library for you, your department, or your entire organization. You pay only for what you need.

As an InformIT partner, Que has shared the knowledge and hands-on advice of our authors with you online. Visit InformIT.com to see what you are missing.

Get your first 14 days **FREE!**

InformIT Online Books is offering its members a 10-book subscription risk free for 14 days.
Visit **http://www.informit.com/onlinebooks** for details.

POWERED BY Safari

informit.com/onlinebooks

informIT

www.informit.com

Your Guide to Information Technology Training and Reference

Que has partnered with **InformIT.com** to bring technical information to your desktop. Drawing on Que authors and reviewers to provide additional information on topics you're interested in, **InformIT.com** has free, in-depth information you won't find anywhere else.

Articles

Keep your edge with thousands of free articles, in-depth features, interviews, and information technology reference recommendations – all written by experts you know and trust.

Online Books

Answers in an instant from **InformIT Online Books'** 600+ fully searchable online books. Sign up now and get your first 14 days **free**.

POWERED BY

Catalog

Review online sample chapters and author biographies to choose exactly the right book from a selection of more than 5,000 titles.

As an **InformIT** partner, **Que** has shared the knowledge and hands-on advice of our authors with you online.
Visit **InformIT.com** to see what you are missing.

www.quepublishing.com

Get Certified!

You have the experience and the training — now demonstrate your expertise and get the recognition your skills deserve. An IT certification increases your credibility in the marketplace and is tangible evidence that you have the know-how to provide top-notch support to your employer.

Why Test with VUE?

Using the speed and reliability of the Internet, the most advanced technology and our commitment to unparalleled service, VUE provides a quick, flexible way to meet your testing needs.

Three easy ways to register for your next exam, all in real time:

- Register online at www.vue.com
- Contact your local VUE testing center. There are over 3000 quality VUE testing centers in more than 130 countries. Visit www.vue.com for the location of a center near you.
- Call a VUE call center. In North America, call toll-free 800-TEST-NOW (800-837-8734). For a complete listing of worldwide call center telephone numbers, visit www.vue.com.

Call your local VUE testing center and ask about TESTNOW!™ same-day exam registration!

The VUE testing system is built with the best technology and backed by even better service. Your exam will be ready when you expect it and your results will be quickly and accurately transmitted to the testing sponsor. Test with confidence!

Visit www.vue.com for a complete listing of IT certification exams offered by VUE

When IT really matters... Test with VUE!

CramSession
– the difference between Pass ... or Fail

"On top of everything else, I find the best deals on training products and services for our CramSession members".

Jami Costin,
Product Specialist

CramSession.com is #1 for IT Certification on the 'Net.

There's no better way to prepare for success in the IT Industry. Find the best IT certification study materials and technical information at CramSession. Find a community of hundreds of thousands of IT Pros just like you who help each other pass exams, solve real-world problems, and discover friends and peers across the globe.

CramSession – #1 Rated Certification Site!

- #1 by TechRepublic.com
- #1 by TechTarget.com
- #1 by CertMag's Guide to Web Resources.

CramSession has IT all!

- **The #1 study guides on the 'Net.** With over 250 study guides for IT certification exams, we are the web site every techie visits before passing an IT certification exam.

- **Practice questions.** Get the answers and explanations with our CramChallenge practice questions delivered to you daily.

- **The most popular IT Forums.** Cramsession has over 400 discussion boards loaded with certification infomation where our subscribers study hard, work hard, and play harder.

- **e-Newsletters.** Our IT e-Newsletters are written by techs for techs: IT certification, technology, humor, career and more.

- **Technical Papers and Product Reviews.** Find thousands of technical articles and whitepapers written by industry leaders, trainers, and IT veterans.

- **Exam reviews.** Get the inside scoop before you take that expensive certification exam.

- **And so much more!**

Visit Cramsession.com today!
...and take advantage of the best IT learning resources.

www.cramsession.com